# Flink 核心技术

## 源码剖析与特性开发

黄伟哲◎著

人民邮电出版社

北京

图书在版编目（CIP）数据

Flink核心技术：源码剖析与特性开发 / 黄伟哲著 . — 北京：人民邮电出版社，2022.7（2024.5重印）
ISBN 978-7-115-58447-2

Ⅰ．①F… Ⅱ．①黄… Ⅲ．①数据处理软件 Ⅳ．①TP274

中国版本图书馆CIP数据核字(2022)第027619号

## 内 容 提 要

本书以核心概念和基本应用为脉络，介绍 Flink 的核心特性（如检查点机制、时间与窗口、混洗机制等）、任务部署、DataStream API、DataSet API、Table API 的应用以及运行时原理等内容。每章先对概念进行基本介绍，然后基于应用实例详细分析 Flink 的设计思想和源码实现，逐步引领读者掌握定制化的开发特性并提升性能，让读者对 Flink 的理解有质的飞跃。本书内容是作者多年工作实践的总结，能够帮助读者实现真实的企业级需求。

本书适合想要学习 Flink 设计原理并希望对 Flink 进行定制化开发的平台开发工程师，需要进行架构设计和技术选型的架构师与项目经理，以及计算机相关专业的学生阅读。

◆ 著　　黄伟哲
　　责任编辑　刘雅思
　　责任印制　王　郁　胡　南
◆ 人民邮电出版社出版发行　北京市丰台区成寿寺路 11 号
　　邮编 100164　电子邮件 315@ptpress.com.cn
　　网址　https://www.ptpress.com.cn
　　北京七彩京通数码快印有限公司印刷
◆ 开本：800×1000　1/16
　　印张：26.75　　　　　　2022 年 7 月第 1 版
　　字数：631 千字　　　　　2024 年 5 月北京第 4 次印刷

定价：109.90 元

读者服务热线：(010)81055410　印装质量热线：(010)81055316
反盗版热线：(010)81055315
广告经营许可证：京东市监广登字 20170147 号

# 序 一

2018年，在北京国家会议中心举办的Flink Forward峰会上，来自阿里巴巴、京东、字节跳动、美团等公司的大数据技术负责人向众多参会者介绍了如何使用Flink解决组织内部的业务问题，做到大规模实践。每个人都在热烈地讨论Flink相关的技术实践和应用场景，很难想象Flink是2014年才正式发布的大数据技术，但在场的所有人都深刻感受到实时计算浪潮的到来。现在看来，Flink的口号"实时即未来"（Real-time is the Future）正一步步地变成现实。

数据的快速产生与快速流动对实时分析提出了很高的要求。如果数据不能被很好地实时处理，那么数据本身所蕴含的实时价值会迅速消失。在Flink之前，流处理的解决方案不尽如人意，Storm、Spark Streaming都只能说是过渡方案。

在大家对流处理的需求日益迫切的同时，Google发表的关于MillWheel和Dataflow的两篇论文让我们依稀看到了下一代流处理技术的模样。恰逢其时，Flink作为新一代流处理计算引擎进入了人们的视野。它吸收了上一代流处理技术的经验，融合了学术界和工业界对下一代流处理技术的理论探索与实践，甫一问世，就迅速成为流处理技术的事实标准。

经过几年的飞速发展和社区的共同努力，Flink通过了众多严苛场景的检验，架构逐渐趋于稳定。本书的出版也可以说是恰逢其时，读者可以从中了解到Flink何以在流处理领域"独步天下"。本书从源码出发，对核心技术进行讲解，有助于读者更好地理解Flink，理解开源技术，实现从学习者、使用者到贡献者的身份转变。

范东来
华为云MVP，《Spark海量数据处理》作者

# 序 二

Flink 自发布以来，在学术界与工业界都得到了极广泛的应用，可以说 Flink 目前已经是全球最为知名的流处理系统框架了，甚至没有之一。但是 Flink 的使用场景远不止流处理，它在批处理、图、机器学习等场景中依旧可以工作得很好。Flink 遵循 MapReduce 的设计，为用户提供了一组低级编程 API（Java 语言），同时设计了更高的抽象 Table API，支持以 Flink SQL 的方式进行数据处理。

目前，湖仓一体架构正逐步得到应用。Flink 的批流一体设计理念在湖仓一体架构下有着非常大的架构优势，基本上已经成为湖仓一体架构下计算引擎的首选，这也得益于 Flink 背后的一些关键设计思想。

- Flink 定位于流计算引擎。这并不意味着 Flink 只能做流计算，而是 Flink 用流处理的设计思想更高维度地抽象了数据处理的过程。Flink 不提供持久化能力，而是专注于提供跨数千台机器的实时数据处理能力，同时提供多种编程接口，从设计理念到 API 提供，统一了批流处理过程。这种设计理念极大程度上降低了不同场景下的数据开发成本以及多服务的维护成本，提供了更友好的架构形态。
- Flink 与大数据生态无缝集成。Flink 与 Zookeeper、HDFS、Yarn 等大数据组件完美集成，可以很好地融入大数据生态，发挥出 1+1＞2 的效果。

对于大数据生态圈或者从事大数据相关工作的从业者，Flink 是一个必学的框架，几乎所有架构下都会出现它的身影。目前 Flink 也在逐步发展，例如实现 Flink Table Store 向存储延伸。Flink 发布多年，其热度依旧不减，因此无论现在还是未来，Flink 都是一个非常值得学习的框架。

从学习难度来说，Flink 的学习门槛会比同类其他框架略高。在 Flink 的设计理念中，批是特殊的流，这种批流一体的架构直接从更高维度抽象了数据处理的理念，比单纯的批处理框架或流处理框架有着更高的理解门槛。Flink 的部署也比其他框架更为复杂，除了自身可以构建一套独立的集群，Flink 也可以与 Yarn 集成。但是 Flink 在 Yarn 下提供了多种不同的 On Yarn 形态，需要花更多时间去研究。

伟哲做事非常认真，他会把自己在工作中遇到的一些知识进行逐一记录并反复琢磨，不断向下探索，直到掌握知识的完整原理。他将自己多年的 Flink 实践经验进行整理，汇聚成书，全面地讲解设计理念、数据传输、计算模型等几乎所有的 Flink 知识模块。

本书由浅入深，从理念到实现，全面地剖析了 Flink 的原理，可以作为深入学习 Flink 整体框架的教程，供读者了解 Flink 的底层知识；也可以作为 Flink 技术参考手册，供读者在日常的数据工作中碰到"疑难杂症"时翻阅，从而解决开发问题。

白发川

字节跳动大数据资深技术专家

Thoughtworks 中国区前数据智能业务线技术负责人

# 前 言

不知不觉，我们已身处大数据时代。即便是一个在学习和工作中完全不会直接接触到此类技术的人，也会频繁听到"大数据"这个词，仿佛一个企业、一个平台或一项服务不与大数据关联，就没有跟上时代的浪潮。前些年，许多组织或公司徒有"大数据"之名，在技术上仍十分守旧，没有任何创新之实。不过，随着近几年大数据技术普及与传统企业向智能化转型，越来越多的企业都已经具备了获取、处理、存储、分析海量数据的能力。通过用户与大数据服务提供方的互动，大数据逐步渗透到人们生产、生活的各个方面。不得不说，大数据时代真的到来了。

对中小企业来说，它们不具备自研框架的能力，因此更愿意拥抱开源，Apache Hadoop 框架及其生态圈的各个框架往往会成为其首选。对大型企业来说，在开发数据工具（计算引擎、存储引擎等）时，也一定会参考这些开源框架的设计思想和代码实现，因为这些框架基本都是从早期研究大数据相关问题的论文中催生的，为后来的框架提供了范式。这些开源框架不仅可以直接运用在真实的生产环境中，解决大数据场景中的一般问题，也可以为个人的学习提供友好的入口。

在我开始参与大数据相关的开发时，Hadoop 已成为一种十分经典（甚至给人感觉有点过时）的批处理框架，其在批处理领域几乎已经成为行业标准。那时 Spark 正处于"新锐巅峰"，不仅在批处理领域能提供比 Hadoop 更高的性能，在流处理领域也有一定的建树，逐步呈现出"取代" Hadoop 的势头。Flink 在那时作为"新一代流处理计算引擎"进入我们团队的视野，经过快速研究和验证，我们使用它作为 Lambda 架构中流处理的一部分，并在生产环境中进行实践。我们在受益于 Flink 流处理的速度的同时，也对其状态存储机制、时间窗口机制、反压机制等技术细节感到困惑。我们对这几个框架的未来发展保持着期待。

后来，由于工作需要和个人兴趣，我对 Flink 等框架的源码实现进行了更加细致的研究。在了解了更多企业的技术选型后，我发现 Hadoop 不但没有被后起之秀淘汰，反而因其稳定性而成为企业处理海量数据的"底牌"；Spark 则因其良好的性能和成功的商业化发展，被更加广泛地应用在生产环境的批处理场景中。由此，关于 Hadoop 和 Spark 的从入门介绍、最佳实践到源码解析的图书层出不穷，给开发人员带来极大的便利。然而，介绍 Flink 的图书相对较少，源码解析类的图书则几乎没有。这主要是因为流处理场景比批处理场景复杂得多，而且 Hadoop 和 Spark 本身就比 Flink 更早进入应用领域，研究和应用 Flink 的团队自然就更少。不过，在批流一体化的方向上，Flink 比 Spark 具有更加先进的设计思想。随着人们对实时性要求的提高，更好的解决方案是将对 Spark 等框架已有的优化移植到 Flink 上，而不是为 Spark 等批处理框架添加更多流处理的特性。实际上，国内许多互联网企业和大数据公司都对流处理服务有广泛的需求，甚至已经开始技术转型。阿里巴巴收购 Flink 背后的商业公司 Data Artisans 这个事件也许就暗示了未来技术的发展方向。

可以预见，以 Flink 为代表的批流一体化的计算引擎会在未来占据更加重要的位置。

在这样的背景下，我想将自己对 Flink 的学习和研究成果成体系地与大家分享。这些内容较为深入，涉及代码的运行流程、底层的数据结构、类与类的依赖关系等，体现了对源码实现细节和相关论文的分析与思考。希望本书对同样热爱技术的同行来说是一本实用之书。

## 本书特色

本书基于 Apache Flink 发行版 1.10.2，从面向开发人员的 API 的使用方式入手，逐步对 Flink 的源码展开分析。本书主要有以下特点。

首先，本书对源码的实现细节进行详细的分析，不仅在逻辑层面展示核心组件之间的交互方式，而且从代码层面揭示它们之间的关系。通过对细节的掌握，读者可以对 Flink 的运行流程、设计思想有更加深刻的认识。

其次，本书内容层次分明、循序渐进，从用户熟悉的 API 开始介绍，基于业务代码的运行流程逐步深入讲解执行图生成、任务调度、数据传输等内容。对于不同的模块，侧重点有所不同。如果想要系统地、全面地了解一个框架，可以将本书的学习路径当作一种参考范例。

最后，本书基于一些社区的讨论和论文，提供了几种 Flink 的特性优化方案。这些优化方案抽象自真实的生产场景，用于解决十分复杂的业务问题。

## 面向读者

本书从基础知识开始讲解，内容由浅入深、循序渐进，特别适合对技术执着和有热情的人阅读。本书的读者对象包括：

- 想要学习 Flink 设计原理的应用开发工程师；
- 希望对 Flink 进行定制化开发的平台开发工程师；
- 需要进行架构设计、技术选型的架构师和项目经理；
- 熟悉其他大数据计算引擎，想要进一步学习流处理计算框架的工程师；
- 计算机相关专业对 Flink 技术感兴趣的学生。

## 阅读方法

本书不仅会介绍 Flink 核心 API 的使用方式，而且会对核心流程与重要组件的源码实现细节进行详细的分析，建议读者在阅读本书的同时运行开发工具，对相关代码进行调试，这样将事半功倍。

本书包含两个部分和两个电子附录。

第一部分为第 1～10 章。第 1 章是序篇，对 Flink 的历史、应用场景、架构等进行总体介绍。第 2 章介绍 Flink 的应用，主要包含核心 API 的使用方式。第 3～10 章分模块介绍 Flink 的源码实现及其设计思想，主要包括执行图生成、任务调度和执行、数据传输、时间与窗口、状态与容错等。

第二部分为第 11～14 章。这些章讲解的是针对 Flink 核心功能的特性开发。在阅读这些章时，可以回顾第一部分的相关内容，这样更能加深理解。这些增强的特性均可运用在生产环境中，相信读者可以从中得到启发，解决棘手的技术难题。

本书还提供两个电子附录，分别介绍 Flink 中的资源管理和类型系统。

## 致谢

感谢我的良师益友范东来先生，从我初入职场至今，是你给了我最多、最有益的帮助。在写作本书的过程中，你多次给予我技术、写作方面的建议，由衷感谢你无私的经验分享，从你的身上我不仅学到了对技术的执着，也学到了待人的真诚。

感谢人民邮电出版社杨海玲编辑与刘雅思编辑在本书出版过程中给予我的信任与帮助，是你们在选题、审阅、排版等工作上的辛勤付出，才使本书得以顺利出版。

感谢我的好友贺鹏数次帮助我解决技术上的难题，与你交流时总有拨云见日之感，是你让我感受到了技术人的纯粹。

最后，感谢我的父母一直以来对我的支持、关怀与宽容，你们是最棒的。感谢代依珊在我写作过程中给我的陪伴与鼓励，你就像一束温暖的光，让我在漫长的写作过程中倍感温暖。

# 资源与支持

本书由异步社区出品，社区（https://www.epubit.com/）为您提供相关资源和后续服务。

## 配套资源

本书提供两个电子附录。要获得相关配套资源，请在异步社区本书页面中点击 ，跳转到下载界面，按提示进行操作即可。注意：为保证购书读者的权益，该操作会给出相关提示，要求输入提取码进行验证。

## 提交勘误

作者和编辑尽最大努力来确保书中内容的准确性，但难免会存在疏漏。欢迎您将发现的问题反馈给我们，帮助我们提升图书的质量。

当您发现错误时，请登录异步社区，按书名搜索，进入本书页面，点击"提交勘误"，输入勘误信息，点击"提交"按钮即可。本书的作者和编辑会对您提交的勘误信息进行审核，确认并接受您的建议后，您将获赠异步社区的 100 积分。积分可用于在异步社区兑换优惠券、样书或奖品。

## 扫码关注本书

扫描下方二维码，您将会在异步社区微信服务号中看到本书信息及相关的服务提示。

## 与我们联系

我们的联系邮箱是 contact@epubit.com.cn。

如果您对本书有任何疑问或建议，请您发邮件给我们，并请在邮件标题中注明本书书名，以便我们更高效地做出反馈。

如果您有兴趣出版图书、录制教学视频，或者参与图书技术审校等工作，可以发邮件给本书的责任编辑（liuyasi@ptpress.com.cn）。

如果您来自学校、培训机构或企业，想批量购买本书或异步社区出版的其他图书，也可以发邮件给我们。

如果您在网上发现有针对异步社区出品图书的各种形式的盗版行为，包括对图书全部或部分内容的非授权传播，请您将怀疑有侵权行为的链接通过邮件发给我们。您的这一举动是对作者权益的保护，也是我们持续为您提供有价值的内容的动力之源。

## 关于异步社区和异步图书

"异步社区"是人民邮电出版社旗下IT专业图书社区，致力于出版精品IT图书和相关学习产品，为作译者提供优质出版服务。异步社区创办于2015年8月，提供大量精品IT图书和电子书，以及高品质技术文章和视频课程。更多详情请访问异步社区官网 https://www.epubit.com。

"异步图书"是由异步社区编辑团队策划出版的精品IT专业图书的品牌，依托于人民邮电出版社的计算机图书出版积累和专业编辑团队，相关图书在封面上印有异步图书的LOGO。异步图书的出版领域包括软件开发、大数据、AI、测试、前端和网络技术等。

异步社区

微信服务号

# 目 录

## 第一部分 设计思想篇

### 第1章 序篇 ... 3
- 1.1 Flink 的诞生与发展 ... 3
  - 1.1.1 Stratosphere 项目 ... 3
  - 1.1.2 Apache Flink 的发展 ... 4
- 1.2 Flink 的应用场景 ... 5
  - 1.2.1 事件驱动型应用 ... 5
  - 1.2.2 数据分析型应用 ... 5
  - 1.2.3 数据管道型应用 ... 9
- 1.3 Flink 的核心特性与架构 ... 9
  - 1.3.1 核心特性 ... 9
  - 1.3.2 架构 ... 10
- 1.4 准备工作 ... 11
- 1.5 总结 ... 12

### 第2章 Flink 编程 ... 13
- 2.1 API 层级 ... 13
- 2.2 DataStream API ... 14
  - 2.2.1 DataStream 版本的 WordCount ... 14
  - 2.2.2 数据源 ... 16
  - 2.2.3 数据的转换操作 ... 16
  - 2.2.4 数据的输出 ... 20
  - 2.2.5 重分区 ... 21
- 2.3 DataSet API ... 21
  - 2.3.1 DataSet 版本的 WordCount ... 21
  - 2.3.2 数据源 ... 22
  - 2.3.3 数据的转换操作 ... 22
  - 2.3.4 数据的输出 ... 24
  - 2.3.5 重分区 ... 25
- 2.4 Table API ... 25
  - 2.4.1 Table API 版本的 WordCount ... 25
  - 2.4.2 初始化执行环境 ... 26
  - 2.4.3 获取 Table 对象 ... 28
  - 2.4.4 Table API 中的转换操作及输出 ... 28
- 2.5 SQL ... 34
- 2.6 总结 ... 34

### 第3章 Flink API 层的实现原理 ... 36
- 3.1 DataStream API ... 37
  - 3.1.1 StreamExecutionEnvironment 执行环境 ... 37
  - 3.1.2 Function 接口分析 ... 42
  - 3.1.3 StreamOperator 算子分析 ... 45
  - 3.1.4 转换操作分析 ... 48
  - 3.1.5 数据流相关类分析 ... 53
- 3.2 DataSet API ... 59
  - 3.2.1 ExecutionEnvironment 执行环境 ... 59
  - 3.2.2 InputFormat 和 OutputFormat ... 62
  - 3.2.3 数据集相关类分析 ... 63
- 3.3 Table API 和 SQL ... 68
- 3.4 总结 ... 71

### 第4章 Flink 的执行图 ... 72
- 4.1 StreamGraph 的生成 ... 73
  - 4.1.1 StreamGraphGenerator 分析 ... 73
  - 4.1.2 StreamGraph 分析 ... 77

4.1.3 StreamNode 和 StreamEdge ...... 80
4.2 Plan 的生成 ...... 81
　　4.2.1 OperatorTranslation 分析 ...... 82
　　4.2.2 Plan 分析 ...... 84
4.3 从 StreamGraph 到 JobGraph ...... 85
　　4.3.1 StreamingJobGraphGenerator 分析 ...... 87
　　4.3.2 JobGraph 分析 ...... 93
　　4.3.3 JobVertex、JobEdge 和 IntermediateDataSet ...... 94
4.4 从 Plan 到 JobGraph ...... 95
4.5 从 JobGraph 到 ExecutionGraph ...... 96
　　4.5.1 ExecutionGraphBuilder 分析 ...... 98
　　4.5.2 ExecutionGraph 分析 ...... 99
　　4.5.3 ExecutionJobVertex、ExecutionVertex 和 Execution 分析 ...... 102
　　4.5.4 IntermediateResult、IntermediateResultPartition 和 ExecutionEdge ...... 106
4.6 总结 ...... 108

## 第 5 章 Flink 的运行时架构 ...... 109
5.1 客户端代码的运行 ...... 110
5.2 高可用相关组件 ...... 115
　　5.2.1 EmbeddedHaServices ...... 115
　　5.2.2 EmbeddedLeaderService ...... 117
5.3 派发器的初始化与启动 ...... 122
5.4 资源管理器的初始化与启动 ...... 128
5.5 TaskExecutor 的初始化与启动 ...... 131
5.6 JobMaster 的初始化与启动 ...... 134
5.7 总结 ...... 137

## 第 6 章 任务调度 ...... 138
6.1 调度器 ...... 138
　　6.1.1 调度器的基本构成与初始化 ...... 139
　　6.1.2 构造 ExecutionGraph ...... 142
6.2 调度拓扑 ...... 143
6.3 调度策略 ...... 147
　　6.3.1 EagerSchedulingStrategy ...... 147
　　6.3.2 LazyFromSourcesSchedulingStrategy ...... 149
　　6.3.3 InputDependencyConstraintChecker ...... 152
6.4 调度过程的实现 ...... 157
　　6.4.1 开始调度 ...... 157
　　6.4.2 更新任务状态 ...... 159
　　6.4.3 调度或更新消费者 ...... 163
6.5 任务的部署 ...... 163
6.6 Execution 对象在调度过程中的行为 ...... 166
6.7 总结 ...... 173

## 第 7 章 任务的生命周期 ...... 174
7.1 任务的提交 ...... 174
　　7.1.1 TaskDeploymentDescriptor ...... 176
　　7.1.2 ResultPartitionDeploymentDescriptor ...... 178
　　7.1.3 InputGateDeploymentDescriptor ...... 180
　　7.1.4 ShuffleDescriptor ...... 181
　　7.1.5 ProducerDescriptor 和 PartitionDescriptor ...... 185
　　7.1.6 TaskDeploymentDescriptor 的提交 ...... 188
7.2 任务的初始化 ...... 189
　　7.2.1 Task 的初始化 ...... 189
　　7.2.2 ResultPartition 的初始化 ...... 191
　　7.2.3 InputGate 的初始化 ...... 194
7.3 任务的执行 ...... 197
　　7.3.1 StreamTask 的初始化 ...... 202
　　7.3.2 StreamTask 中的重要概念 ...... 204

7.3.3 StreamTask 的实现类 ...... 219
7.3.4 StreamTask 的生命周期 ...... 222
7.3.5 DataSourceTask、BatchTask 和 DataSinkTask ...... 227
7.4 总结 ...... 237

## 第 8 章 数据传输 ...... 238
8.1 基本概念与设计思想 ...... 238
　8.1.1 从逻辑执行图到物理执行图 ...... 239
　8.1.2 用同一套模型应对批处理和流处理 ...... 242
　8.1.3 混洗 ...... 242
　8.1.4 流量控制 ...... 245
8.2 数据的输出 ...... 252
　8.2.1 ResultPartitionType ...... 253
　8.2.2 ResultPartitionWriter ...... 256
　8.2.3 ResultSubpartition ...... 262
8.3 数据的读取 ...... 265
　8.3.1 ResultSubpartitionView ...... 266
　8.3.2 InputGate ...... 269
　8.3.3 InputChannel ...... 273
8.4 反压机制的原理 ...... 278
8.5 总结 ...... 283

## 第 9 章 时间与窗口 ...... 284
9.1 基本概念和设计思想 ...... 284
　9.1.1 从批处理到流处理 ...... 284
　9.1.2 数据流模型的设计思想 ...... 287
　9.1.3 Flink 中与窗口操作相关的核心概念 ...... 289
9.2 WindowedStream ...... 290
9.3 窗口相关模型的实现 ...... 292
　9.3.1 Window 类 ...... 292
　9.3.2 WindowAssigner 类 ...... 293
　9.3.3 Trigger 类 ...... 294
　9.3.4 Evictor 类 ...... 296

9.4 WindowOperator ...... 297
9.5 水位线 ...... 299
　9.5.1 产生水位线 ...... 300
　9.5.2 多个数据流传来的水位 ...... 303
9.6 定时器 ...... 304
9.7 总结 ...... 307

## 第 10 章 状态与容错 ...... 308
10.1 基本概念与设计思想 ...... 308
　10.1.1 状态与容错的基本概念 ...... 308
　10.1.2 Hadoop 与 Spark 如何设计容错机制 ...... 311
　10.1.3 Flink 中容错机制的设计思想 ...... 311
　10.1.4 Flink 的状态与容错机制的核心概念 ...... 313
10.2 状态存储 ...... 315
　10.2.1 检查点的触发 ...... 316
　10.2.2 栅栏的传输 ...... 323
　10.2.3 状态数据的更新和存储 ...... 331
　10.2.4 元信息的存储 ...... 336
10.3 状态恢复 ...... 341
　10.3.1 元信息的读取 ...... 342
　10.3.2 状态的重分配 ...... 344
　10.3.3 状态数据的恢复 ...... 347
10.4 状态的重分配策略 ...... 349
　10.4.1 操作符状态的重分配 ...... 350
　10.4.2 键控状态的重分配 ...... 352
10.5 总结 ...... 353

# 第二部分　特性开发篇

## 第 11 章 动态调整并行度 ...... 357
11.1 模型设计 ...... 357
　11.1.1 传统模型的局限 ...... 357
　11.1.2 DS2 模型的核心概念 ...... 358
　11.1.3 算法原理 ...... 359
　11.1.4 架构设计 ...... 360

11.1.5 使用 DS2 模型的注意事项 ……… 361
11.2 指标收集 ……………………… 361
11.3 指标管理 ……………………… 364
11.4 总结 ………………………… 366

## 第 12 章 自适应查询执行 ………… 367
12.1 Flink 框架下的自适应查询执行 …………………… 368
   12.1.1 执行阶段的划分 ……… 368
   12.1.2 优化流程 ……………… 368
   12.1.3 优化策略 ……………… 370
12.2 统计信息的收集 ……………… 373
12.3 执行图与调度拓扑的修改 …… 374
12.4 上下游关系的建立 …………… 377
12.5 总结 ………………………… 378

## 第 13 章 Flink Sort-Merge Shuffle ……… 379
13.1 混洗机制的对比 ……………… 379
13.2 Flink 混洗机制 ……………… 381
13.3 Blink 混洗的数据流转 ……… 382
   13.3.1 ExternalResultPartition ……… 383
   13.3.2 PartitionMergeFileWriter … 384
13.4 Blink 混洗的 Sort-Merge 过程 … 386

13.4.1 PushedUnilateralSortMerger ……………… 387
13.4.2 NormalizedKeySorter …… 390
13.4.3 排序线程 ……………… 393
13.4.4 溢写线程 ……………… 393
13.4.5 合并线程 ……………… 395
13.5 文件的读取和元信息管理 …… 398
   13.5.1 ExternalBlockResultPartitionManager …………… 398
   13.5.2 ExternalBlockResultPartitionMeta ………………… 399
   13.5.3 ExternalBlockSubpartitionView …………………… 400
13.6 总结 ………………………… 402

## 第 14 章 修改检查点的状态 ……… 403
14.1 状态修改的原理 ……………… 403
   14.1.1 状态元信息的读取 …… 404
   14.1.2 状态数据的读取 ……… 405
14.2 状态处理器 API ……………… 407
   14.2.1 数据的读取 …………… 409
   14.2.2 数据的写出 …………… 413
14.3 总结 ………………………… 414

# 第一部分　设计思想篇

# 第 1 章
# 序篇

目前在生产环境中可供选择的大数据分布式计算框架层出不穷,其中较为流行、稳定且性能最良好的开源框架包括 MapReduce 和 Spark,它们几乎分别成为批量计算和微批计算的行业标准。而且 Spark 经过多年的发展,在流式计算领域也有着优异的表现。

虽然 Spark 在多种计算场景下有着不俗的性能表现,但是由于它早期是用于批量计算的,因此在底层架构、数据抽象等方面仍不可避免地保留了许多批量计算的概念。Flink 的诞生并不比 Spark 晚,不过由于它在设计之初就真正将数据当作数据流而非数据集来处理,因此可以说在流式计算方面,Flink 拥有比 Spark 更加先进的计算模型。此外,Flink 以流处理为基础,对批处理也有很好的支持。许多人说"Flink 是(继 Spark 之后的)下一代大数据处理引擎",这并非噱头。本书会从 Flink 的计算封装逻辑、执行图的生成、数据的交换、状态容错、窗口计算等各方面对这一说法进行全面且深入的探讨。

本章旨在让读者对 Flink 有总体的了解,以为接下来的学习打好基础。

希望在学习完本章后,读者能够了解:

- Flink 的发展历史;
- Flink 支持的应用场景;
- Flink 的核心特性;
- Flink 的架构;
- 实现 Flink 源码的基本技巧。

## 1.1 Flink 的诞生与发展

Flink 最早是德国一些大学中的研究项目,经过多年的发展,其已在现实的生产场景中得到了越来越多的应用。近些年,由于社区的推动和商业上的支持,Flink 在流式计算领域相比其他大数据分布式计算引擎有着明显的优势,并且在批量计算、批流一体化的发展上有着令人期待的前景。

### 1.1.1 Stratosphere 项目

在 Flink 官网,有一栏为 Flink Blog(Flink 博客),其中会定期发布一些文章来记录 Flink 发展过程中的重大事件或介绍新引入的重要概念等。Flink Blog 中的第一篇文章宣布了 Flink 0.6 的发布。在 Flink 官网中可供下载的第一个版本是 0.6 版本，因为在 0.5 版本及之前的版本中，项目名称为 Stratosphere，即 Flink 的前身。

Stratosphere 项目起源于德国柏林工业大学（Technische Universität Berlin）Volker Markl 教授于 2008 年提出的构想。由于数据库是 Volker Markl 教授的主要研究方向之一，因此创建该项目的初衷是构建一个以数据库概念为基础、以大规模并行处理（massively parallel processing，MPP）架构为支撑、以 MapReduce 计算模型为逻辑框架的分布式数据计算引擎。在此构想之上，在该项目中还专门引入了流处理，为后来 Flink 的发展"添加"了良好的"基因"。

Volker Markl 教授联络德国柏林工业大学、德国柏林洪堡大学（Humboldt–Universität zu Berlin）和德国哈索·普拉特纳研究所（Hasso Plattner Institute）的多名科研人员，共同开始 Stratosphere 项目的研发。在 2010 年前后，第一版 Stratosphere 以开源的形式发布。在获得初步的应用和一定范围的关注后，该项目组在 2010 年至 2014 年又持续改进并陆续发布了多个版本。从此期间项目组发表的论文可以观察到，该项目已经具备了后来的 Flink 的雏形，可以看到如 JobGraph、ExecutionGraph 等执行图的概念以及作业管理器（JobManager）、任务管理器（TaskManager）等组件的架构设计。

随着知名度的提高，Stratosphere 项目遇到了命名的困扰。项目组成员发现 Stratosphere 这个名字早已由一家商业实体注册，他们不得不对项目重新命名。最终，在 2014 年申请成为 Apache 软件基金会的孵化器项目后，经过项目组成员投票，项目正式更名为 Flink。Flink 在德语中意为"敏捷、快速"。同时，项目组决定使用松鼠形象作为商标，也是为了强调"敏捷、快速"的特性（如图 1-1 所示）。

图 1-1　Flink 商标

### 1.1.2　Apache Flink 的发展

Flink 自从加入 Apache 后发展十分迅猛。自 2014 年 8 月发布 0.6 版本后，Flink 仅用了 3 个月左右的时间，在 2014 年 11 月就发布了 0.7 版本，该版本包含 Flink 目前为止最重要的特性之一——Flink Streaming。Flink 于 2014 年年底顺利从孵化器"毕业"，成为 Apache 顶级项目。随后，Flink 逐步添加了在今天看来都属于其核心功能的特性，如一致性语义、事件时间和 Table API 等。

随着 Flink 受到社区越来越多的关注，其功能和稳定性也不断得到完善。一方面是因为它的功能特性受到了商学两界的广泛认可；另一方面也是因为要应对其他已经商业化的大数据计算引擎的竞争，越来越多的公司开始将 Flink 应用在它们真实的现网环境中，并在技术和商业上共同推动 Flink 的发展。我国很多公司都已经大规模使用 Flink 作为其分布式计算场景的解决方案，如阿里巴巴、华为、小米等。其中，阿里巴巴已经基于 Flink 实时计算平台实现了对淘宝、天猫、支付宝等的数据业务的支持。

早期 Stratosphere 项目的核心成员曾共同创办了一家名叫 Data Artisans 的公司，其多年来一直致力于 Flink 的技术发展和商业化。2019 年，阿里巴巴收购了 Data Artisans 公司，并将其开发的分支 Blink 开源。相信在未来的发展中，凭借阿里巴巴强大的技术储备和商业支持，以及庞大的数据量和丰富的业务场景，Flink 的发展一定会迎来新的机遇。

## 1.2 Flink 的应用场景

Flink 的应用场景十分广泛，下面介绍 3 种常见的应用。

### 1.2.1 事件驱动型应用

在许多场景中，需要处理的数据往往来自事件。小到一些交互式的用户行为，大到一些复杂的业务操作，它们都会被转化成一条条数据，进而形成数据流（事件流）。事件驱动型应用的特点在于，一些事件会触发特定的计算或其他外部行为，其典型场景有异常检测、反欺诈等。

在传统架构下，数据流通常会流入数据库，随后应用系统读取数据库中的数据，根据数据触发计算。在这种架构下，数据和计算分离，而且在存取数据时需要进行远程访问。与传统架构不同，Flink 利用有状态的数据流来完成整个过程的处理，无须将数据先存入数据库再读取出来。数据流的状态数据在本地（local）维护，并且会周期性地持久化以实现容错。图 1-2 展示了传统事务型应用架构与 Flink 事件驱动型应用架构的区别。

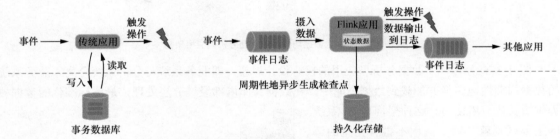

图 1-2 传统事务型应用架构与 Flink 事件驱动型应用架构的区别

Flink 事件驱动型应用架构的优势在于，它的状态数据在本地维护，不需要远程访问数据库，由此获得了更低的延迟、更高的吞吐量。同时，不像传统架构那样多个应用共享同一个数据库，任何对数据库的修改都需要谨慎协调，Flink 事件驱动型应用架构中的每一个应用都独立地维护状态，可以灵活地进行扩/缩容。

从上面的介绍可以了解到，实现事件驱动型应用的关键在于支持"有状态的数据流"及容错机制。这是 Flink 最优秀的设计之一。这部分内容会在后文详细分析。

### 1.2.2 数据分析型应用

从历史发展的角度来看，企业要处理的数据量是由小到大变化的，因此不妨从传统企业的角度来看待数据分析型应用的演变。

过去，传统企业的数据分析型应用往往就是商务智能（business intelligence，BI）系统。一个成熟的 BI 产品是一套集数据清洗、数据分析、数据挖掘、报表展示等功能于一体的完整解决方案。不过，当数据量过大时传统的 BI 系统会出现性能瓶颈，而且它的底层是基于关系数据库的，处理非结构化数据时会十分乏力。因此，当今企业在进行技术选型和架构设计时，更倾向于选择 Hadoop 生态系统组件及其相关架构。

早期大数据场景下的数据分析型应用架构如图 1-3 所示。

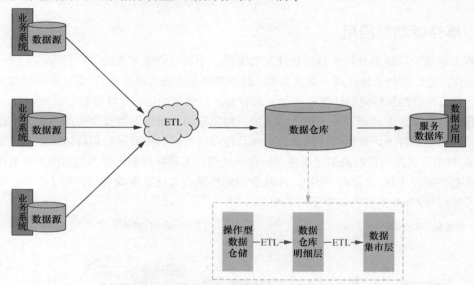

图 1-3 早期大数据场景下的数据分析型应用架构

图 1-3 充分体现了数据分析型应用的核心设计思想，即业务系统与分析系统分离。业务系统的数据周期性地转换并加载到数据仓库中，在数据仓库内部经过分层处理，最终标准化的数据被提供给其他应用使用。这种架构与 BI 系统的主要区别就是整个流程不再有完整的解决方案，而需要技术人员自己选择工具进行开发和组合。

从传统的 BI 系统到早期大数据场景下的数据分析型应用架构，始终存在着一个问题，那就是整个过程中所有的抽取、转换、加载（Extract-Transform-Load，ETL）逻辑都是离线进行的，导致整个分析流程具有较高的延迟。由此，流式架构便应运而生。

流式架构的目的是在不丢失数据的前提下保证整个分析流程的低延迟，如图 1-4 所示。

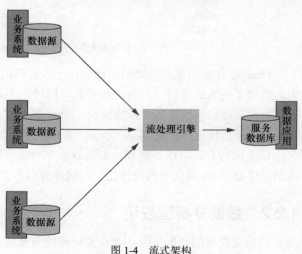

图 1-4 流式架构

图 1-4 所示的整个流程少了 ETL，直接将数据摄入流处理引擎，经过业务处理后输出给其他应用使用。在早期的技术储备条件下，想要保证低延迟，通常就难以保证结果的准确性，因此流式架构仅适用于那些对数据准确性要求不高，而对数据实时性要求极高的场景。

那么，在早期的技术储备条件下，能否通过架构的演进，既保证数据的实时性，又兼顾数据的准确性呢？开源框架 Storm 的创始人 Nathan Marz 提出了 Lambda 架构，有效地解决了这一问题。

Lambda 架构的核心思想是实时处理和离线处理共存，实时处理如流处理一般保证数据的实时性，离线处理通过周期性地合并数据来保证数据的最终一致性。Lambda 架构如图 1-5 所示。

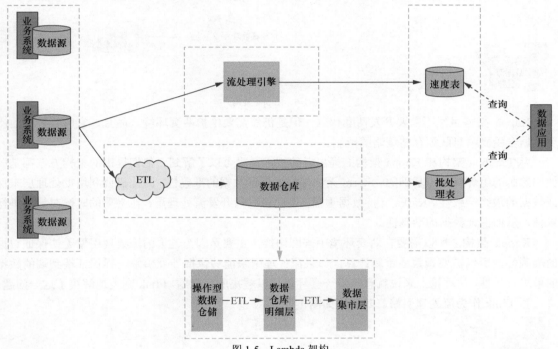

图 1-5　Lambda 架构

在 Lambda 架构下，批处理层将准确结果写入批处理表，流处理层则将数据实时地写入速度表，批处理表的结果会定期与速度表中的数据合并以保证其准确性。数据应用则根据需求进行查询。

显而易见，虽然 Lambda 架构在一定程度上同时保证了数据的准确性与实时性，但它需要开发和维护两套系统，这实在是一笔不小的开销。由此，Kafka 的核心成员之一 Jay Kreps 在 Lambda 架构的基础上提出了 Kappa 架构，解决了"两套代码实现一套业务逻辑"的问题。Kappa 架构舍弃了批处理层，只保留了流处理层。与流式架构不同的是，Kappa 架构需要让业务数据先进入支持数据重播的消息队列（如 Kafka）。如果数据出现错误，那么再执行一个流处理作业，以对历史数据进行重新计算。当新启动的作业消费到最新的数据时，让外部应用访问新的服务数据库，完成服务的切换。Kappa 架构如图 1-6 所示。

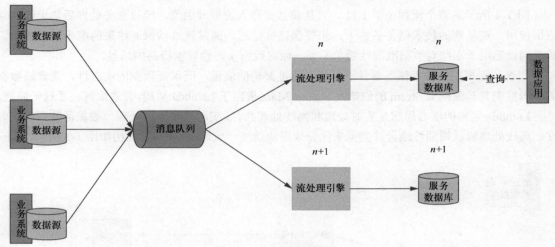

图 1-6 Kappa 架构

Kappa 架构虽然不需要开发两套代码，但是仍然需要维护两套环境。而且，它所能处理的历史数据会受到消息队列存储策略的限制。

从 Lambda 架构和 Kappa 架构的提出者的技术背景可以了解到，他们提出的架构方案都是以他们熟悉的组件特性为基础的。Storm 无法很好地保证数据的准确性，因此需要利用批处理层来保证数据的最终一致性。Kafka 支持数据重播，因此可以只开发流处理层，在必要的时候对数据进行重播，从而保证数据的准确性。

Kappa 架构之所以需要在两套环境中来回切换，主要是因为过去的流处理引擎无法保证数据的准确性，所以需要频繁地重新计算。如果流处理引擎能够像批处理引擎一样保证端到端的数据的最终一致性，从理论上来说就意味着一套环境可以解决所有问题。Flink 完美地解决了这一问题。

以 Flink 作为流处理引擎，其架构如图 1-7 所示。

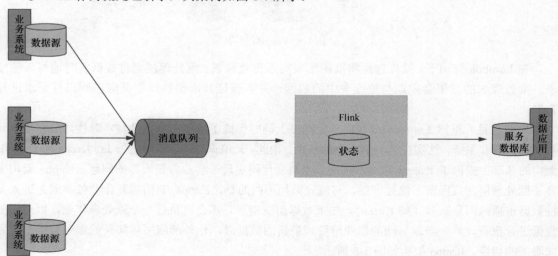

图 1-7 Flink 流式分析架构

Flink 内部维护了数据流的状态,并以容错机制、窗口机制等特性作为支持,可以保证精确地实现端到端的数据的最终一致性。同时,Flink 提供了从 SQL 到底层 API 的多层接口,使分析工作变得十分容易。因为 Flink 本身也能够进行批处理,所以 Flink 流式分析架构可以很容易地被转换成批处理架构。

对 Kappa 架构来说,可以直接选用 Flink 作为其中的流处理引擎,但此时设计两套环境的主要目的不再是保证数据的准确性,而是当 Flink 业务代码发生变动时可以执行新的作业,待数据消费到相同位置时及时完成服务的切换。

### 1.2.3 数据管道型应用

数据管道型应用也常常作为传统 ETL 流程的替代流程,与传统的 ETL 流程相比,其优势在于实时性高。Flink 以流式的方式处理数据,无须像传统 ETL 流程一样进行周期性的离线处理。

数据分析型应用实际上包含数据管道型应用,与数据分析型应用不同的是,数据管道型应用的侧重点在于数据的流转。在数据管道型应用中,数据可能仅仅是从一个消息队列流转到另一个消息队列。

## 1.3 Flink 的核心特性与架构

前文介绍了 Flink 的应用场景,我们已经了解到,正是由于 Flink 拥有一些特性,某些应用和数据架构的实现才成为可能。本节将简单介绍 Flink 的核心特性和架构。

### 1.3.1 核心特性

了解 Flink 的核心特性有助于阅读源码时把握重点。Flink 包括以下核心特性。
- 批流一体化。Flink 可以在底层用同样的数据抽象和计算模型来进行批处理和流处理。事实上,Flink 在设计理念上没有刻意强调批处理或流处理,而更多地强调数据的有界或无界。这个特性在企业技术选型中具有举足轻重的作用,因为这意味着如果 Flink 能够满足业务需求,就无须用两种甚至多种框架分别实现批处理和流处理,这大大降低了架构设计、开发、运维的复杂度,可以节省大量人力成本。
- 支持有状态计算。从产品的角度来看,Flink 最大的"卖点"就是它支持有状态计算,这是实现前文介绍的事件驱动型应用、数据分析型应用等的基础。正如 Flink 官网首页上介绍的那样——Flink 是数据流上的有状态计算。
- 提供多种时间语义。在流处理中,数据到达 Flink 系统的顺序很可能与事件本身发生的顺序不同,这就是流处理中常见的数据乱序现象。针对这个问题,Flink 中区分了事件时间和处理时间:前者表示事件发生的时间,一般从数据自带的时间戳字段提取;后者表示数据被 Flink 处理的系统时间。当 Flink 选用事件时间对数据进行处理时,可以对数据进行排序等操作,从而得到准确的结果;当 Flink 选用处理时间对数据进行处理时,虽然不一定能得到准确的结果,但可以满足低延迟需求。多种时间语义使 Flink 可以在不同的需求实现间达到平衡。
- 轻量级分布式快照。既然 Flink 支持有状态计算,那么同时提供对状态的持久化功能就能实现容错机制。Flink 提供了检查点(checkpoint)机制和相关组件,可实现状态存储与恢复,其最大的特点是,存储状态的操作过程是十分轻量的分布式过程。
- 支持多种一致性语义。Flink 可以精确地满足系统内部的"至少一次"(at least once)语义和"恰好一次"(exactly once)语义。在外部系统的配合下,Flink 也可以比较容易地实现端到端的"恰好一次"行为。
- 多层级 API。Flink 为用户提供了多个层级的 API,用户可以根据自身对于表达力和易用性的需求来选择。不同层级的 API 可以混用,以实现复杂的业务逻辑。

Flink 的特性远不止上面介绍的这些,还包括丰富的连接器、多平台部署等。但主要是上面介绍的这些特性让 Flink 实现了相对于其他框架的差异化,并深刻地影响了 Flink 未来的发展方向。

### 1.3.2 架构

图 1-8 所示为 Flink 对应各层级结构的组件架构。

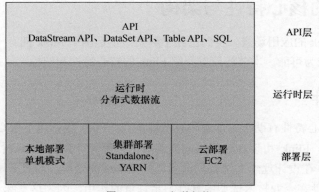

图 1-8　Flink 组件架构

如果想利用 Flink 进行业务开发,那么将重点放在 API 层即可。要想学习 Flink 的底层原理,对源码进行定制化开发,则必须深入学习 API 层下面的层级。

图 1-8 中,最下层为部署层,Flink 提供了多种部署模式,在不同的部署模式之上提供的是相同的运行时架构。

运行时层可以说是 Flink 组件架构中最重要的一层,大部分概念和核心操作定义都在这一层,包括执行图的生成、作业的调度与部署、数据的处理和交换等。这里先对运行时架构和其中的基本概念进行简单介绍,本书后文会对运行时层的各个环节展开分析。

与大部分分布式架构一样,Flink 采用的也是"主从架构"。其中,"主"指作业管理器,负责执行图的生成、作业的调度与部署等;"从"指任务管理器,负责任务的执行、数据的交换等。运行时架构如图 1-9 所示。

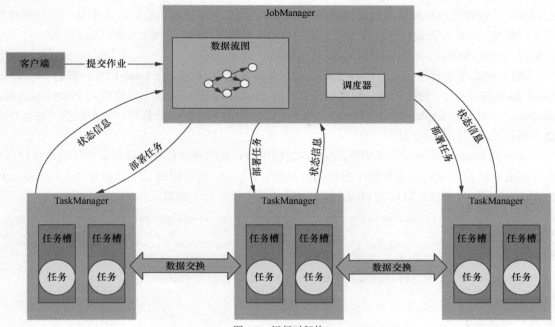

图 1-9 运行时架构

在该架构中，客户端（Client）负责向作业管理器（JobManager）提交作业，从而生成执行图；JobManager 负责任务（Task）的调度和部署，与任务管理器（TaskManager）通信，将任务派发到 TaskManager 中；TaskManager 根据资源情况将任务放在各个任务槽（TaskSlot）中执行，并向 JobManager 汇报任务状态等信息。任务执行过程中会涉及上下游任务的数据交换，这个过程发生在 TaskManager 内部或 TaskManager 之间。

图 1-9 所示的只是一个简化版的运行时架构，但它也基本涵盖 Flink 运行时的核心组件和流程。学习 Flink 源码的过程实际上就是逐步理解该架构的过程。比如，可以针对图 1-9 提出如下问题。

- 执行图是如何生成的？
- 任务是如何拆分的？
- 调度器（scheduler）是如何调度和部署这些任务的？
- TaskManager 是如何划分资源的？
- 任务是如何在 TaskManager 中执行的？
- 任务的数据是如何交换的？

JobManager 与 TaskManager 之间还存在其他交互，如有关检查点机制的流程中涉及的通信等。这些问题会在阅读后文后一一得到解决。

## 1.4 准备工作

本书不仅会对 Flink 各方面的内容进行概念性的介绍，还会更多地就源码的逻辑、设计思想等进行分析,读者可以在阅读本书的同时用调试源码的方式增强理解,也可以基于自己的理解对源码进行修改。因此,我强烈建议读者用 IntelliJ IDEA 来调试代码,并用 Git 进行版本控制,甚至可以在自己的代码仓库中不断地对 Flink 进行定制化开发。

通过 git 命令从代码仓库把 Flink 源码工程下载到本地后,切换到 release-1.10.2,执行 mvn clean install-DskipTests 命令进行构建。构建完成后,在 Flink-examples 模块下分别找到 Flink-examples-streaming、Flink-examples-batch、Flink-examples-table 中的 WordCount 程序,如果程序能够运行成功,则可以开始后面的学习。

考虑到 Flink 是分布式计算引擎,在学习过程中很可能需要将作业部署到集群环境中执行,因此可以通过远程调试对各个进程中的代码进行追踪。Flink 的远程调试主要指对 JobManager 和 TaskManager 进行调试,可以在 Flink-conf.yaml 文件中进行如下配置:

```
env.java.opts.jobmanager: "-agentlib:jdwp=transport=dt_socket,server=y,suspend=n,address=5005"
env.java.opts.taskmanager: "-agentlib:jdwp=transport=dt_socket,server=y,suspend=n,address=5006"
```

在 IntelliJ IDEA 中,可以通过相应的 IP 地址和端口号进行调试。

## 1.5 总结

本章首先从发展历史的角度介绍了 Flink 的发展历程;随后列举了几个典型的应用场景,让读者对使用 Flink 的场景有直观的感受;接着对 Flink 的核心特性与架构进行了简单介绍,主要目的是帮助读者建立学习框架;最后介绍了调试源码的工具和技巧。

第 2 章将参考大多数 Flink 入门教程的学习路径,列举在 Flink 应用开发中常用的 API,从 API 层级、编程模型等业务开发角度对 Flink 的使用进行介绍(不过不会对这部分内容讲解太多)。从第 3 章起,会从 API 层的设计思想和代码实现入手进行讲解,以逐步加深读者对源码的学习和理解。

# 第 2 章 Flink 编程

当学习一门新的编程语言时,往往会从"hello world"程序开始,而接触一套新的大数据计算框架时,则一般会从 WordCount 用例入手。可千万不要小看 WordCount,这个用例除了有简洁、易懂的优点,还包含对数据的映射处理和聚合操作,这正是 MapReduce 编程模型。在分布式架构中,要聚合不同物理节点上的数据,这意味着需要进行网络传输、数据的重分区等。可以说,如果完全了解了 WordCount 用例在分布式计算框架中的运行原理,基本上就掌握了该框架的核心设计思想。

本章从 Flink 源码工程中的 WordCount 程序入手进行讲解,"开门见山"地讲解何谓"Flink 编程"。随后会在介绍各个 API 的语义时对该用例进行简单的修改,引入新的转换操作对数据进行处理。本章会依次介绍 DataStream API、DataSet API、Table API 和 SQL。

希望在学习本章后,读者能够了解:

- Flink 编程的基本模式;
- 常用 API 的语义和用法。

## 2.1 API 层级

对 Flink 业务代码开发人员来说,非常重要的是了解如何使用 Flink 编程模型中的 API。Flink 将 API 抽象成了图 2-1 所示的层级结构。

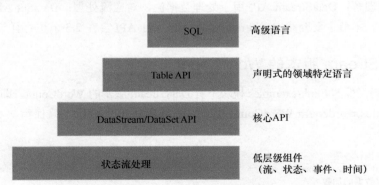

图 2-1 Flink 中的 API 层级结构

在图 2-1 所示的层级结构中,上层并不是下层的高级封装,该结构展示抽象程度和易用性程度。

底层的状态流处理 API 的抽象程度最低，而且它只能用于流处理。不过它提供了非常灵活的接口，可以用于自定义底层与状态、时间相关的操作。

DataStream/DataSet API 这一层级的 API 是 Flink 中的核心 API。这一层级中要处理的数据会被抽象成数据流（DataStream）或数据集（DataSet），然后在其上通过定义转换操作实现业务逻辑。这一层级的 API 的使用风格与 Java 8 中的 Stream 编程风格十分类似。

在 DataStream/DataSet API 之上是 Table API。Table API 和 DataStream/DataSet API 不同，不是用复杂的函数定义业务过程的，而是用陈述性的语言加以描述。这样就可大大地降低编程难度，增强描述性。这种语言来自 SQL 语法，只不过以 API 的形式呈现出来。既然有了 Table API，那么自然可以直接使用 SQL 来进行描述。这就是最上层的 SQL。

总而言之，越上层的 API，其描述性和可读性越强；越下层的 API，其灵活度越高、表达力越强。多数时候上层 API 能做到的事，下层 API 也能做到，反过来却未必。不过，这些 API 的底层模型是一致的，可以混合使用。

敏锐的读者或许可以从中得到一个推论：当我们用上层 API 开发业务代码时，在 Flink 内部会有一个将其转化为底层 API 的"翻译"过程（比如用 SQL 开发时，会有各种规则对执行计划进行优化），而这个过程不是业务开发人员可以介入的，得到的结果很有可能不是最优的。这就好比我们用高级语言进行开发，虽然开发效率得到了提高，但是系统的性能往往不如使用底层语言进行开发时那么高。这也提示了我们学习底层实现原理的必要性——从理论上来讲，要想写出性能最佳的 Flink 业务代码，就应该在理解其转换规则和运行的原理后，用底层的 API 进行开发。若有余力，则可以思考如何优化转换规则，使框架能够自动生成最佳执行计划。

一般来说，DataStream/DataSet API 及其上层 API 已经能够描述清楚整个业务场景，需要用到底层 API 的场景较少，因此下面会略过底层的 API，直接从核心 API 层开始介绍。

## 2.2 DataStream API

2.1 节提到，DataStream API 和 DataSet API 属于核心 API。在探究各种 API 的底层原理之前，可以这样简单地理解：DataStream API 用于处理数据流，对应流处理；DataSet API 用于处理数据集，对应批处理。本节主要介绍 DataStream API，DataSet API 会在 2.3 节进行详细介绍。

### 2.2.1 DataStream 版本的 WordCount

在 Flink 源码工程的 Flink-examples 模块下有 DataStream 版本的 WordCount，即 org.apache.Flink.streaming.examples.wordcount.WordCount 类。这里将其进一步简化，只保留核心的步骤，代码如下：

```
// （1）获取执行环境
StreamExecutionEnvironment env = StreamExecutionEnvironment.getExecutionEnvironment();
// （2）指定数据的加载方式
DataStream<String> text = env.fromElements(WordCountData.WORDS);
// （3）指定数据的转换方式
DataStream<Tuple2<String, Integer>> counts =
```

```
            text.flatMap(new Tokenizer())
                .keyBy(0).sum(1);
// (4) 指定如何输出
counts.print();
// (5) 触发程序运行
env.execute();
```

上述示例中的 5 个步骤几乎可代表所有 Flink 业务代码的编程模式（不仅仅是 DataStream API）。读者如果接触过 Spark 等类似 Flink 的框架，那么应该对这种编程模式并不陌生。下面对这 5 个步骤稍加解释。

（1）获取执行环境。要想编写一个流处理任务，则需要获取 StreamExecutionEnvironment 对象，并通过它调用 DataStream API。对于其他 API，也需要分别获取对应的对象作为入口。在后面介绍其他 API 时会依次说明对应的类。

（2）指定数据的加载方式。通过执行环境对象 env 指定数据的加载方式，获取 DataStream 对象。因为任何一个分布式计算框架对数据的加载、转换、输出操作都是懒执行的，所以这一步并没有真正加载数据，DataStream 对象也不会保存数据。它只是封装了加载逻辑，以便继续调用其他接口来定义数据的转换方式。

（3）指定数据的转换方式。这一步与第 2 步类似，同样是懒执行的，即仅定义数据的转换方式，并不会真正对数据进行处理。flatMap()方法与其他框架中的语义相同。Tokenizer 类表示一个分词的处理逻辑，返回一个二元组。keyBy()方法表示对数据进行重分区，按照二元组的第一位（索引为 0）来分区，随后对重分区后的数据进行聚合操作。整个逻辑与 Hadoop 或 Spark 中的 WordCount 示例的逻辑完全一致，只是方法名和参数可能有些许差别。

（4）指定如何输出。这一步可以与 Spark 的编程模式进行对比。在 Spark 中，算子被分为转换（Transformation）算子和行动（Action）算子。上述代码中的 print()看起来对应 Spark 中的行动算子，但其实两者完全不同。在 Spark 中，行动算子不仅要定义如何输出，还要肩负生成数据库可用性组（Database Availability Group，DAG）图、划分阶段（stage）、提交作业等多个任务。在 DataStream API 中，这一步仅定义了数据的输出逻辑，作业的触发执行是在第 5 步中完成的。

（5）触发程序运行。前面几步中指定了计算逻辑，这些逻辑被封装在 env 对象中，调用其 execute()方法即可完成任务的拆分、提交等工作。这个方法有一个 JobExecutionResult 类型的返回值，通过该返回值，可以获取任务在执行中返回的结果。

Tokenizer 类的代码如下：

```
public static final class Tokenizer implements FlatMapFunction<String, Tuple2<String, Integer>> {
    @Override
    public void flatMap(String value, Collector<Tuple2<String, Integer>> out) {
        String[] tokens = value.toLowerCase().split("\\W+");
        for (String token : tokens) {
            if (token.length() > 0) {
                out.collect(new Tuple2<>(token, 1));
            }
        }
    }
}
```

## 2.2.2 数据源

Flink 针对常用的数据源提供了一些现成的方法,也提供了多个接口让用户自定义数据源。大体上,数据源的读取方式分为 4 类——从文件读取、从套接字读取、从集合读取和自定义数据源。下面介绍这些分类下的方法示例。

(1) 从文件读取。

如果想从文本文件读取数据源,则可以调用 readTextFile()方法。方法签名如下:

```
public DataStreamSource<String> readTextFile(String filePath)
```

该方法是从文件读取数据的最简单的方法。参数为文本文件的路径。

如果调用更下层的方法,则可以更加灵活地读取其他格式的文件,并且可以监控是否有新的文件生成,以决定是否周期性地读取数据源。读取数据源后会对文件进行分片,并分发给后面的任务进行处理。

(2) 从套接字读取。

如果想从套接字读取数据源,则可以调用 socketTextStream()方法。方法签名如下:

```
public DataStreamSource<String> socketTextStream(String hostname, int port)
```

通过指定主机名与端口号就可以从套接字读取数据源。

(3) 从集合读取。

从集合读取数据源一般只在演示代码中存在。常用的方法是 fromElements():

```
public final <OUT> DataStreamSource<OUT> fromElements(OUT... data)
```

前文介绍的 WordCount 程序中使用的就是该方法。

(4) 自定义数据源。

自定义数据源常用的接口是 addSource()方法:

```
public <OUT> DataStreamSource<OUT> addSource(SourceFunction<OUT> function)
```

通过实现 SourceFunction 接口可以自定义数据源的读取方式。SourceFunction 接口如下:

```
public interface SourceFunction<T> extends Function, Serializable {
    void run(SourceContext<T> ctx) throws Exception;
    void cancel();
}
```

其中最主要的是实现 run()方法。上述几个内置的 source 在底层也实现了 SourceFunction 接口。在生产环境中常用抽象类 FlinkKafkaConsumerBase 的实现类(如 FlinkKafkaConsumer010 类)来读取 Kafka 的消息,FlinkKafkaConsumerBase 就是一个自定义的 SourceFunction。它在 run()方法中会循环读取 Kafka 中的数据。

## 2.2.3 数据的转换操作

数据从数据源读取出来后,需要经过转换处理。Flink 定义了相当多的转换方法,因此大多数情况下开发人员无须调用底层的 API。

（1）Map/FlatMap/Filter。

这 3 种转换操作的语义与其他框架的一致，可以通过自定义函数将数据流转换成 SingleOutputStreamOperator，如：

```
public <R> SingleOutputStreamOperator<R> map(MapFunction<T, R> mapper)
public <R> SingleOutputStreamOperator<R> flatMap(FlatMapFunction<T, R> flatMapper)
public SingleOutputStreamOperator<T> filter(FilterFunction<T> filter)
```

这 3 个方法的返回值类型 SingleOutputStreamOperator 类继承自 DataStream 类。由于 SingleOutputStreamOperator 类在行为上与 DataStream 类没有太大的不同，因此若官方文档或一些图书中称这 3 个方法完成了 DataStream 到 DataStream 的转换是没有问题的。这里特此说明，以免读者在阅读不同资料时产生歧义。

（2）KeyBy。

keyBy()方法可以通过指定 KeySelector()等方法对数据流进行重分区，如：

```
public <K> KeyedStream<T, K> keyBy(KeySelector<T, K> key)
```

DataStream 经过 keyBy()方法的转换后会变成 KeyedStream。KeyedStream 也继承自 DataStream，因此它拥有所有 DataStream 的属性与行为。

（3）Reduce/Fold/Aggregations。

这 3 种类型的操作必须作用在 KeyedStream 上，其含义是对键（key）相同的数据进行某种聚合操作。分区后进行聚合操作在业务上是非常普遍的需求，典型的应用是 WordCount。

Reduce 和 Fold 操作通过指定自定义函数规定聚合的逻辑，其中 Fold 操作需要指定初始值，如：

```
public SingleOutputStreamOperator<T> reduce(ReduceFunction<T> reducer)
public <R> SingleOutputStreamOperator<R> fold(R initialValue, FoldFunction<T, R> folder)
```

Aggregations 操作用于表示一些内置的聚合操作，如：

```
public SingleOutputStreamOperator<T> sum(int positionToSum)
public SingleOutputStreamOperator<T> min(int positionToMin)
public SingleOutputStreamOperator<T> max(int positionToMax)
```

经过聚合操作后，数据流又变回了 DataStream。

（4）Window/WindowAll 与 Window Apply/Window Reduce/ Window Fold/Window Aggregations。

Window/WindowAll 这两种操作都用于对数据流进行加窗，它们的区别是：前者对 KeyedStream 加窗，即对单个分区内的数据加窗；后者对 DataStream 加窗，即对所有数据加窗。两者的返回值类型也不同，分别是 WindowedStream 和 AllWindowedStream。

在 WindowedStream 和 AllWindowedStream 上可以定义窗口内的数据转换操作，即 Window Apply/Window Reduce/Window Fold/Window Aggregations 操作，返回值类型为 DataStream。

关于窗口的应用和底层设计原理在第 9 章会详细介绍。

（5）Union。

利用 Union 操作可以将多个数据流合并成一个数据流。对前文介绍的示例 WordCount 进行简单的修改：

```
StreamExecutionEnvironment env = StreamExecutionEnvironment.getExecutionEnvironment();
DataStream<String> text = env.fromElements(WordCountData.WORDS);
```

```
DataStream<Tuple2<String, Integer>> counts =
        text.flatMap(new Tokenizer())
        .keyBy(0).sum(1);
DataStream<Tuple2<String, Integer>> filter1 = counts.filter(value -> value.f1 == 1);
DataStream<Tuple2<String, Integer>> filter2 = counts.filter(value -> value.f1 == 2);
DataStream<Tuple2<String, Integer>> union = filter1.union(filter2);
union.print();
env.execute();
```

将统计个数为 1 和 2 的单词分别过滤出来，再利用 Union 操作将两个数据流合并，最终的输出结果就是所有统计个数为 1 或 2 的单词。

（6）Window Join/Window CoGroup/IntervalJoin。

对两个数据集进行连接（join）是容易理解的，但是对两个无界的数据流进行连接，就需要在相同的时间窗口上进行。Window Join 和 Window CoGroup 操作是在相同时间窗口上对两个数据流进行连接，需要分别指定连接键（join key）。通过观察底层源码能够发现，前者可以理解为后者的一个特例，后者能够更加灵活地对同一分区的数据进行操作。

代码示例如下：

```
dataStream.join(otherStream)
    .where(<key selector>).equalTo(<key selector>)
    .window(TumblingEventTimeWindows.of(Time.seconds(3)))
    .apply (new JoinFunction () {...});
dataStream.coGroup(otherStream)
    .where(0).equalTo(1)
    .window(TumblingEventTimeWindows.of(Time.seconds(3)))
    .apply (new CoGroupFunction () {...});
```

IntervalJoin 是对两个数据流在某一指定时间区间内进行连接。输入流均为 KeyedStream，即连接键已提前指定好。

代码示例如下：

```
keyedStream.intervalJoin(otherKeyedStream)
    .between(Time.milliseconds(-2), Time.milliseconds(2))
    .upperBoundExclusive(true)
    .lowerBoundExclusive(true)
    .process(new IntervalJoinFunction() {...});
```

（7）Connect 与 CoMap/CoFlatMap。

Connect 操作可将两个数据流合并为一个 ConnectedStreams。与 Union 操作不同的是，使用 Connect 操作合并数据流后，两个数据流还是独立地进行操作的，常用的转换操作是 CoMap/CoFlatMap，转换后汇聚成一个数据流输出。返回值类型为 DataStream。对 WordCount 进行简单的修改：

```
StreamExecutionEnvironment env = StreamExecutionEnvironment.getExecutionEnvironment();
DataStream<String> text = env.fromElements(WordCountData.WORDS);
DataStream<Tuple2<String, Integer>> counts =
        text.flatMap(new Tokenizer())
        .keyBy(0).sum(1);
DataStream<Tuple2<String, Integer>> filter1 = counts.filter(value -> value.f1 == 1);
DataStream<Tuple2<String, Integer>> filter2 = counts.filter(value -> value.f1 == 2);
ConnectedStreams<Tuple2<String, Integer>, Tuple2<String, Integer>> connectedStreams =
filter1.connect(filter2);
```

```java
DataStream<Tuple2<String, Integer>> result = connectedStreams.map(new CoMapFunction
<Tuple2<String, Integer>, Tuple2<String, Integer>, Tuple2<String, Integer>>() {
    @Override
    public Tuple2<String, Integer> map1(Tuple2<String, Integer> value) throws Exception {
        value.f1 += 1;
        return value;
    }
    @Override
    public Tuple2<String, Integer> map2(Tuple2<String, Integer> value) throws Exception {
        return value;
    }
});
result.print();
env.execute();
```

在上面的代码中，先分别过滤出了统计个数为 1 和 2 的单词，对两个数据流执行 Connect 操作，再将一个 CoMap 作用在数据流上。注意，Connect 操作是有顺序的，对于前面的数据流会用 map1() 方法进行转换，对于后面的数据流会用 map2() 方法进行转换。上面代码的逻辑是，将统计个数为 1 的单词的个数再加 1，那么最后输出结果中二元组的第二位全部为 2。

（8）Split 与 Select。

Split 操作相当于给满足某些条件的数据贴一个标签，标签相同的数据会被输出到同一个位置，其通常与 Select 操作一起使用。Split 操作的返回值类型为 SplitStream，执行 Select 操作后再返回 DataStream。对 WordCount 进行简单的修改：

```java
StreamExecutionEnvironment env = StreamExecutionEnvironment.getExecutionEnvironment();
DataStream<String> text = env.fromElements(WordCountData.WORDS);
DataStream<Tuple2<String, Integer>> counts =
        text.flatMap(new Tokenizer())
            .keyBy(0).sum(1);
SplitStream<Tuple2<String, Integer>> split = counts.split(new OutputSelector<Tuple2
<String, Integer>>() {
    @Override
    public Iterable<String> select(Tuple2<String, Integer> value) {
        List<String> list = new ArrayList<>();
        if (value.f1 % 2 == 0) {
            list.add("even"); // 相当于给数据贴标签。一条数据可以有多个标签，即在 list 中添加多个字符串
        } else {
            list.add("odd");
        }
        return list;
    }
});
// 选择某一标签下的数据。参数可以有多个，表示将多个标签下的数据选择出来并形成数据流
DataStream<Tuple2<String, Integer>> even = split.select("even");
even.print();
env.execute();
```

（9）Project。

Project 操作只能作用在 Tuple 类型的数据流上。它会选择其中一部分索引对应的数据形成新的 Tuple 类型的数据流。对 WordCount 进行简单的修改：

```java
StreamExecutionEnvironment env = StreamExecutionEnvironment.getExecutionEnvironment();
DataStream<String> text = env.fromElements(WordCountData.WORDS);
```

```
DataStream<Tuple2<String, Integer>> counts =
        text.flatMap(new Tokenizer())
        .keyBy(0).sum(1);
DataStream<Tuple2<String, Integer>> filter = counts.filter((FilterFunction<Tuple2<String,
Integer>>) value -> value.f1 == 1);
DataStream<Tuple> project = filter.project(0); // 选择索引为0对应的数据形成新的Tuple类型
                                                // 的数据流
project.print();
env.execute();
```

因为在 Flink 中 Tuple 的实现类最多只能到 Tuple25，所以要注意不能超出 25 这个值。

### 2.2.4 数据的输出

对数据进行转换操作后，往往需要定义输出方式，以将数据输出到外部系统进行存储。大体上，数据输出方式也有 4 类——输出到文件、输出到套接字、标准输出或标准错误输出和自定义输出。下面介绍这些分类下的方法示例。

（1）输出到文件。

如果想将数据输出到文本文件，则可以调用 writeAsText() 方法：

```
public DataStreamSink<T> writeAsText(String path)
```

调用该方法时需指定文件路径。类似的方法还有 writeAsCsv()，其可以用于将数据输出为 CSV 文件。通常底层调用的是 writeUsingOutputFormat() 方法，该方法可以用于自定义输出格式（OutputFormat）等。

（2）输出到套接字。

如果想将数据输出到套接字，则可以调用 writeToSocket() 方法：

```
public DataStreamSink<T> writeToSocket(String hostName, int port, SerializationSchema<T>
schema)
```

通过指定主机名和端口号输出到套接字。这种输出方式的并行度强制为 1。

（3）标准输出或标准错误输出。

如果想实现标准输出或标准错误输出，则可以调用 print() 方法或 printToErr() 方法：

```
public DataStreamSink<T> print()
public DataStreamSink<T> printToErr()
```

（4）自定义输出。

自定义输出时，可以调用 addSink() 方法：

```
public DataStreamSink<T> addSink(SinkFunction<T> sinkFunction)
```

利用这个方法可以实现 SinkFunction 接口的自定义输出。SinkFunction 接口如下：

```
public interface SinkFunction<IN> extends Function, Serializable {
    default void invoke(IN value) throws Exception {}
    default void invoke(IN value, Context context) throws Exception {
        invoke(value);
    }
}
```

其中主要是要实现 invoke() 方法。上述几个内置的 sink 在底层也实现了 SinkFunction 接口。

在生产环境中常用的用于输出数据到 Kafka 的 sink 的基类 FlinkKafkaProducerBase 也是一个自定义的 SinkFunction，它在 invoke() 方法中实现了将数据发往 Kafka 的逻辑。

### 2.2.5 重分区

重分区操作实际上是在底层指定一种分区策略，当消息从上游发往下游时，会根据这种分区策略决定发往哪个任务实例。

前面介绍过的 KeyBy 也实现对数据进行重分区，只不过经过 keyBy() 方法的处理后，数据流变成了 KeyedStream，可以进行一些分区的聚合操作等。与 KeyBy 操作不同的是，这里要介绍的重分区操作的返回值类型仍是 DataStream。

重分区操作可分为如下几种。

- 随机重分区（random partitioning）：顾名思义，即随机地对数据进行重分区，通过调用 shuffle() 方法实现。
- 重新平衡（rebalance）分区：通过重新平衡的方式循环地将数据分发到各个分区，通过调用 rebalance() 方法实现。
- 本地重新平衡（local rebalance）分区：将下游的并行实例平均分配给每个上游的并行实例，然后上游的并行实例将数据在本地重新平衡地分配给它对应的下游的并行实例，如图 2-2 所示。重新平衡分区可实现全局平衡，而本地重新平衡分区实现的是上游的每个并行实例级的平衡，通过调用 rescale() 方法实现。

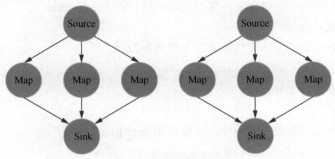

图 2-2　本地重新平衡分区

- 广播（broadcasting）：将数据广播到下游的每个并行实例上，通过调用 broadcast() 方法实现。
- 自定义重分区（custom partitioning）：通过调用 partitionCustom() 方法实现。

## 2.3　DataSet API

DataSet API 与 DataStream API 一样，处于核心 API 层，用于批处理的场景。

### 2.3.1　DataSet 版本的 WordCount

在 Flink 源码工程的 Flink-examples 模块下有 DataSet 版本的 WordCount，即 org.apache.Flink.

examples.java.wordcount.WordCount 类。这里将其进一步简化,只保留核心的步骤,代码如下:

```java
// (1) 获取执行环境
ExecutionEnvironment env = ExecutionEnvironment.getExecutionEnvironment();
// (2) 指定数据的加载方式
DataSet<String> text = env.fromElements(WordCountData.WORDS);
// (3) 指定数据的转换方式
DataSet<Tuple2<String, Integer>> counts =
        text.flatMap(new Tokenizer())
                .groupBy(0).sum(1);
// (4) 指定如何输出并触发执行
counts.print();
```

将 DataStream 版本的 WordCount 与 DataSet 版本的 WordCount 的核心步骤进行对比可以发现,二者的核心逻辑一致,主要的区别在于:一是执行环境的类不同,DataStream 版本中是 StreamExecutionEnvironment 类,DataSet 版本中是 ExecutionEnvironment 类;二是执行转换操作后的返回值类型不同,分别为 DataStream 和 DataSet。

在上面代码的第 4 步中,print()方法既指定了输出,又触发了作业的执行。这看起来与 DataStream 版本不同,但实际上是把触发作业执行的 execute()方法封装在了 print()方法中,因此这与 DataStream 版本仍是一致的。

### 2.3.2 数据源

在 DataSet API 中读取数据源时需要指定输入格式(InputFormat)。Flink 提供了一些现成的实现方法,如从文件或集合中读取数据,其中内置了对应的 InputFormat。

(1) 从文件读取。

可以调用 readTextFile()方法实现从文本文件读取数据:

```java
public DataSource<String> readTextFile(String filePath)
```

调用该方法需要指定文件路径。

(2) 从集合读取。

从集合读取数据常用在演示代码中,可以调用 fromElements()方法:

```java
public final <X> DataSource<X> fromElements(X... data)
```

(3) 自定义输入源。

调用 createInput()方法可以自定义 InputFormat:

```java
public <X> DataSource<X> createInput(InputFormat<X, ?> inputFormat)
```

### 2.3.3 数据的转换操作

如果将 DataSet API 中的转换操作与 DataStream API 中的进行对比,则会发现它们无法一一对应。这主要是因为人们看待数据流和数据集的方式不一样。对于数据流,更多考虑的是如何处理数据流中的单条数据;而对于数据集,更多考虑的是如何对完整数据集进行处理。

下面对相关操作进行介绍。

(1) Map/FlatMap/Filter/MapPartition/Project。

前 3 种转换操作的语义与其他框架的一致。MapPartition 与 Map 类似，只不过 Map 操作是对迭代器中的每个元素单独进行操作，而 MapPartition 是通过传入迭代器来定义转换操作。相关代码如下：

```
ExecutionEnvironment env = ExecutionEnvironment.getExecutionEnvironment();
DataSet<String> text = env.fromElements(WordCountData.WORDS);
DataSet<Tuple2<String, Integer>> counts =
        text.flatMap(new Tokenizer())
                .mapPartition(
                        (MapPartitionFunction<Tuple2<String, Integer>, Tuple2<String,
                                Integer>>)
                                (values, out) -> values.forEach(out::collect))
                .groupBy(0)
                .sum(1);
counts.print();
```

这里 Project 操作的语义和用法与 DataStream API 中一致，用于从 Tuple 类型的数据中提取出元素并返回，形成新的 Tuple 类型的数据。

(2) 分组数据集上的转换操作。

对数据集调用 groupBy()方法后，其返回值类型为 UnsortedGrouping；对 UnsortedGrouping 调用 sortGroup()方法后，其返回值类型为 SortedGrouping，它们都继承自 Grouping 类。对于 UnsortedGrouping，可以执行 Reduce/GroupReduce/GroupCombine/Aggregate 操作；对于 SortedGrouping，可以执行 GroupReduce/GroupCombine 操作。

```
ExecutionEnvironment env = ExecutionEnvironment.getExecutionEnvironment();
DataSet<String> text = env.fromElements(WordCountData.WORDS);
DataSet<Tuple2<String, Integer>> counts =
        text.flatMap(new Tokenizer())
                .groupBy(0)
                .sum(1);
DataSet first = counts.groupBy(1).sortGroup(0, Order.DESCENDING).first(5);
first.print();
```

(3) 完整数据集上的聚合操作。

DataSet API 中的聚合操作可以直接作用在未分区的完整数据集上，这些操作包括 Reduce/ReduceGroup/ ReduceCombine/Aggregate 操作。

```
ExecutionEnvironment env = ExecutionEnvironment.getExecutionEnvironment();
DataSet<String> text = env.fromElements(WordCountData.WORDS);
DataSet<Tuple2<String, Integer>> counts =
        text.flatMap(new Tokenizer())
                .first(10)
                .groupBy(0)
                .sum(1);
counts.print();
```

(4) Distinct。

Distinct 的语义与其他框架中的一致，用于去重。

```
ExecutionEnvironment env = ExecutionEnvironment.getExecutionEnvironment();
DataSet<String> text = env.fromElements(WordCountData.WORDS);
```

```
DataSet<Tuple2<String, Integer>> counts =
        text.flatMap(new Tokenizer()).distinct()
        .groupBy(0).sum(1);
counts.print();
```

输入数据经过分词处理后,通过 distinct()方法可实现对相同的单词去重,因此所有单词的统计结果为 1。

(5) Join/Cross/CoGroup。

相比 DataStream API 中的 Join 操作,DataSet API 中的 Join 操作更好理解一些,因为不需要考虑无界的数据流,也不需要考虑窗口与时间,其与传统数据库中的 Join 语义类似。当两个数据集连接到一起后,通过 where()和 equalTo()方法指定连接键,还可以选择性地调用 with()方法添加 JoinFunction,用于定义两个数据集连接到一起后的操作。

在 Join 操作中可以指定 JoinHint,如 BROADCAST_HASH_FIRST、REPARTITION_SORT_MERGE 等,表示两个数据集以什么方式连接。

```
ExecutionEnvironment env = ExecutionEnvironment.getExecutionEnvironment();
DataSet<String> text = env.fromElements(WordCountData.WORDS);
// 大数据集
DataSet<Tuple2<String, Integer>> counts =
        text.flatMap(new Tokenizer())
        .groupBy(0).sum(1);
// 小数据集
DataSet<Tuple2<String, Integer>> small = text
        .flatMap(new Tokenizer())
        .map(new MapFunction<Tuple2<String, Integer>, Tuple2<String, Integer>>() {
    @Override
    public Tuple2<String, Integer> map(Tuple2<String, Integer> value) throws Exception {
        return new Tuple2<>(value.f0, value.f0.length());
    }
}).first(10);
// 对两个数据集进行连接操作,并指定连接方式
DataSet<Tuple2<Tuple2<String, Integer>, Tuple2<String, Integer>>> result
        = counts.join(small, JoinHint.BROADCAST_HASH_SECOND).where(0).equalTo(0);
result.print();
```

第一个数据集仍是 WordCount 的结果,第二个数据集中的每个元素是一个二元组,二元组中的第一位表示单词,第二位表示单词长度,并且只取结果中的前 10 个,以形成一个小数据集。

Cross 操作是对两个数据集计算笛卡儿积。CoGroup 操作的语义与 DataStream API 中的一致。

(6) Union。

DataSet API 中的 Union 操作的语义与 DataStream API 中的一致,表示将两个数据集合并。可以连续执行 Union 操作合并多个数据集,这些数据集的数据类型必须相同。

### 2.3.4 数据的输出

下面介绍指定数据输出的方式。

(1) 输出到文件。

如果想将数据输出到文本文件,则可以调用 writeAsText()方法:

```
public DataSink<T> writeAsText(String filePath)
```
调用该方法时需要指定文件路径。类似的方法还有 writeAsCsv()等。

（2）标准输出和标准错误输出。

调用 print()或 printToErr()方法可以实现标准输出或标准错误输出：
```
public void print() throws Exception
public void printToErr() throws Exception
```
（3）自定义输出。

调用 output()方法可以自定义 OutputFormat（输出格式）：
```
public DataSink<T> output(OutputFormat<T> outputFormat)
```

### 2.3.5 重分区

DataSet API 中的重分区操作的返回值类型是 DataSet。重分区操作可分为如下几种。
- 重新平衡（rebalance）分区：数据平均分配到每个下游的任务实例上，通过调用 rebalance() 方法实现。
- 哈希重分区（hash-partitioning）：根据哈希值进行重分区，通过调用 partitionByHash()方法实现。
- 范围重分区（range-partitioning）：根据范围进行重分区，通过调用 partitionByRange()方法实现。
- 自定义重分区（custom partitioning）：通过调用 partitionCustom()方法实现。

## 2.4 Table API

Table API 在 Flink 中的 API 层级结构中位于核心 API 的上层，它在底层可转换成数据流或数据集。它的核心思想是对要处理的数据流或数据集注册为一个表（Table），然后用一种直观的方式对其进行关系型操作，如 Join、Select、Filter 等。由于它是靠上层的 API，因此它的灵活度不如下层 API 的高。目前这套 API 正在完善中，有些语义尚不支持。

### 2.4.1 Table API 版本的 WordCount

在 Flink 源码工程的 Flink-examples/Flink-examples-table 模块下有多个 Table API 版本的 WordCount。不过这些用例将 Table API 与 SQL 混合在一起。下面的示例对 WordCount 进行了修改，其中只包含 Table API。
```
// （1）获取执行环境
ExecutionEnvironment env = ExecutionEnvironment.getExecutionEnvironment();
BatchTableEnvironment tEnv = BatchTableEnvironment.create(env);
// （2）指定数据的加载方式
DataSet<WC> input = env.fromElements(
        new WC("Hello", 1),
        new WC("Ciao", 1),
        new WC("Hello", 1));
// （3）获取 Table 对象，定义对表的转换操作
tEnv.createTemporaryView("WordCount", input, "word, frequency");
Table table = tEnv
```

```
            .from("WordCount")
            .select("word, frequency")
            .groupBy("word")
            .select("word, frequency.sum as frequency");
// (4) 指定如何输出并触发程序运行
DataSet<WC> result = tEnv.toDataSet(table, WC.class);
result.print();
```

虽然 Table API 常常与 SQL 或核心 API 混合在一起使用，但编程模式与核心 API 的没有区别。

首先仍然需要获取执行环境。这里的关键是要获取 Table API 所需的执行环境，即 TableEnvironment。

接着是要获取 Table 对象，这是最重要的一步。获取 Table 对象有多种方式，上例中是将 DataSet 对象注册成临时表，再通过临时表创建表。得到 Table 对象后就可以利用 Table API 定义转换操作。这部分操作的语义是本节的主要内容。

最后，可以选择性地指定如何输出并触发程序运行。

在整个过程中，Table API 可以与其他层级的 API 混合使用，因此可以在 Table 对象和 DataStream/DataSet 对象间自由转换。

上例中的 WC 类如下：

```
public static class WC {
   public String word;
   public long frequency;
   public WC() {}
   public WC(String word, long frequency) {
      this.word = word;
      this.frequency = frequency;
   }
   @Override
   public String toString() {
      return "WC " + word + " " + frequency;
   }
}
```

### 2.4.2 初始化执行环境

使用 DataStream API 或 DataSet API 时，需要在程序开始处指定执行环境。对 DataStream API 来说，需要初始化 StreamExecutionEnvironment 对象；对 DataSet API 来说，需要初始化 ExecutionEnvironment 对象。因为 Table API 在底层要转换成数据流或数据集，所以使用这套 API 时也需要指定执行环境。

从上层 API 转换到下层 API，可以理解为"翻译"的过程。该"翻译"过程有多种实现方式。目前 Flink 提供两种"翻译"方式——Old Planner 和 Blink Planner。但因为 Blink Planner 做了批流统一，所以在创建执行环境时可以直接指定流模式（StreamingMode）或批模式（BatchMode），在这两种模式下 API 在底层都会转换成数据流；而 Old Planner 只能进行流处理，即只能指定流模式，如果想要用 Flink 原有方式将 Table API "翻译"成数据库，则需要初始化 BatchTableEnvironment。

总而言之，在初始化执行环境时，存在下面 4 种情况。

## 2.4 Table API

（1）Flink 流查询。

```
StreamExecutionEnvironment fsEnv = StreamExecutionEnvironment.getExecutionEnvironment();
StreamTableEnvironment fsTableEnv = StreamTableEnvironment.create(fsEnv);
```

（2）Flink 批查询。

```
ExecutionEnvironment fbEnv = ExecutionEnvironment.getExecutionEnvironment();
BatchTableEnvironment fbTableEnv = BatchTableEnvironment.create(fbEnv);
```

（3）Blink 流查询。

```
EnvironmentSettings bsSettings = EnvironmentSettings.newInstance().useBlinkPlanner().inStreamingMode().build();
TableEnvironment bsTableEnv = TableEnvironment.create(bsSettings);
```

（4）Blink 批查询。

```
EnvironmentSettings bbSettings = EnvironmentSettings.newInstance().useBlinkPlanner().inBatchMode().build();
TableEnvironment bbTableEnv = TableEnvironment.create(bbSettings);
```

对第一种情况而言，**StreamTableEnvironment.create()** 方法的内部实际上是这样的：

```
static StreamTableEnvironment create(StreamExecutionEnvironment executionEnvironment) {
    return create(
        executionEnvironment,
        EnvironmentSettings.newInstance().build());
}
```

这里利用默认值实例化了 EnvironmentSettings 对象。因此也可以显式地指定默认值，利用重载的方法完成初始化，如下所示：

```
EnvironmentSettings fsSettings = EnvironmentSettings.newInstance().useOldPlanner().inStreamingMode().build();
StreamExecutionEnvironment fsEnv = StreamExecutionEnvironment.getExecutionEnvironment();
StreamTableEnvironment fsTableEnv = StreamTableEnvironment.create(fsEnv, fsSettings);
```

或者用如下方式完成初始化：

```
EnvironmentSettings fsSettings = EnvironmentSettings.newInstance().useOldPlanner().inStreamingMode().build();
TableEnvironment fsTableEnv = TableEnvironment.create(fsSettings);
```

这样就与第三种情况的初始化方式形成了对比。

StreamTableEnvironment 接口继承自 TableEnvironment 接口，提供了更多方法。在流处理中一般直接使用 StreamTableEnvironment 接口。因而对于上述第三种情况也可以写成如下形式：

```
StreamExecutionEnvironment bsEnv = StreamExecutionEnvironment.getExecutionEnvironment();
EnvironmentSettings bsSettings = EnvironmentSettings.newInstance().useBlinkPlanner().inStreamingMode().build();
StreamTableEnvironment bsTableEnv = StreamTableEnvironment.create(bsEnv, bsSettings);
```

对于初始化内容可以做如下总结。

- 原生的 Flink 和 Blink 分别提供了一套初始化方式。原生的 Flink 可以直接通过 StreamTableEnvironment 或 BatchTableEnvironment 的 create() 方法进行初始化；Blink 则要实例化 EnvironmentSettings 对象，需要指定 Planner 以及是否为流模式。
- 原生的 Flink 会根据执行环境的不同，将 Table API 分别转换成 DataStream API 或 DataSet

API；Blink 只需要指定流模式，在底层都会转换成 DataStream API，能做到批流统一。
- 因为 Blink 无论如何都会将 Table API 转换成 DataStream API，所以在底层，原生的 Flink 对于流处理执行环境的初始化方式与 Blink 对于执行环境的初始化方式在形式上达成了一致，通过指定 Planner 就可以进行区分。需要注意的是，如果用 Old Planner，则 StreamingMode 必须为 true。

### 2.4.3　获取 Table 对象

要想用 Table API 定义数据的转换操作，首先需要获取 Table 对象。一般在这个过程中就会定义数据的加载方式。如 2.4.1 节的示例中，通过 env.fromElements()方法获取了 DataSet 对象，并将其注册在执行环境中。

除此以外，还可以定义 TableSource，通过 TableEnvironment 的 registerTableSource()方法对其进行注册：

```
ExecutionEnvironment env = ExecutionEnvironment.getExecutionEnvironment();
BatchTableEnvironment tEnv = BatchTableEnvironment.create(env);
tEnv.registerTableSource("WordCount", new CsvTableSource(...));
Table table = tEnv.from("WordCount")
        .select("word, frequency")
        .groupBy("word")
        .select("word, frequency.sum as frequency");
DataSet<WC> result = tEnv.toDataSet(table, WC.class);
result.print();
```

CsvTableSource 类中已经定义好了数据的加载方式。

如果已经获取了 Table 对象，那么可将其直接注册在执行环境中：

```
tEnv.registerTable("WordCount", table);
```

这里所说的注册，主要是指在内存中将临时表的信息维护起来，这样就可以利用 TableEnvironment 的 from()等方法通过指定表名的方式来获取 Table 对象。如果连接了外部数据源的元信息，那么可以通过指定外部数据源的表名来直接获取 Table 对象。

总而言之，这种方式是通过内存中或者外部数据源的元信息来构造 Table 对象的。另外，还可以将 DataStream/DataSet 直接转换成 Table 对象。

```
DataSet<WC> input = env.fromElements(
        new WC("Hello", 1),
        new WC("Ciao", 1),
        new WC("Hello", 1));
Table table = tEnv.fromDataSet(input);
```

这样通过省略注册的步骤，可直接得到 Table 对象。实际上在注册过程中就是先调用 fromDataStream()/fromDataSet()方法构造 Table 对象，再将其注册到执行环境中。

### 2.4.4　Table API 中的转换操作及输出

Table 算子的语义来源于 SQL，可以看到 Table API 的方法名和其中的参数都保留了大量 SQL

语法的"痕迹"。Table API 中大多数转换操作同时支持数据流和数据集,不过由于其语义来源于 SQL,而 SQL 的操作对象主要是数据集而非数据流,因此许多语义对数据流并不适用,这些操作包括 Union、Intersect、OrderBy 等。针对数据流的处理,Table API 专门设计了一些转换操作以满足业务需求,这些操作不能作用在数据集上。

Table API 中的转换操作相对复杂,下面介绍一些常用的转换操作。在介绍这些转换操作时,如果没有特殊说明,则表示该操作同时支持批处理和流处理。

(1) Select/As。

select()操作已在 2.2.3 节中介绍过。as 操作用于给某一列取一个别名,相关代码如下:

```
Table table = tEnv.from("WordCount")
    .select("word, frequency")
    .as("word_alias, frequency_alias")
    .groupBy("word_alias")
    .select("word_alias as word, frequency_alias.sum as frequency");
```

一旦给某一列取了别名,后面的操作就都要以这个别名为准。最后如果要把 Table 对象转换为 DataStream/DataSet,则需要使列名与 DataStream/DataSet 中数据类型的属性名一致。从上例也可以看出,As 操作可以通过 as()方法实现,也可以在 select()方法的参数中直接使用 AS 关键字。

(2) Filter/Where。

这两个算子在 Table API 中的语义完全相同。实际上 where()方法在底层调用的就是 filter()方法,相关代码如下:

```
Table table = tEnv.from("WordCount")
    .select("word, frequency")
    .where("word == 'Ciao'")
    .groupBy("word")
    .select("word, frequency.sum as frequency");
```

(3) InnerJoin/OuterJoin。

Join 操作的语义与 SQL 中的一致,相关代码如下。

```
ExecutionEnvironment env = ExecutionEnvironment.getExecutionEnvironment();
BatchTableEnvironment tEnv = BatchTableEnvironment.create(env);
DataSet<WC> input = env.fromElements(
    new WC("Hello", 1),
    new WC("Ciao", 1),
    new WC("Hello", 1));
DataSet<Dict> dict = env.fromElements(
    new Dict("Hello", "English"),
    new Dict("Ciao", "Italian"));
tEnv.createTemporaryView("WordCount", input, "word, frequency");
Table table = tEnv.from("WordCount")
    .select("word, frequency")
    .groupBy("word")
    .select("word as word_alias, frequency.sum as frequency");
Table dictTable = tEnv.fromDataSet(dict);
Table result = table.join(dictTable).where("word_alias == word").select("word, frequency, language");
    tEnv.toDataSet(result, Row.class).print();
```

其中 Dict 类如下：

```
public static class Dict {
  public String word;
  public String language;
  public Dict() {}
  public Dict(String word, String language) {
      this.word = word;
      this.language = language;
  }
}
```

（4）Union/UnionAll/Intersect/IntersectAll/Minus/MinusAll。

这部分操作属于集合操作，分别对应 SQL 中的关键字 UNION、UNION ALL、INTERSECT、INTERSECT ALL、EXCEPT 和 EXCEPT ALL。除了 UNION ALL，其余操作均只能作用在数据集上。以 UnionAll 操作为例：

```
ExecutionEnvironment env = ExecutionEnvironment.getExecutionEnvironment();
BatchTableEnvironment tEnv = BatchTableEnvironment.create(env);
DataSet<WC> input = env.fromElements(
      new WC("Hello", 1),
      new WC("Ciao", 1),
      new WC("Hello", 1));
DataSet<WC> input2 = env.fromElements(
      new WC("Hello", 1),
      new WC("Hello", 1),
      new WC("Hello", 1));
tEnv.createTemporaryView("WordCount", input, "word, frequency");
Table table = tEnv.from("WordCount")
      .select("word, frequency")
      .unionAll(tEnv.fromDataSet(input2).select("word, frequency"))
      .groupBy("word")
      .select("word, frequency.sum as frequency");
DataSet<WC> result = tEnv.toDataSet(table, WC.class);
result.print();
```

（5）In。

In 操作对应 SQL 中的 IN 关键字。其一般与 filter()/where() 方法一起使用，相关代码如下：

```
ExecutionEnvironment env = ExecutionEnvironment.getExecutionEnvironment();
BatchTableEnvironment tEnv = BatchTableEnvironment.create(env);
DataSet<WC> input = env.fromElements(
      new WC("Hello", 1),
      new WC("Ciao", 1),
      new WC("Hello", 1));
DataSet<String> wordFilter = env.fromElements("Hello");
tEnv.createTemporaryView("WordCount", input, "word, frequency");
tEnv.createTemporaryView("WordFilter", wordFilter, "word");
Table table = tEnv.from("WordCount")
      .select("word, frequency")
      .where("word.in(WordFilter)")
      .groupBy("word")
      .select("word, frequency.sum as frequency");
DataSet<WC> result = tEnv.toDataSet(table, WC.class);
result.print();
```

（6）OrderBy/Offset/Fetch。

OrderBy 操作用于排序，语义与 SQL 中的一致。Offset 和 Fetch 操作必须用在 OrderBy 操作的后面，表示从某个偏移量开始取值以及取值数量。这一套转换操作仅针对批处理。

```
Table table = tEnv.from("WordCount")
        .select("word, frequency")
        .groupBy("word")
        .select("word, frequency.sum as frequency")
        .orderBy("frequency.asc")
        .offset(0)
        .fetch(1);
```

上例表示根据词频按升序排序，从第一位开始取一个值。

（7）Distinct。

Distinct 的语义与 SQL 中的一致，用于去重。

```
Table table = tEnv.from("WordCount")
        .select("word, frequency")
        .distinct()
        .groupBy("word")
        .select("word, frequency.sum as frequency");
```

（8）GroupBy Aggregation/GroupBy Window Aggregation。

GroupBy Aggregation 操作就是 WordCount 中的 groupBy()方法加上后续 select()中对某一列的聚合操作（frequency.sum）。GroupBy Window Aggregation 就是给数据加窗后，在每个窗口内调用 groupBy()方法。

```
ExecutionEnvironment env = ExecutionEnvironment.getExecutionEnvironment();
BatchTableEnvironment tEnv = BatchTableEnvironment.create(env);
DataSet<WC> input = env.fromElements(
        new WC("Hello", 1, new Timestamp(1600489315000L)),
        new WC("Ciao", 1, new Timestamp(1600489374000L)),
        new WC("Hello", 1, new Timestamp(1600489854000L)));
tEnv.createTemporaryView("WordCount", input, "word, frequency, rowtime");
Table table = tEnv.from("WordCount")
        .select("word, frequency, rowtime")
        .window(Tumble.over("5.minutes").on("rowtime").as("w"))
        .groupBy("word, w")
        .select("word, frequency.sum as frequency, w.start, w.end, w.rowtime");
DataSet<Row> result = tEnv.toDataSet(table, Row.class);
result.print();
```

既然要加窗，那么必须有一列用于表示时间。此时的 WC 类被修改为如下形式：

```
public static class WC {
   public String word;
   public long frequency;
   public Timestamp rowtime;
   public WC() {}
   public WC(String word, long frequency, Timestamp rowtime) {
      this.word = word;
      this.frequency = frequency;
      this.rowtime = rowtime;
   }
```

```
    @Override
    public String toString() {
        return "WC " + word + " " + frequency + " " + rowtime;
    }
}
```

上例中第一个 Hello 与 Ciao 在一个窗口中，第二个 Hello 在另一个窗口中。

（9）Row-based Operations。

这种类型的操作需要定义一个函数，用于对整行进行操作。常用的函数有 map()/flatMap()/aggregate()。

1）map()。

```
ExecutionEnvironment env = ExecutionEnvironment.getExecutionEnvironment();
BatchTableEnvironment tEnv = BatchTableEnvironment.create(env);
DataSet<String> input = env.fromElements("Hello", "Ciao", "Hello");
ScalarFunction func = new MyMapFunction();
tEnv.registerFunction("func", func);
tEnv.createTemporaryView("Word", input, "word");
Table table = tEnv.from("Word")
            .select("word")
            .map("func(word)")
            .as("word, frequency")
            .groupBy("word")
            .select("word, frequency.sum as frequency");
DataSet<WC> result = tEnv.toDataSet(table, WC.class);
result.print();
```

其中，**MyMapFunction** 类如下：

```
public static class MyMapFunction extends ScalarFunction {
    public Row eval(String a) {
        return Row.of(a, 1L);
    }
    @Override
    public TypeInformation<?> getResultType(Class<?>[] signature) {
        return Types.ROW(Types.STRING, Types.LONG);
    }
}
```

2）flatMap()。

```
ExecutionEnvironment env = ExecutionEnvironment.getExecutionEnvironment();
BatchTableEnvironment tEnv = BatchTableEnvironment.create(env);
DataSet<String> input = env.fromElements("Hello", "Ciao", "Hello");
TableFunction func = new MyFlatMapFunction();
tEnv.registerFunction("func", func);
tEnv.createTemporaryView("Text", input, "line");
Table table = tEnv.from("Text")
            .select("line")
            .flatMap("func(line)")
            .as("word, frequency")
            .groupBy("word")
            .select("word, frequency.sum as frequency");
DataSet<WC> result = tEnv.toDataSet(table, WC.class);
result.print();
```

其中，**MyFlatMapFunction** 类如下：

```java
public static class MyFlatMapFunction extends TableFunction<Row> {
    public void eval(String str) {
        String[] tokens = str.toLowerCase().split("\\W+");
        for (String token : tokens) {
            if (token.length() > 0) {
                collector.collect(Row.of(token, 1L));
            }
        }
    }
    @Override
    public TypeInformation<Row> getResultType() {
        return Types.ROW(Types.STRING, Types.LONG);
    }
}
```

3）aggregate()。

```java
ExecutionEnvironment env = ExecutionEnvironment.getExecutionEnvironment();
BatchTableEnvironment tEnv = BatchTableEnvironment.create(env);
DataSet<WC> input = env.fromElements(
        new WC("Hello", 1),
        new WC("Ciao", 1),
        new WC("Hello", 1));
AggregateFunction myAggFunc = new MySum();
tEnv.registerFunction("myAggFunc", myAggFunc);
tEnv.createTemporaryView("WordCount", input, "word, frequency");
Table table = tEnv.from("WordCount")
        .select("word, frequency")
        .groupBy("word")
        .aggregate("myAggFunc(frequency) as frequency")
        .select("word, frequency");
DataSet<WC> result = tEnv.toDataSet(table, WC.class);
result.print();
```

其中，**MySum** 类如下：

```java
public static class MySumAcc {
    public Long sum = 0L;
}
public static class MySum extends AggregateFunction<Row, MySumAcc> {
    public void accumulate(MySumAcc acc, Long value) {
        acc.sum += value;
    }
    @Override
    public MySumAcc createAccumulator() {
        return new MySumAcc();
    }
    public void resetAccumulator(MySumAcc acc) {
        acc.sum = 0L;
    }
    @Override
    public Row getValue(MySumAcc acc) {
        return Row.of(acc.sum);
    }
    @Override
    public TypeInformation<Row> getResultType() {
```

```
            return new RowTypeInfo(Types.LONG);
    }
}
```

（10）列操作。

这种类型的操作有 AddColumns、AddOrReplaceColumns、DropColumns 和 RenameColumns 等。

（11）表的输出。

Table API 可以与其他层级的 API 混合使用，因此有多种方式可用于定义其输出，比如 Table 对象在执行环境中注册，再由 SQL 定义输出；或者将其转换为 DataStream/DataSet，再由对应的 API 定义输出。Table 对象本身可以由 insertInto()方法定义输出，其参数为一个注册过的 TableSink。

## 2.5 SQL

SQL 位于 Flink 中的 API 层级结构中的顶层。它的使用离不开 Table API 和 Table API 所需的执行环境 TableEnvironment。TableEnvironment 主要提供了 sqlQuery()和 sqlUpdate()方法。前者用于查询，后者用于表、函数的创建、删除操作以及数据的输入操作等。

下面是 SQL 版本的 WordCount：

```
ExecutionEnvironment env = ExecutionEnvironment.getExecutionEnvironment();
BatchTableEnvironment tEnv = BatchTableEnvironment.create(env);
DataSet<WC> input = env.fromElements(
        new WC("Hello", 1),
        new WC("Ciao", 1),
        new WC("Hello", 1));
tEnv.createTemporaryView("WordCount", input, "word, frequency");
Table table = tEnv.sqlQuery("SELECT word, SUM(frequency) as frequency FROM WordCount GROUP BY word");
DataSet<WC> result = tEnv.toDataSet(table, WC.class);
result.print();
```

SQL 语句中查询的表必须是在执行环境中可以查询到的表。上例还有另一种写法，如下：

```
ExecutionEnvironment env = ExecutionEnvironment.getExecutionEnvironment();
BatchTableEnvironment tEnv = BatchTableEnvironment.create(env);
DataSet<WC> input = env.fromElements(
        new WC("Hello", 1),
        new WC("Ciao", 1),
        new WC("Hello", 1));
Table wordCount = tEnv.fromDataSet(input);
Table table = tEnv.sqlQuery("SELECT word, SUM(frequency) as frequency FROM " + wordCount + " GROUP BY word");
DataSet<WC> result = tEnv.toDataSet(table, WC.class);
result.print();
```

其中没有显式的注册步骤，但其实 Table 对象的 toString()方法会将其自身注册在执行环境中。

## 2.6 总结

本章首先展示了 Flink 应用程序开发中常用 API 的层级结构，并从核心 API 层开始对每一层的编程模式和常用接口等进行了介绍。越下层的 API 灵活度和复杂性越高，越上层的 API 易用性和表达力越强。这些 API 的底层实现一致，因而可以互相转换，混合使用。

每一层的编程模式基本相同，都从获取执行环境开始，然后定义数据源、添加转换操作，最终定义数据输出方式并执行。对于核心 API 层，操作的对象为数据流或数据集，读者学习的重点在于理解各种转换操作的语义。Table API 层之上的层操作的对象则是表，许多转换操作的语义与下层的基本相同，读者学习的重点在于了解获取表的方式以及表与数据流/数据集的相互转换。

# 第 3 章
# Flink API 层的实现原理

第 2 章详细介绍了 DataStream/DataSet API、Table API 和 SQL 的使用方法。这些内容虽不足以实现复杂的业务逻辑，但已经能够使人完整地了解 Flink 的编程模式，并且学习这些内容能使后续的学习变得轻松。本章会以第 2 章的内容为基础，逐步深入介绍方法的内部实现。

对分布式计算框架来说，它需要将关于分布式的实现细节隐藏起来，让开发人员无须过多地考虑数据的正确性、一致性、机器的负载、网络上的数据交换等问题，从而将开发的重点放在业务逻辑的实现上。这其实对框架的可复用性提出了极高的要求。可复用性并不仅仅指对某一段代码的重复使用，而是指需要抽象出一种编程模式，使之适合多种具体的业务场景。关于这一点，MapReduce 编程模式为后来绝大多数分布式计算框架提供了一种标准范式。Flink 遵循了这一范式，将定义计算逻辑的接口暴露给开发人员，这些接口包括 map()、reduce() 等方法。

读者可以就下面这个 WordCount 用例的代码片段思考几个简单的问题：

```
DataStream<String> text = env.fromElements(WordCountData.WORDS);
DataStream<Tuple2<String, Integer>> counts =
        text.flatMap(new Tokenizer())
            .keyBy(0).sum(1);
```

当调用 fromElements() 方法时，数据被读取到内存了吗？text 对象中封装了具体的数据吗？当调用 flatMap() 方法时，数据立即发生转换了吗？返回值 counts 对象中封装了最终的计算结果吗？

这些问题的答案显然都是否定的。如果这些对象中封装了数据，并且程序在运行到这几行时数据发生了转换，就失去了分布式计算的意义。对于任何一个分布式框架暴露给开发人员的接口，都不会真正地进行计算，而只是提供定义计算逻辑的功能。当开发人员定义好计算逻辑后，会再调用另一个方法去提交整个作业，这个方法在 Spark 中是行动算子，在 Flink 中是 execute() 方法。随后的工作就与开发人员无关了，框架本身会解析这些计算逻辑，将其拆分成多个任务，并提交到集群中的各个物理节点去执行。在真正执行任务时调用先前定义的计算逻辑，这时才会进行真正的计算。

本章会从业务代码获取执行环境对象开始介绍，逐步分析到调用 execute() 方法之前。因为每种 API 的封装逻辑并不完全相同，所以还是会分开介绍 DataStream API、DataSet API、Table API 和 SQL。

希望在学习完本章后，读者能够了解：
- 执行环境的初始化原理；
- Flink 提供的 API 如何封装业务逻辑。

## 3.1 DataStream API

本节仍以 WordCount 为基础，对常用的几个 API 进行分析。

```
// （1）获取执行环境
StreamExecutionEnvironment env = StreamExecutionEnvironment.getExecutionEnvironment();
// （2）指定数据的加载方式
DataStream<String> text = env.fromElements(WordCountData.WORDS);
// （3）指定数据的转换方式
DataStream<Tuple2<String, Integer>> counts =
        text.flatMap(new Tokenizer())
            .keyBy(0).sum(1);
// （4）指定如何输出
counts.print();
// （5）触发程序运行
env.execute();
```

### 3.1.1 StreamExecutionEnvironment 执行环境

执行环境主要用于封装配置、定义和业务逻辑并提供作业启动的入口。StreamExecutionEnvironment 类位于 Flink-streaming-java 的 org.apache.Flink.streaming.api.environment 包中。对于 StreamExecutionEnvironment 类，可以从执行环境的初始化、定义数据的加载方式、添加计算逻辑、触发程序运行这 4 个方面来把握。

1. 执行环境的初始化

对于 StreamExecutionEnvironment，至少有下面两个要点。

- 由于任何程序最后都需要调用 env.execute()方法来触发运行，因此可以推测出所有的计算逻辑最后都会被封装在执行环境对象中。
- 同样的代码既可以在 IDE 里直接运行，又可以用 Flink 的 run 命令以不同的方式去部署运行，这说明 StreamExecutionEnvironment.getExecutionEnvironment()方法会在不同的执行模式下返回不同类型的对象（或者是其中的属性值不同），导致调用 execute()方法时会产生不同的结果。

图 3-1 展示了 StreamExecutionEnvironment 类与相关常用类的继承关系。

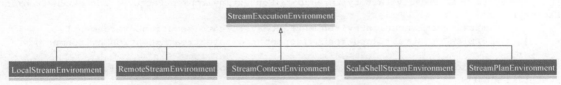

图 3-1 StreamExecutionEnvironment 类与相关常用类的继承关系

StreamExecutionEnvironment 类本身并不是接口或抽象类，可以直接被实例化。其他类中，有些提供了公有的构造方法，可以直接被实例化，有些则需要将 StreamExecutionEnvironment 类作为"入口"才能被实例化。这些类会在不同的执行模式下被构造出来。比如 ScalaShellStreamEnvironment 类的对象是在执行 start-scala-shell.sh 脚本时通过公有的构造方法直接被实例化；LocalStreamEnvironment

类的对象和 StreamContextEnvironment 类的对象可以通过 StreamExecutionEnvironment.getExecutionEnvironment()方法在不同条件下进行构造。

简单观察一下 StreamExecutionEnvironment 类的子类中的方法就会发现，这些子类的核心逻辑与其父类的是完全相同的，因此可以把重点放在 StreamExecutionEnvironment 类上。

下面介绍 StreamExecutionEnvironment 类的如下重要字段。

- contextEnvironmentFactory：StreamExecutionEnvironmentFactory 类型，用于创建执行环境的工厂类。
- threadLocalContextEnvironmentFactory：ThreadLocal 类型，用于保存 StreamExecutionEnvironmentFactory 对象。
- defaultLocalParallelism：用于设置本地模式下默认的并行度。
- config：ExecutionConfig 类型，用于维护各种配置，如并行度、最大并行度等。
- checkpointCfg：CheckpointConfig 类型，用于维护各种与检查点有关的配置，如检查点的时间间隔等。关于检查点机制会在第 10 章详细介绍。
- transformations：Transformation 类型，所有的计算逻辑都封装在这个对象中。
- defaultStateBackend：StateBackend 类型，用于作为状态存储的状态后端（StateBackend）。状态后端的相关内容会在第 10 章详细介绍。
- timeCharacteristic：TimeCharacteristic 类型，表示该作业使用 ProcessingTime、IngestionTime 或 EventTime。关于时间与窗口的内容会在第 9 章详细介绍。
- configuration：Configuration 类型，从配置文件中读取出的配置会保存在其中。

上述字段中，contextEnvironmentFactory 和 threadLocalContextEnvironmentFactory 与创建执行环境有关，它们是静态字段。相关代码如下：

```
private static StreamExecutionEnvironmentFactory contextEnvironmentFactory = null;
private static final ThreadLocal<StreamExecutionEnvironmentFactory> threadLocalContextEnvironmentFactory = new ThreadLocal<>();
```

正如 WordCount 用例所展示的那样，在业务代码中，常会调用 StreamExecutionEnvironment.getExecutionEnvironment()方法来初始化执行环境。该方法的内容如下：

```
public static StreamExecutionEnvironment getExecutionEnvironment() {
    return Utils.resolveFactory(threadLocalContextEnvironmentFactory, contextEnvironmentFactory)
        .map(StreamExecutionEnvironmentFactory::createExecutionEnvironment)
        .orElseGet(StreamExecutionEnvironment::createStreamExecutionEnvironment);
}
public static <T> Optional<T> resolveFactory(ThreadLocal<T> threadLocalFactory, @Nullable T staticFactory) {
    final T localFactory = threadLocalFactory.get();
    final T factory = localFactory == null ? staticFactory : localFactory;
    return Optional.ofNullable(factory);
}
```

getExecutionEnvironment()方法的逻辑是：如果 threadLocalContextEnvironmentFactory 中设置了工厂类对象，则利用该对象初始化执行环境；如果没有设置，则使用 contextEnvironmentFactory

对象来初始化执行环境；如果该对象也没有设置，则利用 StreamExecutionEnvironment.createStreamExecutionEnvironment()方法来初始化执行环境。

正是因为 contextEnvironmentFactory 和 threodLocalContextEnvironmentFactory 字段为静态字段，所以在利用 Flink 客户端提交任务时，在执行到 main()方法前可以对这两个字段提前赋值，由此，在 getExecutionEnvironment()方法中就可能会返回不同类型的执行环境。

如果没有提前赋值，那么会执行到 StreamExecutionEnvironment.createStreamExecutionEnvironment()方法。

```
private static StreamExecutionEnvironment createStreamExecutionEnvironment() {
    // 获取 DataSet 的执行环境
    ExecutionEnvironment env = ExecutionEnvironment.getExecutionEnvironment();
    // 根据 DataSet 执行环境类型的不同，返回不同的 DataStream 执行环境
    if (env instanceof ContextEnvironment) {
      return new StreamContextEnvironment((ContextEnvironment) env);
    } else if (env instanceof OptimizerPlanEnvironment) {
      return new StreamPlanEnvironment(env);
    } else {
      return createLocalEnvironment();
    }
}
```

这是非常令人费解的设计——初始化 DataStream 的执行环境竟然需要先初始化 DataSet 的执行环境。

其实这样的设计实属无奈。前文已经介绍过，在 Flink 执行到 main()方法前，也就是在客户端代码中，可以对上述两个静态字段进行设置。由于 Flink 的客户端代码在 Flink-clients 模块中，而 StreamExecutionEnvironment 类在 Flink-streaming-java 模块中，因此如果想在客户端代码中直接设置 StreamExecutionEnvironment 类中的这两个字段，那么需要在 Flink-clients 模块中引入 Flink-streaming-java 模块的依赖，而在 Flink 之前的设计中，后者已经依赖了前者（图 3-2 展示了目前几个模块间的依赖关系），为了避免循环依赖的问题，只能另择他法。所幸的是，DataSet 的执行环境 ExecutionEnvironment 类所在的模块 Flink-java 不存在这样的问题，于是可以在客户端中对 ExecutionEnvironment 类中的静态字段赋值，该类的字段与 StreamExecutionEnvironment 类的类似。这样，在 StreamExecutionEnvironment.createStreamExecutionEnvironment()方法中根据 ExecutionEnvironment 类对象的实际类型，就能判断出 DataStream 中应该具体使用哪种执行环境。在以后的版本中，这部分代码会得到重构。

其中，ExecutionEnvironment.getExecutionEnvironment()方法如下：

```
public static ExecutionEnvironment getExecutionEnvironment() {
    return Utils.resolveFactory(threadLocalContextEnvironmentFactory,contextEnvironmentFactory)
        .map(ExecutionEnvironmentFactory::createExecutionEnvironment)
        .orElseGet(ExecutionEnvironment::createLocalEnvironment);
}
```

与 StreamExecutionEnvironment 的设计思路完全一致。这里的 threadLocalContextEnvironmentFactory 和 contextEnvironmentFactory 为 ExecutionEnvironment 类中的静态字段。

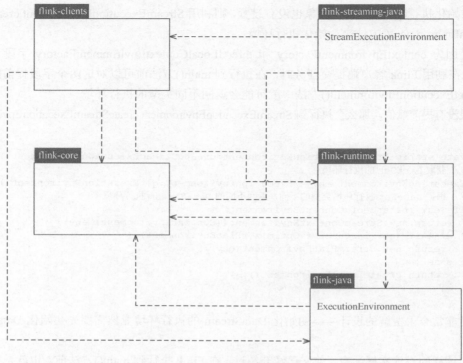

图 3-2　模块间的依赖关系

### 2. 定义数据的加载方式

获取执行环境 StreamExecutionEnvironment 对象后,可以做的重要事情之一就是定义数据的加载方式。在第 2 章中对各种方法的语义进行了介绍,这里以 fromElements()方法为例,清晰地展示添加数据源的整个过程。

```
public final <OUT> DataStreamSource<OUT> fromElements(OUT...
data) {
    ...
    TypeInformation<OUT> typeInfo;
    try {
        // 获取输入的数据类型
        typeInfo = TypeExtractor.getForObject(data[0]);
    }
    catch (Exception e) {
        ...
    }
    return fromCollection(Arrays.asList(data), typeInfo);
}
public <OUT> DataStreamSource<OUT> fromCollection(Collection<OUT> data, TypeInformation
<OUT> typeInfo) {
    ...
    SourceFunction<OUT> function;
    try {
        // 构造 SourceFunction 对象,定义数据的加载方式
        function = new FromElementsFunction<>(typeInfo.createSerializer(getConfig()), data);
```

```
    }
    catch (IOException e) {
        ...
    }
    // 调用 addSource()方法添加 SourceFunction 对象
    return addSource(function, "Collection Source", typeInfo).setParallelism(1);
}
```

上例中省略了校验参数、处理异常的代码,并在关键步骤处给出了注释。其中核心的逻辑就是构造 SourceFunction 对象(FromElementsFunction 类实现了 SourceFunction 接口),定义数据的加载方式,然后调用 addSource()方法对 SourceFunction 对象进行添加。分析其他定义输入源的方法时会发现也是类似的步骤,如 socketTextStream()方法:

```
public DataStreamSource<String> socketTextStream(String hostname, int port, String delimiter, long maxRetry) {
    return addSource(new SocketTextStreamFunction(hostname, port, delimiter, maxRetry),
        "Socket Stream");
}
```

这里的 SocketTextStreamFunction 类同样实现了 SourceFunction 接口。总之,无论定义何种数据源,最终都会构造一个 SourceFunction 对象并调用 addSource()方法。

在实际场景中,通常会直接调用 addSource()方法添加自定义的或者外部模块定义的 SourceFunction,比如用于读取 Kafka 数据的 FlinkKafkaConsumer,其实现如下:

```
public <OUT> DataStreamSource<OUT> addSource(SourceFunction<OUT> function, String sourceName, TypeInformation<OUT> typeInfo) {
    // (1) 获取输入的数据类型
    if (typeInfo == null && function instanceof ResultTypeQueryable) {
        typeInfo = ((ResultTypeQueryable<OUT>) function).getProducedType();
    }
    if (typeInfo == null) {
        try {
            typeInfo = TypeExtractor.createTypeInfo(
                SourceFunction.class,
                function.getClass(), 0, null, null);
        } catch (final InvalidTypesException e) {
            typeInfo = (TypeInformation<OUT>) new MissingTypeInfo(sourceName, e);
        }
    }
    ...
    // (2) 将 SourceFunction 对象封装到 StreamSource 对象中
    final StreamSource<OUT, ?> sourceOperator = new StreamSource<>(function);
    // (3) 返回 DataStreamSource 对象
    return new DataStreamSource<>(this, typeInfo, sourceOperator, isParallel, sourceName);
}
```

上述代码省略了校验 function 对象能否序列化等步骤。代码主要的逻辑可分为以下 3 方面。

(1)获取输入的数据类型。虽然 fromElements()方法中也涉及获取输入的数据类型的步骤,因为该方法的参数就是输入,可以直接获取数据类型,而其他输入源并没有这样的信息,因此最终需要在 addSource()方法中获取。

(2)将 SourceFunction 对象封装到 StreamSource 对象中。

(3)将 StreamSource 对象以及输入的数据类型、并行度等信息一同封装到 DataStreamSource

对象中，然后返回该对象。DataStreamSource 是 DataStream 的子类，它们都表示数据流对象。

根据对上述源码的分析可以了解到，计算逻辑以 SourceFunction 的形式，通过层层封装，最终被封装到 DataStreamSource 对象中。这个 DataStreamSource 对象就是最终的返回值。

3. 添加计算逻辑

执行环境提供一个方法用于添加计算逻辑：

```
public void addOperator(Transformation<?> transformation) {
    this.transformations.add(transformation);
}
```

**Transformation** 类型表示转换操作，无论是定义的数据输入输出方式还是 Map、KeyBy 等操作，在 Flink 中都被抽象成 Transformation。

4. 触发程序运行

执行环境调用 execute() 方法触发程序运行。从 execute() 方法开始，作业就开始执行生成执行图的流程。这部分内容将在第 4 章详细分析。

### 3.1.2 Function 接口分析

任何一个分布式计算框架都需要提供接口让开发人员定义计算逻辑。Flink 中提供了 addSource() 方法以定义数据的加载方式，还提供了 map()、filter() 等接口来定义数据的转换逻辑。这些方法的参数类型都实现了 Flink-core 模块下的 org.apache.Flink.api.common.functions.Function 接口。相关代码如下：

```
public interface Function extends java.io.Serializable {
}
```

这个顶层接口并没有定义任何方法，只是继承了 Serializable 接口。Function 接口的实现类极多，这里介绍几个常用的具有代表性的子类或子接口。

1. SourceFunction

SourceFunction 用于定义数据的加载方式，它是 Function 接口的子接口。相关代码如下：

```
public interface SourceFunction<T> extends Function, Serializable {
    void run(SourceContext<T> ctx) throws Exception;
    void cancel();
}
```

调用该接口的 run() 方法可实现数据加载逻辑。下面以 FromElementsFunction 这个实现类为例进行分析，该类的构造方法如下：

```
public FromElementsFunction(TypeSerializer<T> serializer, Iterable<T> elements) throws
IOException {
    ByteArrayOutputStream baos = new ByteArrayOutputStream();
    DataOutputViewStreamWrapper wrapper = new DataOutputViewStreamWrapper(baos);
    int count = 0;
    try {
        for (T element : elements) {
            serializer.serialize(element, wrapper);
            count++;
        }
```

```
    }
    catch (Exception e) {
        ...
    }
    this.serializer = serializer;
    this.elementsSerialized = baos.toByteArray();
    this.numElements = count;
}
```

结合上文的分析可以知道,这里传入的第一个参数是由输入的数据类型得到的,表示一个序列化器。关于 Flink 的类型系统和序列化器的内容会在附录 B 详细分析。

在构造方法中,输入的数据被序列化到 elementsSerialized 字段中。

该类的 run() 方法如下:

```
public void run(SourceContext<T> ctx) throws Exception {
    ByteArrayInputStream bais = new ByteArrayInputStream(elementsSerialized);
    final DataInputView input = new DataInputViewStreamWrapper(bais);
    ...
    while (isRunning && numElementsEmitted < numElements) {
        T next;
        try {
            next = serializer.deserialize(input);
        }
        catch (Exception e) {
            ...
        }
        synchronized (lock) {
            ctx.collect(next);
            numElementsEmitted++;
        }
    }
}
```

一般在 SourceFunction 的 run() 中会出现 while 循环,通过循环不断获取数据,并向下游发送数据。尤其是 DataStream 中用到的 SourceFunction,常常会是 while(true)形式的死循环,以产生无界的数据流。这里在 while 循环中将 elementsSerialized 中的数据进行了反序列化,通过 SourceContext 的 collect()方法向下游发送数据。

这里之所以对数据先进行序列化后进行反序列化,是因为构造方法在客户端执行,而 run()方法是在任务被调度和部署后在任务管理器中执行的,序列化是为了方便数据传输。

2. FlatMapFunction

数据被加载并形成数据流后,需要对数据流执行转换操作。转换操作的逻辑是通过 Function 接口来定义的,其中 FlatMapFunction 是常用的转换接口,也是 Function 接口的子接口。

```
public interface FlatMapFunction<T, O> extends Function, Serializable {
    void flatMap(T value, Collector<O> out) throws Exception;
}
```

回顾 WordCount 用例,得到 DataStream 对象后,调用 flatMap()方法便可添加转换逻辑,该方法需要传入 FlatMapFunction 实例。FlatMapFunction 中定义了 flatMap()方法,调用该方法即可实现 FlatMap 语义。

下面以 WordCount 用例中出现的 Tokenizer 类为例进行分析：

```
public void flatMap(String value, Collector<Tuple2<String, Integer>> out) {
    String[] tokens = value.toLowerCase().split("\\W+");
    for (String token : tokens) {
        if (token.length() > 0) {
            out.collect(new Tuple2<>(token, 1));
        }
    }
}
```

最后，FlatMapFunction 调用 Collector 的 collect()方法将数据发送至下游。

### 3. RichFunction

RichFunction 是 Function 的子接口，可以将其理解为 Function 接口的增强版，比如 FlatMapFunction 继承自 Function 接口，可以用来实现 FlatMap 语义，与之相对应，会存在 RichFlatMapFunction 用于对 FlatMapFunction 的功能进行增强。

RichFunction 接口的定义如下：

```
public interface RichFunction extends Function {
    void open(Configuration parameters) throws Exception;
    void close() throws Exception;
    RuntimeContext getRuntimeContext();
    IterationRuntimeContext getIterationRuntimeContext();
    void setRuntimeContext(RuntimeContext t);
}
```

该接口不仅定义了 open()方法和 close()方法分别，用于在 Function 初始化和关闭时完成一些任务，还定义了运行时上下文对象的 get/set 方法，可以用于获取任务信息。

AbstractRichFunction 是实现了 RichFunction 接口的抽象类，该接口定义好了 getRuntimeContext()方法、getIterationRuntimeContext()方法和 setRuntimeContext()方法，而 open()方法和 close()方法的实现为空。

类似 RichFlatMapFunction 这样的类通常会继承 AbstractRichFunction 类并实现 FlatMapFunction 接口，这种关系在 Function 体系中是十分典型的继承关系。图 3-3 展示了 Function 体系中典型的继承关系。

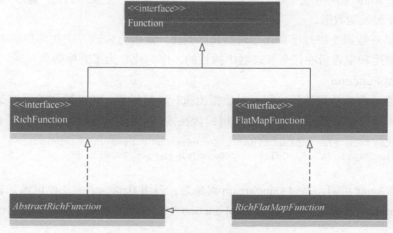

图 3-3 Function 体系中典型的继承关系

RichFlatMapFunction 的定义如下：

```
public abstract class RichFlatMapFunction<IN, OUT> extends AbstractRichFunction
implements FlatMapFunction<IN, OUT> {
    private static final long serialVersionUID = 1L;
    @Override
    public abstract void flatMap(IN value, Collector<OUT> out) throws Exception;
}
```

用户通常首先需要实现 flatMap()方法，然后可能会根据业务需求重写 AbstractRichFunction 中的 open()方法和 close()方法。

### 3.1.3　StreamOperator 算子分析

在 Flink 流处理中用 Flink-streaming-java 模块的 org.apache.Flink.streaming.api.operators.StreamOperator 表示一个算子，它可以被理解为"获取数据-处理数据-输出数据"这个过程的最小计算逻辑单元。一个任务中可以有多个算子。算子的数据可能来自外部数据源，也可能来自上一个算子；算子的输出可能是外部系统、中间文件，也可能是同一个任务中的下一个算子。当任务在 TaskManager 中被执行时，是通过调用 StreamOperator 的方法进而调用 Function 的相关方法的。

本书后文只要提到算子，均指 StreamOperator。

StreamOperator 接口的定义如下：

```
public interface StreamOperator<OUT> extends CheckpointListener, KeyContext, Disposable,
Serializable {
    // -----------------------------------------------------------------------
    //   与生命周期相关的方法
    // -----------------------------------------------------------------------
    void open() throws Exception;
    void close() throws Exception;
    void dispose() throws Exception;
    // -----------------------------------------------------------------------
    //   与状态快照相关的方法
    // -----------------------------------------------------------------------
    void prepareSnapshotPreBarrier(long checkpointId) throws Exception;
    OperatorSnapshotFutures snapshotState(
        long checkpointId,
        long timestamp,
        CheckpointOptions checkpointOptions,
        CheckpointStreamFactory storageLocation) throws Exception;
    void initializeState() throws Exception;
    // -----------------------------------------------------------------------
    //   其他方法
    // -----------------------------------------------------------------------
    void setKeyContextElement1(StreamRecord<?> record) throws Exception;
    void setKeyContextElement2(StreamRecord<?> record) throws Exception;
    ChainingStrategy getChainingStrategy();
    void setChainingStrategy(ChainingStrategy strategy);
    MetricGroup getMetricGroup();
    OperatorID getOperatorID();
}
```

该接口定义的方法大致可以分为 3 类，已在注释中说明。

抽象类 AbstractUdfStreamOperator 实现了该接口并封装了 Function 接口，该类实现了一些方法，如 open()方法和 close()方法，实质上就是调用 Function 对象的对应方法：

```
public void open() throws Exception {
    super.open();
    FunctionUtils.openFunction(userFunction, new Configuration());
}
public void close() throws Exception {
    super.close();
    functionsClosed = true;
    FunctionUtils.closeFunction(userFunction);
}
public static void openFunction(Function function, Configuration parameters) throws Exception{
    if (function instanceof RichFunction) {
      RichFunction richFunction = (RichFunction) function;
      richFunction.open(parameters);
    }
}
public static void closeFunction(Function function) throws Exception{
    if (function instanceof RichFunction) {
      RichFunction richFunction = (RichFunction) function;
      richFunction.close();
    }
}
```

常用的实现类基本继承自 AbstractUdfStreamOperator。下面举几个典型的例子。

1. StreamSource

StreamSource 类的定义如下：

```
public class StreamSource<OUT, SRC extends SourceFunction<OUT>> extends AbstractUdfStreamOperator<OUT, SRC>
```

在构造方法中主要是对 Function 字段赋值：

```
public StreamSource(SRC sourceFunction) {
    super(sourceFunction);
    ...
}
public AbstractUdfStreamOperator(F userFunction) {
    this.userFunction = requireNonNull(userFunction);
    ...
}
```

StreamSource 类中实现了 run()方法，该方法会在任务被部署到 TaskManager 后被触发执行。

```
public void run(final Object lockingObject,
    final StreamStatusMaintainer streamStatusMaintainer,
    final Output<StreamRecord<OUT>> collector,
    final OperatorChain<?, ?> operatorChain) throws Exception {
    ...
    // 实例化 SourceFunction.SourceContext 对象
    this.ctx = StreamSourceContexts.getSourceContext(
      timeCharacteristic,
      getProcessingTimeService(),
```

```
            lockingObject,
            streamStatusMaintainer,
            collector,
            watermarkInterval,
            -1);
    try {
        // 调用 SourceFunction 对象中的 run()方法
        userFunction.run(ctx);
        ...
    } finally {
        ...
    }
}
```

回顾上文介绍过的 SourceFunction 中的 run()方法正是在这里被调用的。

2. StreamFlatMap

StreamFlatMap 类也继承自 AbstractUdfStreamOperator 抽象类,同时它还实现了 OneInputStreamOperator 接口。OneInputStreamOperator 接口是 StreamOperator 接口的子接口,其定义如下:

```
public interface OneInputStreamOperator<IN, OUT> extends StreamOperator<OUT> {
    void processElement(StreamRecord<IN> element) throws Exception;
    void processWatermark(Watermark mark) throws Exception;
    void processLatencyMarker(LatencyMarker latencyMarker) throws Exception;
}
```

这里读者只需要理解 processElement()方法。该方法定义了单条数据该如何处理。从上文可以看出,StreamSource 类是没有该方法的,这是因为源的任务需要定义如何加载数据,后面的任务则只需要定义单条数据的处理方式。该方法同样是在 TaskManager 处理数据时被执行。

StreamFlatMap 的 processElement()方法如下:

```
public void processElement(StreamRecord<IN> element) throws Exception {
    ...
    userFunction.flatMap(element.getValue(), collector);
}
```

在该方法中调用了 FlatMapFunction 的 flatMap()方法。

3. CoStreamMap/CoStreamFlatMap

第 2 章介绍了 CoMap/CoFlatMap 转换操作。该转换操作会将自定义的 Function 对象封装到 CoStreamMap/CoStreamFlatMap 中,它们也继承自 AbstractUdfStreamOperator 抽象类且实现了 TwoInputStreamOperator 接口。TwoInputStreamOperator 接口也是 StreamOperator 的子接口。

```
public interface TwoInputStreamOperator<IN1, IN2, OUT> extends StreamOperator<OUT> {
    void processElement1(StreamRecord<IN1> element) throws Exception;
    void processElement2(StreamRecord<IN2> element) throws Exception;
    void processWatermark1(Watermark mark) throws Exception;
    void processWatermark2(Watermark mark) throws Exception;
    void processLatencyMarker1(LatencyMarker latencyMarker) throws Exception;
    void processLatencyMarker2(LatencyMarker latencyMarker) throws Exception;
}
```

由于是将两个数据流连接在一起,因此在 TwoInputStreamOperator 中分别为每个数据流定义了一套方法。CoStreamMap/CoStreamFlatMap 对 processElement1()和 processElement2()的实现分别

如下。

（1）CoStreamMap。

```
public void processElement1(StreamRecord<IN1> element) throws Exception {
    output.collect(element.replace(userFunction.map1(element.getValue())));
}
public void processElement2(StreamRecord<IN2> element) throws Exception {
    output.collect(element.replace(userFunction.map2(element.getValue())));
}
```

（2）CoStreamFlatMap。

```
public void processElement1(StreamRecord<IN1> element) throws Exception {
    ...
    userFunction.flatMap1(element.getValue(), collector);
}
public void processElement2(StreamRecord<IN2> element) throws Exception {
    ...
    userFunction.flatMap2(element.getValue(), collector);
}
```

它们都分别调用了 Function 对象中自定义的方法。

上面提及的接口和类的继承关系如图 3-4 所示，其在 StreamOperator 体系中具有代表性。

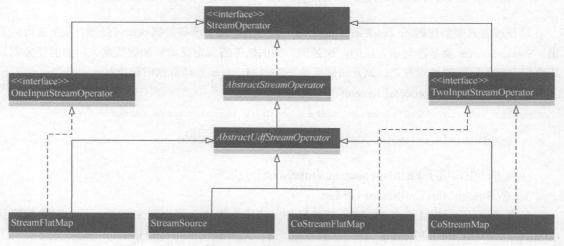

图 3-4  StreamOperator 继承关系

### 3.1.4  转换操作分析

Flink-core 模块下的 org.apache.Flink.api.dag.Transformation 是一个抽象类，是 DataStream API 中的重要概念之一。在 DataStream API 中，Flink 将每一种数据转换操作（包括数据的加载、输出、重分区等）都抽象成了一种 Transformation 对象。后添加的 Transformation 对象中封装了前面的 Transformation 对象，由此产生树状的结构，在生成执行图时直接从 Transformation 树进行转换。不过，并不是每一种 Transformation 对象都会被转换成具体的算子，从下文也可以看出，一些实现类并没有封装 StreamOperator。图 3-5 所示为 Transformation 树。

在 Transformation 树中，会包含重分区、合并等操作的 Transformation 对象，但是在运行时树中却并非如此，运行时树如图 3-6 所示。

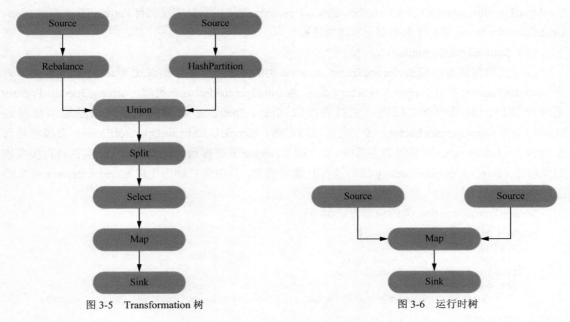

图 3-5　Transformation 树　　　　　　　　　图 3-6　运行时树

在图 3-6 所示的运行时树中，Source、Map、Sink 这些操作都被转换成了具体的算子，而 HashPartition、Union 这些转换操作却没有出现。实际上，在后续生成执行图时，Source、Map、Sink 这些操作会被转换成执行图中的节点（node），而 HashPartition、Union 这些操作的相关信息会被保存在连接节点的边（edge）上。基于这样的区别，Transformation 被分成了物理 Transformation 和虚拟 Transformation。在代码中则体现为物理 Transformation 都继承自 PhysicalTransformation 抽象类（该抽象类也继承自 Transformation），而虚拟 Transformation 都直接继承自 Transformation。

下面是 Transformation 类的一些重要字段。
- id：Transformation 对象的唯一标识，由一个递增的静态变量生成。
- name：转换操作的名称。对于常用的转换操作，有其默认的名称，也提供了 set 方法对其进行设置。
- outputType：TypeInformation 类型，表示输出的类型。
- parallelism：该转换操作的并行度。
- maxParallelism：该转换操作的最大并行度。
- uid：用户给该转换操作指定的标识。该标识一方面与 name 字段作用相同，可以给转换操作标记一些业务含义。另一方面，如果用户指定了 uid，那么该任务节点会利用 uid 生成 OperatorID，以确保作业重启时该节点的 OperatorID 不变；如果用户没有指定 uid，任务节点的 OperatorID 就会根据当时作业的拓扑结构生成，这意味着停止作业后若添加或减少一些转换操作，则 Flink 会找不到原来的节点。

- slotSharingGroup:共享的任务槽组。

PhysicalTransformation 的实现类有 SourceTransformation、OneInputTransformation、TwoInputTransformation 和 SinkTransformation。Transformation 的其他实现类有 PartitionTransformation、UnionTransformation 等。下面对部分实现类进行介绍。

(1) SourceTransformation。

在定义数据源时会构造 SourceTransformation 类的实例。该实现类仅比 Transformation 类多一个 operatorFactory 字段。operatorFactory 表示 StreamOperatorFactory 类型。StreamOperatorFactory 是一个接口,表示一个工厂类,它具有生成 StreamOperator 实例的功能。这里通常会得到 SimpleUdfStreamOperatorFactory 这个实现类的实例。SimpleUdfStreamOperatorFactory 直接将先前生成的 StreamOperator 对象封装在其中,在 TaskManager 需要得到 StreamOperator 实例时直接将该对象取出。StreamOperatorFactory 接口还有其他实现类,其中包括动态生成 StreamOperator 对象的工厂类。

SourceTransformation 类的构造方法如下:

```
public SourceTransformation(
    String name,
    StreamSource<T, ?> operator,
    TypeInformation<T> outputType,
    int parallelism) {
        this(name, SimpleOperatorFactory.of(operator), outputType, parallelism);
}
public SourceTransformation(
    String name,
    StreamOperatorFactory<T> operatorFactory,
    TypeInformation<T> outputType,
    int parallelism) {
        super(name, outputType, parallelism);
        this.operatorFactory = operatorFactory;
}
```

(2) OneInputTransformation。

在进行 Map、FlatMap、Filter 等单输入源的转换操作时,会生成 OneInputTransformation 实现类的实例。比如在调用 flatMap() 方法时,会进入下面的调用链:

```
public <R> SingleOutputStreamOperator<R> flatMap(FlatMapFunction<T, R> flatMapper,
TypeInformation<R> outputType) {
    return transform("Flat Map", outputType, new StreamFlatMap<>(clean(flatMapper)));

}
public <R> SingleOutputStreamOperator<R> transform(
    String operatorName,
    TypeInformation<R> outTypeInfo,
    OneInputStreamOperator<T, R> operator) {
        return doTransform(operatorName, outTypeInfo, SimpleOperatorFactory.of(operator));
}
protected <R> SingleOutputStreamOperator<R> doTransform(
    String operatorName,
    TypeInformation<R> outTypeInfo,
```

```
        StreamOperatorFactory<R> operatorFactory) {
    transformation.getOutputType();
    // 构造一个 Transformation 对象，将上一个 Transformation 对象、operatorFactory 对象封装进去
    OneInputTransformation<T, R> resultTransform = new OneInputTransformation<>(
        this.transformation,
        operatorName,
        operatorFactory,
        outTypeInfo,
        environment.getParallelism());
    // 构造一个 DataStream 对象并返回
    @SuppressWarnings({"unchecked", "rawtypes"})
    SingleOutputStreamOperator<R> returnStream = new SingleOutputStreamOperator(environment, resultTransform);
    // 将该 Transformation 对象添加到执行环境中
    getExecutionEnvironment().addOperator(resultTransform);
    return returnStream;
}
```

该实现类定义了如下字段。

- input：Transformation 类型，表示上一个转换操作。
- operatorFactory：StreamOperatorFactoryl 类型，可生成 StreamOperator 实例的工厂类。
- stateKeySelector：KeySelector 类型，表示键的提取器，默认值为 null；在 keyBy()方法后添加 OneInputTransformation 对象时，就会对该对象设置 stateKeySelector。
- stateKeyType：TypeInformation 类型，表示键的数据类型，默认值为 null；在 keyBy()方法后添加 OneInputTransformation 对象时，就会对该对象设置 stateKeyType。

OneInputTransformation 类的构造方法如下：

```
public OneInputTransformation(
    Transformation<IN> input,
    String name,
    StreamOperatorFactory<OUT> operatorFactory,
    TypeInformation<OUT> outputType,
    int parallelism) {
    super(name, outputType, parallelism);
    this.input = input;
    this.operatorFactory = operatorFactory;
}
```

在构造 OneInputTransformation 对象时，需要传入前一个 Transformation 对象，并赋值给 input 字段。

（3）PartitionTransformation。

调用 keyBy()方法时，会生成 PartitionTransformation 实现类的实例。这时会调用 KeyedStream 的构造方法，如下：

```
public KeyedStream(DataStream<T> dataStream, KeySelector<T, KEY> keySelector) {
    this(dataStream, keySelector, TypeExtractor.getKeySelectorTypes(keySelector, dataStream.getType()));
}
public KeyedStream(DataStream<T> dataStream, KeySelector<T, KEY> keySelector,
    TypeInformation<KEY> keyType) {
    this(
```

```
        dataStream,
        new PartitionTransformation<>(
            dataStream.getTransformation(),
            new KeyGroupStreamPartitioner<>(keySelector, StreamGraphGenerator.DEFAULT_
LOWER_BOUND_MAX_PARALLELISM)),
        keySelector,
        keyType);
}
```

PartitionTransformation 类定义了如下字段。

- input：Transformation 类型，表示上一个转换操作。
- partitioner：StreamPartitioner 类型，表示分区器，其中封装了 KeySelector 对象。
- shuffleMode：ShuffleMode 类型，表示上下游算子的数据交换方式，它是一个枚举类，其中包括 PIPELINED（上游一生产数据就被下游消费）、BATCH（上游将数据全部生产完毕再由下游消费）和 UNDEFINED（暂时未定义，最终框架会在 PIPELINED 和 BATCH 中选择其一）。

PartitionTransformation 类的构造方法如下：

```
public PartitionTransformation(Transformation<T> input, StreamPartitioner<T> partitioner) {
    this(input, partitioner, ShuffleMode.UNDEFINED);
}
public PartitionTransformation(
    Transformation<T> input,
    StreamPartitioner<T> partitioner,
    ShuffleMode shuffleMode) {
    super("Partition", input.getOutputType(), input.getParallelism());
    this.input = input;
    this.partitioner = partitioner;
    this.shuffleMode = checkNotNull(shuffleMode);
}
```

PartitionTransformation 类没有继承自 PhysicalTransformation 类，它只封装分区器等信息。

（4）UnionTransformation。

调用 union() 方法时，会生成 UnionTransformation 实现类的实例，如下：

```
public final DataStream<T> union(DataStream<T>... streams) {
    // 将每个数据流中的 Transformation 对象都添加到列表中
    List<Transformation<T>> unionedTransforms = new ArrayList<>();
    unionedTransforms.add(this.transformation);
    for (DataStream<T> newStream : streams) {
        if (!getType().equals(newStream.getType())) {
            throw new IllegalArgumentException("Cannot union streams of different types: "
                + getType() + " and " + newStream.getType());
        }
        unionedTransforms.add(newStream.getTransformation());
    }
    // 构造 UnionTransformation 对象，将其封装到 DataStream 中并返回
    return new DataStream<>(this.environment, new UnionTransformation<>(unionedTransforms));
}
```

UnionTransformation 类定义的字段如下。

inputs：Transformation 类型，表示合并在一起的每个数据流中的上一个转换操作。

UnionTransformation 类没有继承自 PhysicalTransformation，它只封装了每个数据流中的 Transformation 对象。其构造方法如下：

```java
public UnionTransformation(List<Transformation<T>> inputs) {
    super("Union", inputs.get(0).getOutputType(), inputs.get(0).getParallelism());
    for (Transformation<T> input: inputs) {
        if (!input.getOutputType().equals(getOutputType())) {
            throw new UnsupportedOperationException("Type mismatch in input " + input);
        }
    }
    this.inputs = Lists.newArrayList(inputs);
}
```

### 3.1.5　数据流相关类分析

DataStream 类位于 Flink-streaming-java 模块的 org.apache.Flink.streaming.api.datastream 包中，它是在 DataStream API 中对数据流的抽象。它不会保存真实的数据，只会封装数据的加载、计算和输出逻辑。它与 Transformation、StreamOperator 和 Function 等类的关系如图 3-7 所示。

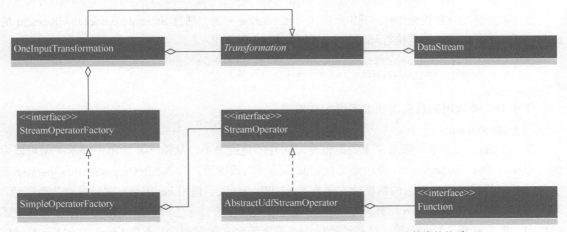

图 3-7　DataStream 与 Transformation、StreamOperator 和 Function 等类的关系

图 3-8 展示了 DataStream 等表示数据流的类的继承关系。

当调用执行环境的 addSource() 方法添加数据源时，返回的是 DataStreamSource 类的对象。调用一般的转换操作返回的是 SingleOutputStreamOperator 类的对象。调用 keyBy() 方法返回的是 KeyedStream 类的对象。

除了 DataStream 类及其子类，Flink 的 Flink-streaming-java 模块中还有很多类可用于表示不同类型的数据流。比如，调用 join() 方法会返回 JoinedStreams 类的对象，表示两个数据流连接到一起后形成的数据流，可以在该数据流上执行一些与连接相关的操作；调用 window() 方法会返回 WindowedStream/AllWindowedStream 类的对象，表示加窗的数据流，可以在该数据流上执行一些与加窗相关的操作；定义输出方式时会返回 DataStreamSink 类的对象，该类没有继承自 DataStream

类，因为在它之后不能再定义其他转换操作。此外还有 CoGroupedStreams、ConnectedStreams 等类。所有的转换操作都是直接定义在数据流的对象之上的。

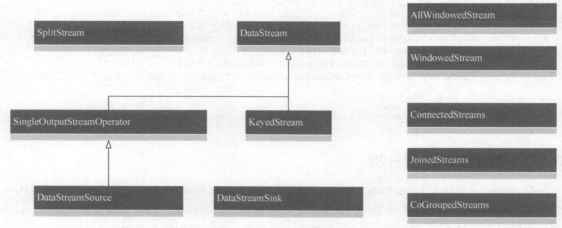

图 3-8 DataStream 等表示数据流的类的继承关系

在对数据流定义转换操作时，其中一些 Transformation 对象会通过 StreamExecutionEnvironment 的 addOperator()方法被添加到执行环境的 transformations 列表中：

```
public void addOperator(Transformation<?> transformation) {
    this.transformations.add(transformation);
}
```

下面对几个常用的数据流相关类进行分析。

（1）DataStream。

DataStream 类表示一般意义上的数据流，它并不是抽象类。从图 3-8 所示的继承关系可以看到，已经有 DataStreamSource 表示定义数据源后返回的数据流，有 SingleOutputStreamOperator 表示执行一般的转换操作返回的数据流，还有 KeyedStream 表示调用 keyBy()方法后返回的数据流，那么为何还需要 DataStream 这样一个实现类呢？这是因为还有一些转换操作并没有被包含在上述操作中，如 Shuffle、Union 等，这些转换操作既不属于定义数据源，又不像 Map、FlatMap 等操作一样是具体的任务，也不是用键的方式进行重分区，而且调用这些方法返回的数据流也不像 JoinedStreams 或 WindowedStream 一样可以用于执行一些特殊的操作。因此，这些转换操作返回的数据流就用 DataStream 这一最一般意义上的数据流来表示。

DataStream 中的重要字段如下。

- environment：StreamExecutionEnvironment 类型，表示执行环境。正是因为有该对象，才使得对 DataStream 定义的转换操作都被封装到了执行环境中。
- transformation：Transformation 类型，表示数据流经过这个 transformation 操作才形成该 DataStream 对象。

在构造方法中对这两个字段赋值：

```
public DataStream(StreamExecutionEnvironment environment, Transformation<T> transformation) {
```

```java
    this.environment = Preconditions.checkNotNull(environment, "Execution Environment
must not be null.");
    this.transformation = Preconditions.checkNotNull(transformation, "Stream Transformation
must not be null.");
}
```

因为封装了上述两个字段,所以 DataStream 针对这两个字段提供了很多获取其中的属性值的方法,如:

```java
public int getId() {
    return transformation.getId();
}
public int getParallelism() {
    return transformation.getParallelism();
}
public TypeInformation<T> getType() {
    return transformation.getOutputType();
}
public ExecutionConfig getExecutionConfig() {
    return environment.getConfig();
}
```

对数据流类来说,重要的是定义转换操作的方法。可以在 DataStream 上执行的转换操作主要如下:

```java
public <K> KeyedStream<T, K> keyBy(KeySelector<T, K> key) {
    return new KeyedStream<>(this, clean(key));
}
public <R> SingleOutputStreamOperator<R> map(MapFunction<T, R> mapper) {
    TypeInformation<R> outType = TypeExtractor.getMapReturnTypes(clean(mapper), getType(),
        Utils.getCallLocationName(), true);
    return map(mapper, outType);
}
public <R> SingleOutputStreamOperator<R> flatMap(FlatMapFunction<T, R> flatMapper) {
    TypeInformation<R> outType = TypeExtractor.getFlatMapReturnTypes(clean(flatMapper),
        getType(), Utils.getCallLocationName(), true);
    return flatMap(flatMapper, outType);
}
public SingleOutputStreamOperator<T> filter(FilterFunction<T> filter) {
    return transform("Filter", getType(), new StreamFilter<>(clean(filter)));
}
public DataStreamSink<T> addSink(SinkFunction<T> sinkFunction) {
    ...
    StreamSink<T> sinkOperator = new StreamSink<>(clean(sinkFunction));
    DataStreamSink<T> sink = new DataStreamSink<>(this, sinkOperator);
    getExecutionEnvironment().addOperator(sink.getTransformation());
    return sink;
}
```

此外还有重分区、加窗等操作,在此不一一列出。大部分转换操作的语义在第 2 章介绍过。
在执行转换操作时,底层会调用 doTransform()方法:

```java
protected <R> SingleOutputStreamOperator<R> doTransform(
    String operatorName,
    TypeInformation<R> outTypeInfo,
    StreamOperatorFactory<R> operatorFactory) {
```

```
    ...
    // 将计算逻辑封装成 Transformation 对象
    OneInputTransformation<T, R> resultTransform = new OneInputTransformation<>(
        this.transformation,
        operatorName,
        operatorFactory,
        outTypeInfo,
        environment.getParallelism());
    // 构造一个数据流，将 Transformation 对象封装进去
    @SuppressWarnings({"unchecked", "rawtypes"})
    SingleOutputStreamOperator<R> returnStream = new SingleOutputStreamOperator(environment, resultTransform);
    // 将 Transformation 对象添加到执行环境的 transformations 列表中
    getExecutionEnvironment().addOperator(resultTransform);
    return returnStream;
}
```

（2）SingleOutputStreamOperator。

SingleOutputStreamOperator 类是 DataStream 的子类，表示定义一个转换操作后返回的数据流，该转换操作会形成具体的任务在 TaskManager 被执行。总体而言，SingleOutputStreamOperator 与 DataStream 没有太大差异。因为定义了具体的转换操作，所以有下面这些 set 方法：

```
public SingleOutputStreamOperator<T> name(String name){
    transformation.setName(name);
    return this;
}
@PublicEvolving
public SingleOutputStreamOperator<T> uid(String uid) {
    transformation.setUid(uid);
    return this;
}
public SingleOutputStreamOperator<T> setParallelism(int parallelism) {
    OperatorValidationUtils.validateParallelism(parallelism, canBeParallel());
    transformation.setParallelism(parallelism);
    return this;
}
@PublicEvolving
public SingleOutputStreamOperator<T> setMaxParallelism(int maxParallelism) {
    OperatorValidationUtils.validateMaxParallelism(maxParallelism, canBeParallel());
    transformation.setMaxParallelism(maxParallelism);
    return this;
}
public SingleOutputStreamOperator<T> returns(TypeInformation<T> typeInfo) {
    transformation.setOutputType(typeInfo);
    return this;
}
```

（3）KeyedStream。

对 DataStream 调用 keyBy()方法后，会返回 KeyedStream。

KeyedStream 中定义的字段如下。

- keySelector：KeySelector 类型，表示键的提取器。
- keyType：TypeInformation 类型，表示键的数据类型。

在 **KeyedStream** 类的构造方法中，这两个字段会被赋值：

```
public KeyedStream(DataStream<T> dataStream, KeySelector<T, KEY> keySelector) {
    this(dataStream, keySelector, TypeExtractor.getKeySelectorTypes(keySelector, dataStream.getType()));
}
public KeyedStream(DataStream<T> dataStream, KeySelector<T, KEY> keySelector,
TypeInformation<KEY> keyType) {
    this(
        dataStream,
        new PartitionTransformation<>(
            dataStream.getTransformation(),
            new KeyGroupStreamPartitioner<>(keySelector, StreamGraphGenerator.DEFAULT_LOWER_BOUND_MAX_PARALLELISM)),
        keySelector,
        keyType);
}
KeyedStream(
    DataStream<T> stream,
    PartitionTransformation<T> partitionTransformation,
    KeySelector<T, KEY> keySelector,
    TypeInformation<KEY> keyType) {
    super(stream.getExecutionEnvironment(), partitionTransformation);
    this.keySelector = clean(keySelector);
    this.keyType = validateKeyType(keyType);
}
```

从这里可以看到，在构造 KeyedStream 的过程中，同时构造了 PartitionTransformation 对象，并赋值给了 transformation 字段。

因为 KeyedStream 继承自 DataStream，所以对 DataStream 的转换操作对 KeyedStream 同样适用，只不过 KeyedStream 重写了 doTransform()方法和 addSink()方法：

```
protected <R> SingleOutputStreamOperator<R> doTransform(
        final String operatorName,
        final TypeInformation<R> outTypeInfo,
        final StreamOperatorFactory<R> operatorFactory) {
    SingleOutputStreamOperator<R> returnStream = super.doTransform(operatorName, outTypeInfo,operatorFactory);
    // 给 Transformation 对象设置与键相关的字段
    OneInputTransformation<T, R> transform = (OneInputTransformation<T, R>) returnStream.getTransformation();
    transform.setStateKeySelector(keySelector);
    transform.setStateKeyType(keyType);
    return returnStream;
}
public DataStreamSink<T> addSink(SinkFunction<T> sinkFunction) {
    DataStreamSink<T> result = super.addSink(sinkFunction);
    // 给 Transformation 对象设置与键相关的字段
    result.getTransformation().setStateKeySelector(keySelector);
    result.getTransformation().setStateKeyType(keyType);
    return result;
}
```

**OneInputTransformation** 和 **SinkTransformation** 中都有 stateKeySelector 和 stateKeyType 字段，其默认值都为 null，若在 keyBy()方法后添加这些 Transformation 对象，则会给这两个字段

赋值。

此外，还有一些转换操作只能在 keyBy()方法后被执行，这些操作的语义在第 2 章中已经介绍过。下面是其中的一些方法：

```
public SingleOutputStreamOperator<T> max(String field) {
    return aggregate(new ComparableAggregator<>(field, getType(), AggregationFunction.
AggregationType.MAX, false, getExecutionConfig()));
}
protected SingleOutputStreamOperator<T> aggregate(AggregationFunction<T> aggregate) {
    StreamGroupedReduce<T> operator = new StreamGroupedReduce<T>(
        clean(aggregate), getType().createSerializer(getExecutionConfig()));
    return transform("Keyed Aggregation", getType(), operator);
}
public SingleOutputStreamOperator<T> reduce(ReduceFunction<T> reducer) {
    return transform("Keyed Reduce", getType(), new StreamGroupedReduce<T>(
        clean(reducer), getType().createSerializer(getExecutionConfig())));
}
```

经过这些转换操作，数据流类又变回了 SingleOutputStreamOperator。

（4）CoGroupedStreams 与 JoinedStreams。

对两个数据流进行 Window CoGroup/Window Join 转换操作时，会产生 CoGroupedStreams/JoinedStreams。这两种数据流类在字段、方法、内部类方面都极为相似，并且它们都没有继承自 DataStream。可以将 JoinedStreams 理解为 CoGroupedStreams 的一个特例，因为 JoinedStreams 在内部会被转换为 CoGroupedStreams。下面分析 CoGroupedStreams。

CoGroupedStreams 中定义了两个字段。

- input1：DataStream 类型，表示第一个数据流。
- input2：DataStream 类型，表示第二个数据流。

因为 CoGroupedStreams 没有继承自 DataStream，所以对于这样的数据流类，只能调用 where()方法及其重载方法：

```
public <KEY> Where<KEY> where(KeySelector<T1, KEY> keySelector) {
    final TypeInformation<KEY> keyType = TypeExtractor.getKeySelectorTypes(keySelector,
input1.getType());
    return where(keySelector, keyType);
}
```

这里的 Where 类型是内部类，包含两个字段。

- keySelector1：KeySelector 类型，表示第一个数据流的键的提取器。
- keyType：TypeInformation 类型，表示键的数据类型。

对于 Where 对象，只能调用 equalTo()方法及其重载方法：

```
public EqualTo equalTo(KeySelector<T2, KEY> keySelector) {
    final TypeInformation<KEY> otherKey = TypeExtractor.getKeySelectorTypes(keySelector,
input2.getType());
    return equalTo(keySelector, otherKey);
}
```

返回的 EqualTo 类型是 Where 类的内部类。它的字段如下。
keySelector2：KeySelector 类型，表示第二个数据流的键的提取器。

这样，就准备好了两个数据流的键的提取器和数据类型，接下来只能调用 window()方法构建 Window 的相关信息：

```
public <W extends Window> WithWindow<T1, T2, KEY, W> window(WindowAssigner<? super TaggedUnion<T1, T2>, W> assigner) {
    return new WithWindow<>(input1, input2, keySelector1, keySelector2, keyType, assigner, null, null, null);
}
```

WithWindow 是 CoGroupedStreams 的内部类，可以对其进行一些关于窗口的特殊操作。得到一个 WithWindow 对象后，通过调用 apply()方法添加 CoGroupFunction：

```
public <T> DataStream<T> apply(CoGroupFunction<T1, T2, T> function) {
    TypeInformation<T> resultType = TypeExtractor.getCoGroupReturnTypes(
        function,
        input1.getType(),
        input2.getType(),
        "CoGroup",
        false);
    return apply(function, resultType);
}
```

接下来，Flink 会将其转换成 WindowedStream，并对 function 对象进行一些封装，核心思想与前面介绍过的数据流的相关内容一致。

## 3.2 DataSet API

回顾第 2 章中介绍的 DataSet API 版本的 WordCount 用例：

```
// （1）获取执行环境
ExecutionEnvironment env = ExecutionEnvironment.getExecutionEnvironment();
// （2）指定数据的加载方式
DataSet<String> text = env.fromElements(WordCountData.WORDS);
// （3）指定数据的转换方式
DataSet<Tuple2<String, Integer>> counts =
    text.flatMap(new Tokenizer())
        .groupBy(0).sum(1);
// （4）指定如何输出并触发执行
counts.print();
```

下面对上述代码涉及的流程的核心内容进行分析。

### 3.2.1 ExecutionEnvironment 执行环境

ExecutionEnvironment 位于 Flink-java 模块的 org.apache.Flink.api.java 包中，是批处理时的执行环境，作用与 StreamExecutionEnvironment 类似。下面从执行环境的初始化、定义数据的加载方式、注册 DataSink 和触发程序运行这几个方面来进行介绍。

1. 执行环境的初始化

在第 2 章介绍过，虽然上述代码最后调用的是 DataSet 的 print()方法，但是该方法内部还是调用了执行环境的 execute()方法。因此 ExecutionEnvironment 与 StreamExecutionEnvironment 的设计思路是一致的。ExecutionEnvironment 与其常用子类的继承关系如图 3-9 所示。

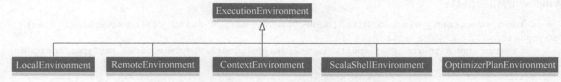

图 3-9　ExecutionEnvironment 与其常用子类的继承关系

与 StreamExecutionEnvironment 及其子类的关系一样，ExecutionEnvironment 的子类与它的核心逻辑完全相同，下面只分析 ExecutionEnvironment 类，其核心字段如下。

- contextEnvironmentFactory：ExecutionEnvironmentFactory 类型，用于创建执行环境的工厂类。
- threadLocalContextEnvironmentFactory：ThreadLocal 类型，用于保存 ExecutionEnvironmentFactory 对象。
- defaultLocalDop：本地模式下默认的并行度。
- config：ExecutionConfig 类型，用于维护各种配置，如并行度等。
- sinks：DataSink<?> 类型，所有计算逻辑都被封装在这个对象中。其可类比为 StreamExecutionEnvironment 中的 transformations 字段。
- lastJobExecutionResult：JobExecutionResult 类型，表示整个作业被执行完以后返回的对象。
- configuration：Configuration 类型，从配置文件中读取出的配置信息会保存在其中。

与 StreamExecutionEnvironment 相比，ExecutionEnvironment 主要有以下几个不同点。

- 没有与时间、窗口相关的字段。由于 DataSet API 处理的是整个数据集，因此不需要加窗操作。
- 没有与检查点有关的字段。这符合批量计算框架下并非必须有状态容错机制的朴素设计思想。
- 多了 lastJobExecutionResult 字段。因为所处理的是有界的数据集，所以当数据处理完后可以获得一些返回值，以用来做下一步的处理。

ExecutionEnvironment 的初始化方式与 StreamExecutionEnvironment 的一致：

```
public static ExecutionEnvironment getExecutionEnvironment() {
    return Utils.resolveFactory(threadLocalContextEnvironmentFactory, contextEnvironmentFactory)
        .map(ExecutionEnvironmentFactory::createExecutionEnvironment)
        .orElseGet(ExecutionEnvironment::createLocalEnvironment);
}
```

静态字段 threadLocalContextEnvironmentFactory、contextEnvironmentFactory 同样是可以在提交作业的过程中被提前设置的。在提交作业的过程中，客户端代码会设置一个 ContextEnvironmentFactory 类的对象，它的 createExecutionEnvironment() 方法如下：

```
public ExecutionEnvironment createExecutionEnvironment() {
    return new ContextEnvironment(
        executorServiceLoader,
        configuration,
        userCodeClassLoader);
}
```

这样就创建了一个 ContextEnvironment 对象。

### 2. 定义数据的加载方式

这里仍以 fromElements()方法为例进行介绍。读者可观察其与 DataStream 添加数据源的不同。

```
public final <X> DataSource<X> fromElements(X... data) {
    ...
    TypeInformation<X> typeInfo;
    try {
        typeInfo = TypeExtractor.getForObject(data[0]);
    }
    catch (Exception e) {
        ...
    }
    return fromCollection(Arrays.asList(data), typeInfo, Utils.getCallLocationName());
}
private <X> DataSource<X> fromCollection(Collection<X> data, TypeInformation<X> type, String callLocationName) {
    CollectionInputFormat.checkCollection(data, type.getTypeClass());
    return new DataSource<>(this, new CollectionInputFormat<>(data, type.createSerializer(config)), type, callLocationName);
}
```

在定义数据源的过程中，构造了 CollectionInputFormat 对象。CollectionInputFormat 是 InputFormat 接口的实现类。最后将该对象封装成了 DataSource 对象并返回。DataSource 的父类是 DataSet。

回顾前文介绍的在 DataStream API 中，需要先构造一个 Function 对象，然后将其封装成 StreamOperator 对象，最后返回一个 DataStream 对象。由此可见，DataStream API 与 DataSet API 的封装逻辑略有不同。这里的 InputFormat 可以被类比成 DataStream API 中的 Function 对象。可以注意到这里暂时没有与 StreamOperator 同级别的概念，InputFormat 对象会被直接放入 DataSource 对象中。可以推测在 TaskManager 流任务和批任务算子的封装逻辑有所不同。

定义数据加载方式的一般方法如下：

```
public <X> DataSource<X> createInput(InputFormat<X, ?> inputFormat, TypeInformation<X> producedType) {
    if (inputFormat == null) {
        throw new IllegalArgumentException("InputFormat must not be null.");
    }
    if (producedType == null) {
        throw new IllegalArgumentException("Produced type information must not be null.");
    }
    return new DataSource<>(this, inputFormat, producedType, Utils.getCallLocationName());
}
```

利用 createInput()方法可自定义 InputFormat。

### 3. 注册 DataSink

在 DataStream API 中，计算逻辑会被添加到 StreamExecutionEnvironment 的 transformations 字段；在 DataSet API 中，计算逻辑被封装在 DataSet 对象中，最终 DataSet 会作为 DataSink 的输入，这样 DataSink 对象就持有了所有的计算逻辑。这个 DataSink 对象会被注册到执行环境中。

```
void registerDataSink(DataSink<?> sink) {
    this.sinks.add(sink);
}
```

这里的 sinks 字段可被类比成 transformations 字段。

4. 触发程序运行

与 StreamExecutionEnvironment 类似，ExecutionEnvironment 也提供了 execute()方法用于触发程序运行。此后的流程与 DataStream API 的一致，只不过生成执行图的具体算法有所不同。

## 3.2.2　InputFormat 和 OutputFormat

在 DataStream API 中，数据的加载方式在 SourceFunction 中定义，并且之后对数据进行具体处理的转换操作都由 Function 接口的实现类来定义。在 DataSet API 中，中间的转换操作仍由 Function 接口定义，而 source 和 sink 的逻辑分别由 Flink-core 模块下的 org.apache.Flink.api.common.io.InputFormat 和 org.apache.Flink.api.common.io.OutputFormat 定义。

InputFormat 接口定义了多个方法，其中不少方法已在实现了该接口的抽象类中实现，需要自定义的方法主要是下面几个：

```
void open(T split) throws IOException;
boolean reachedEnd() throws IOException;
OT nextRecord(OT reuse) throws IOException;
```

以 CollectionInputFormat 类为例，其构造方法如下：

```
public CollectionInputFormat(Collection<T> dataSet, TypeSerializer<T> serializer) {
    if (dataSet == null) {
        throw new NullPointerException();
    }
    this.serializer = serializer;
    this.dataSet = dataSet;
}
```

在构造方法中，将传入的数据集赋值给了其中的 dataSet 字段。于是在 open()方法中定义了如下逻辑：

```
public void open(GenericInputSplit split) throws IOException {
    super.open(split);
    this.iterator = this.dataSet.iterator();
}
```

通过 dataSet 字段构造了一个迭代器对象。接下来，在 reachedEnd()方法和 nextRecord()方法中，通过迭代器不断获取下一条数据：

```
public boolean reachedEnd() throws IOException {
    return !this.iterator.hasNext();
}
public T nextRecord(T record) throws IOException {
    return this.iterator.next();
}
```

OutputFormat 接口中主要需要自定义的方法有下面几个：

```
void open(int taskNumber, int numTasks) throws IOException;
void writeRecord(IT record) throws IOException;
void close() throws IOException;
```

以 CollectHelper 为例，DataSet API 版本的 WordCount 会在 collect()方法中定义 DataSink：

```
public List<T> collect() throws Exception {
    final String id = new AbstractID().toString();
    final TypeSerializer<T> serializer = getType().createSerializer(getExecutionEnvironment().
getConfig());
    // 构造 DataSink。在构造过程中会构造累加器并添加到上下文中
    this.output(new Utils.CollectHelper<>(id, serializer)).name("collect()");
    JobExecutionResult res = getExecutionEnvironment().execute();
    // 获取指定的累加器
    ArrayList<byte[]> accResult = res.getAccumulatorResult(id);
    ...
}
```

上例中的 CollectHelper 就是 OutputFormat 的一个实现类,它实现了上述 3 个方法。

```
public void open(int taskNumber, int numTasks) {
    this.accumulator = new SerializedListAccumulator<>();
}
public void writeRecord(T record) throws IOException {
    accumulator.add(record, serializer);
}
public void close() {
    getRuntimeContext().addAccumulator(id, accumulator);
}
```

首先在 open()方法中构造累加器,然后在 writeRecord()方法中将数据添加进其中,最后在 close()方法中添加累加器到上下文中,这样在整个任务执行完后,就可以通过返回值 JobExecutionResult 对象获取累加器中的数据。

### 3.2.3 数据集相关类分析

正如在 DataStream API 中用 DataStream 等相关类来表示数据流的抽象,在 DataSet API 中可以用 Flink-java 模块下的 org.apache.Flink.api.java.DataSet 类及其相关类来表示有界数据集的抽象。

以 DataSet 为父类的实现类或者不以其为父类的实现类都非常多,DataSet 等类与 Function、InputFormat、OutputFormat 的关系如图 3-10 所示。

本节以几个常用的类为例进行分析。

**1. DataSet**

DataSet 类本身是抽象类,其核心字段是封装在其中的 ExecutionEnvironment 执行环境对象和数据类型 type。DataSet 类提供了几乎所有的转换操作,在每一种转换操作中,都会将计算逻辑封装在一个新的数据集抽象中并返回。如 map()方法:

```
public <R> MapOperator<T, R> map(MapFunction<T, R> mapper) {
    ...
    TypeInformation<R> resultType = TypeExtractor.getMapReturnTypes(mapper, getType(),
callLocation, true);
    return new MapOperator<>(this, resultType, clean(mapper), callLocation);
}
```

这里将计算逻辑 mapper 对象、数据类型 resultType 以及该 DataSet 对象本身封装在 MapOperator 对象中。MapOperator 就是 DataSet 的一个实现类。

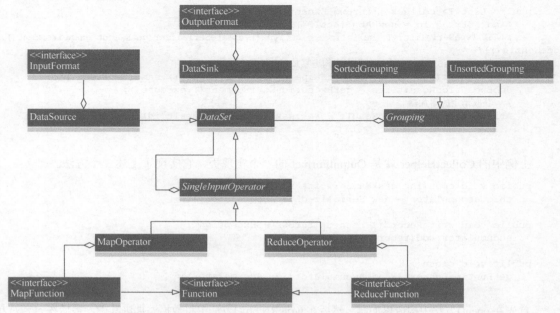

图 3-10　DataSet 等类与 Function、InputFormat、OutputFormat 的关系

再如 groupBy() 方法：

```
public UnsortedGrouping<T> groupBy(int...
fields) {
    return new UnsortedGrouping<>(this, new Keys.ExpressionKeys<>(fields, getType()));
}
```

这里将该 DataSet 对象本身以及分区的键的相关信息封装到 UnsortedGrouping 对象中。UnsortedGrouping 类并不是 DataSet 的子类，但是可以将它理解为按照一定的分区规则分组的数据集。

同时，DataSet 提供了构造 DataSink 的方法：

```
public DataSink<T> output(OutputFormat<T> outputFormat) {
    ...
    DataSink<T> sink = new DataSink<>(this, outputFormat, getType());
    this.context.registerDataSink(sink);
    return sink;
}
```

正是在这个方法中，通过 OutputFormat 对象构造出了 DataSink，并将其注册到了执行环境中。从上述每个数据集对象的构造过程对应的代码可以发现，上一个数据集对象都会作为下一个数据集对象的字段被传递下来，最终在 DataSink 对象中就有了完整的计算逻辑。

2．DataSource

通过执行环境定义数据源的返回值类型就是 DataSource。该类为 DataSet 的一个实现类，继承自 Operator 抽象类。Operator 类中定义了以下几个字段。

- name：算子的名称。
- parallelism：算子的并行度。

- minResources：ResourceSpec 类型，表示该转换操作所需的资源下限。
- preferredResources：ResourceSpec 类型，表示该转换操作所需的资源上限。

Operator 类中提供的方法就是这些字段的 get/set 方法。DataSource 中定义的核心字段就是上文介绍过的 InputFormat 对象。

DataSource 中有一个重要方法 translateToDataFlow()，该方法会在后续生成执行图时被调用：

```
protected GenericDataSourceBase<OUT, ?> translateToDataFlow() {
    ...
    // 将 InputFormat 对象封装成 GenericDataSourceBase 对象
    @SuppressWarnings({"unchecked", "rawtypes"})
    GenericDataSourceBase<OUT, ?> source = new GenericDataSourceBase(this.inputFormat,
        new OperatorInformation<OUT>(getType()), name);
    ...
    return source;
}
```

这里的 GenericDataSourceBase 对象继承自 org.apache.Flink.api.common.operators.Operator 类，这些 Operator 对象共同组成执行计划（Plan），进而转换成 JobGraph。执行计划的转换会在第 4 章介绍。

注意，上面 DataSource 类继承的 Operator 类是 org.apache.Flink.api.java.operators.Operator 类。本节中所有表示数据集抽象的类所继承的 Operator 类都指的是 org.apache.Flink.api.java.operators.Operator 类。

3. SingleInputOperator

抽象类 SingleInputOperator 同样是 Operator 类的子类。该类定义了一个 DataSet 类型的 input 字段，表示作为输入的数据集。以实现类 MapOperator 类为例，MapOperator 类中多了一个 function 字段。SingleInputOperator 类中定义了一个方法 translateToDataFlow()，其含义与 DataSource 中的 translateToDataFlow()方法相同，只是这里需要一个 input 参数。在 MapOperator 中该方法的实现如下：

```
protected MapOperatorBase<IN, OUT, MapFunction<IN, OUT>> translateToDataFlow(Operator<IN> input) {
    ...
    // 将计算逻辑封装在 MapOperatorBase 对象中
    MapOperatorBase<IN, OUT, MapFunction<IN, OUT>> po = new MapOperatorBase<IN, OUT, MapFunction<IN, OUT>>(function,
        new UnaryOperatorInformation<IN, OUT>(getInputType(), getResultType()), name);
    // 设置 po 对象的输入
    po.setInput(input);
    ...
    return po;
}
```

MapOperatorBase 类同样继承自 org.apache.Flink.api.common.operators.Operator 类。

4. DataSink

DataSink 表示数据集的输出。因为在 DataSink 之后不能再定义其他转换操作，所以它没有继承自 DataSet 类。其核心字段如下：

- format：OutputFormat 类型，定义数据的输出格式。
- type：TypeInformation 类型，表示数据集中的数据类型。
- data：DataSet 类型，表示输入的数据集。

- name：表示 sink 的名称。
- parallelism：表示 sink 的并行度。
- minResources：ResourceSpec 类型，表示该转换操作所需的资源下限。
- preferredResources：ResourceSpec 类型，表示该转换操作所需的资源上限。
- parameters：Configuration 类型。
- sortKeyPositions：int 数组类型，表示用于排序的键的位置。
- sortOrders：Order 数组类型，表示排序方式，如升序或降序等。

DataSink 中很重要的方法同样是 translateToDataFlow()方法：

```
protected GenericDataSinkBase<T> translateToDataFlow(Operator<T> input) {
    ...
    // 将计算逻辑封装在 GenericDataSinkBase 对象中
    GenericDataSinkBase<T> sink = new GenericDataSinkBase<>(this.format, new UnaryOperatorInformation<>(this.type, new NothingTypeInfo()), name);
    // 设置 sink 对象的输入
    sink.setInput(input);
    ...
    return sink;
}
```

上述代码省略了对 sink 对象设置并行度等步骤。与其他算子类似，DataSink 的计算逻辑 format 被封装到了 GenericDataSinkBase 对象中。GenericDataSinkBase 类同样继承自 org.apache.Flink.api.common.operators.Operator 类。

### 5. Grouping

Grouping 是一个抽象类，表示根据某个规则进行分区的数据集。该类没有继承自 DataSet 类，其定义了如下字段。

- inputDataSet：DataSet 类型，表示输入的数据集。
- keys：Keys 类型，表示分区的键。
- customPartitioner：Partitioner 类型，表示分区器。

Grouping 的实现类有两个，分别是 UnsortedGrouping 和 SortedGrouping。对数据集调用 groupBy()方法后就会得到 UnsortedGrouping 数据集，再对 UnsortedGrouping 调用 sortGroup()方法就会得到 SortedGrouping 数据集。UnsortedGrouping 类没有定义额外的字段，SortedGrouping 定义了下面几个字段。

- groupSortKeyPositions：int 数组类型，表示用于排序的键的位置。
- groupSortOrders：Order 数组类型，表示排序方式，如升序或降序等。
- groupSortSelectorFunctionKey：Keys.SelectorFunctionKeys 类型，其中封装了 KeySelector 对象。

可见 SortedGrouping 比 UnsortedGrouping 多了一些排序的信息。

对于 UnsortedGrouping，可以执行许多聚合操作，如 max、min、sum 等，这类聚合操作最终都会调用 aggregate()方法。

```
private AggregateOperator<T> aggregate(Aggregations agg, int field, String callLocationName) {
    return new AggregateOperator<T>(this, agg, field, callLocationName);
}
```

计算逻辑 agg 对象以及该数据集本身都被封装到了 AggregateOperator 中。AggregateOperator 也是一种数据集，继承自 DataSet。

另外一些聚合操作的方法如 reduce() 方法如下：

```java
public ReduceOperator<T> reduce(ReduceFunction<T> reducer) {
    if (reducer == null) {
        throw new NullPointerException("Reduce function must not be null.");
    }
    return new ReduceOperator<T>(this, inputDataSet.clean(reducer), Utils.getCallLocationName());
}
```

ReduceOperator 也是一种数据集，继承自 DataSet。

对于 SortedGrouping，主要可以调用 reduceGroup() 方法和 combineGroup() 方法：

```java
public <R> GroupReduceOperator<T, R> reduceGroup(GroupReduceFunction<T, R> reducer) {
    if (reducer == null) {
        throw new NullPointerException("GroupReduce function must not be null.");
    }
    TypeInformation<R> resultType = TypeExtractor.getGroupReduceReturnTypes(reducer,
            inputDataSet.getType(), Utils.getCallLocationName(), true);
    return new GroupReduceOperator<>(this, resultType, inputDataSet.clean(reducer),
Utils.getCallLocationName());
}
public <R> GroupCombineOperator<T, R> combineGroup(GroupCombineFunction<T, R> combiner) {
    if (combiner == null) {
        throw new NullPointerException("GroupCombine function must not be null.");
    }
    TypeInformation<R> resultType = TypeExtractor.getGroupCombineReturnTypes(combiner,
            this.getInputDataSet().getType(), Utils.getCallLocationName(), true);
    return new GroupCombineOperator<>(this, resultType, inputDataSet.clean(combiner),
Utils.getCallLocationName());
}
```

GroupReduceOperator 和 GroupCombineOperator 也是数据集，均继承自 DataSet。

6. TwoInputOperator

TwoInputOperator 抽象类表示两个数据集汇聚到一起形成的数据集，因此它有以下两个字段。

- input1：DataSet 类型，表示第一个输入的数据集。
- input2：DataSet 类型，表示第二个输入的数据集。

TwoInputOperator 的子类主要有 UnionOperator、JoinOperator、CrossOperator、CoGroupOperator 等。以 CrossOperator 为例，在这个实现类中，定义了如下字段。

- function：CrossFunction 类型，即具体的计算逻辑。
- defaultName：表示默认的算子名称。
- hint：CrossHint 类型，这是一个枚举类，提示系统做相应的优化。

```java
public static enum CrossHint {
    OPTIMIZER_CHOOSES,
    FIRST_IS_SMALL,
    SECOND_IS_SMALL
}
```

在 translateToDataFlow() 方法中，完成了以下转换：

```
    protected CrossOperatorBase<I1, I2, OUT, CrossFunction<I1, I2, OUT>> translateToDataFlow
(Operator<I1> input1, Operator<I2> input2) {
    ...
        // 将计算逻辑封装在 CrossOperatorBase 对象中
        CrossOperatorBase<I1, I2, OUT, CrossFunction<I1, I2, OUT>> po =
            new CrossOperatorBase<I1, I2, OUT, CrossFunction<I1, I2, OUT>>(function,
                new BinaryOperatorInformation<I1,I2, OUT>(getInput1Type(), getInput2Type(),
getResultType()),
                name);
        // 设置输入
        po.setFirstInput(input1);
        po.setSecondInput(input2);
        po.setParallelism(getParallelism());
        po.setCrossHint(hint);
        return po;
}
```

CrossOperatorBase 类也是 org.apache.Flink.api.common.operators.Operator 类的子类。其封装逻辑与前文介绍的算子的一致。

## 3.3 Table API 和 SQL

Table 类位于 Flink-table-api-java 模块的 org.apache.Flink.table.api 包中。Table API 和 SQL 在实际场景中常常被混合使用,并且 Table 还会与 DataStream/DataSet 进行互相转换。回顾 Table API 版本的 WordCount:

```
// (1) 获取执行环境
ExecutionEnvironment env = ExecutionEnvironment.getExecutionEnvironment();
BatchTableEnvironment tEnv = BatchTableEnvironment.create(env);
// (2) 指定数据的加载方式
DataSet<WC> input = env.fromElements(
        new WC("Hello", 1),
        new WC("Ciao", 1),
        new WC("Hello", 1));
// (3) 获取 Table 对象,定义对表的转换操作
tEnv.createTemporaryView("WordCount", input, "word, frequency");
Table table = tEnv
        .from("WordCount")
        .select("word, frequency")
        .groupBy("word")
        .select("word, frequency.sum as frequency");
// (4) 指定如何输出并触发程序运行
DataSet<WC> result = tEnv.toDataSet(table, WC.class);
result.print();
```

在这个过程中,主要分析表的转换操作。

正如在 DataStream API 中计算逻辑被封装在 Transformation 对象中,在 DataSet API 中计算逻辑直接被封装在 DataSet、DataSink 等对象中,在 Table API 和 SQL 中也需要用一个概念来封装转换逻辑,这个概念由 Flink-table-api-java 模块的 org.apache.Flink.table.operations.Operation 来表示,每个 Table 对象中都有这个字段,由此把计算逻辑层层封装并传递下去。最终,Operation 会经过逻辑执行计划、物理执行计划的转换和优化,被转换为 DataStream API 中的 Transformation 对象或

DataSet API 中的 DataSet 对象。

下面以几个方法为例介绍如何构造 Table 对象和 Operation 对象。

在 createTemporaryView()方法中，会将 DataSet 对象注册成执行环境中的一张表。

```
override def createTemporaryView[T](
    path: String,
    dataSet: DataSet[T],
    fields: String): Unit = {
  createTemporaryView(path, fromDataSet(dataSet, fields))
}
```

其中，fromDataSet()方法中就生成了一个 Operation 对象，并以该对象为参数构造 Table 对象后返回。

```
override def fromDataSet[T](dataSet: DataSet[T]): Table = {
  createTable(asQueryOperation(dataSet, None))
}
protected def asQueryOperation[T](dataSet: DataSet[T], fields: Option[Array[Expression]])
  : DataSetQueryOperation[T] = {
  ...
  // 构造 Operation 对象
  val tableOperation = new DataSetQueryOperation[T](
    dataSet,
    fieldsInfo.getIndices,
    fieldsInfo.toTableSchema)
  tableOperation
}
protected def createTable(tableOperation: QueryOperation): TableImpl = {
  // 以 Operation 对象为参数构造 Table 对象
  TableImpl.createTable(
    this,
    tableOperation,
    operationTreeBuilder,
    functionCatalog)
}
```

得到 Table 对象后，回到 createTemporaryView()的逻辑中。请观察接下来的注册步骤：

```
private def createTemporaryView(identifier: UnresolvedIdentifier, view: Table): Unit = {
  ...
  // 取出 Operation 对象，将其封装成 QueryOperationCatalogView 对象，并注册到执行环境中
  catalogManager.createTemporaryTable(
    new QueryOperationCatalogView(view.getQueryOperation),
    objectIdentifier,
    false)
}
```

catalogManager 是 CatalogManager 类型，维护了表的信息。

这时已经将表注册到了执行环境中，于是可以调用执行环境的 from()方法扫描这张表。

```
override def from(path: String): Table = {
  val parser = planningConfigurationBuilder.createCalciteParser()
  val unresolvedIdentifier = UnresolvedIdentifier.of(parser.parseIdentifier(path).names: _*)
  scanInternal(unresolvedIdentifier) match {
    case Some(table) => createTable(table)
    case None => throw new TableException(s"Table '$unresolvedIdentifier' was not found.")
  }
}
```

通过表名生成标识，调用 scanInternal()方法找寻这张表：

```
private[Flink] def scanInternal(identifier: UnresolvedIdentifier)
  : Option[CatalogQueryOperation] = {
  val objectIdentifier: ObjectIdentifier = catalogManager.qualifyIdentifier(identifier)
  JavaScalaConversionUtil.toScala(catalogManager.getTable(objectIdentifier))
    .map(t => new CatalogQueryOperation(objectIdentifier, t.getTable.getSchema))
}
```

通过标识在 catalogManager 中找到相关信息，生成 CatalogQueryOperation 对象，即 Operation 对象。于是调用 createTable()方法，将 Operation 对象传入，构造 Table 对象并返回：

```
protected def createTable(tableOperation: QueryOperation): TableImpl = {
  TableImpl.createTable(
    this,
    tableOperation,
    operationTreeBuilder,
    functionCatalog)
}
```

当调用转换操作的方法时采用同样的逻辑。先构造 Operation 对象，再构造 Table 对象。如调用 select()方法：

```
public Table select(Expression... fields) {
  ...
  if (!extracted.getAggregations().isEmpty()) {
     QueryOperation aggregate = operationTreeBuilder.aggregate(
        Collections.emptyList(),
        extracted.getAggregations(),
        operationTree
     );
     return createTable(operationTreeBuilder.project(extracted.getProjections(), aggregate, false));
  } else {
     return createTable(operationTreeBuilder.project(expressionsWithResolvedCalls, operationTree, false));
  }
}
```

通过 operationTreeBuilder.project()方法生成 Operation 对象，并将其作为 createTable()方法的参数传入，构造 Table 对象。

最后，简单分析一下 Operation 对象如何被转换成了 Transformation 对象或 DataSet 对象。

如果执行环境是 BatchTableEnvironment，那么在 toDataSet()方法中会完成这个转换：

```
override def toDataSet[T](table: Table, clazz: Class[T]): DataSet[T] = {
  translate[T](table)(TypeExtractor.createTypeInfo(clazz))
}
protected def translate[A](table: Table)(implicit tpe: TypeInformation[A]): DataSet[A] = {
  // 取出 Operation 对象
  val queryOperation = table.getQueryOperation
  // 生成逻辑执行计划
  val relNode = getRelBuilder.tableOperation(queryOperation).build()
  // 生成物理执行计划
  val dataSetPlan = optimizer.optimize(relNode)
  // 生成 DataSet
```

```
translate(
  dataSetPlan,
  getTableSchema(queryOperation.getTableSchema.getFieldNames, dataSetPlan))
}
```

如果执行环境是 StreamExecutionEnvironment，那么在 toDataStream()方法中可以看到转换的过程：

```
private <T> DataStream<T> toDataStream(Table table, OutputConversionModifyOperation modifyOperation) {
    // 根据指定的 planner，将 Operation 转换成 Transformation
    List<Transformation<?>> transformations = planner.translate(Collections.singletonList(modifyOperation));
    Transformation<T> transformation = getTransformation(table, transformations);
    executionEnvironment.addOperator(transformation);
    return new DataStream<>(executionEnvironment, transformation);
}
```

下面是在 PlannerBase 中实现的 translate()方法：

```
override def translate(
    modifyOperations: util.List[ModifyOperation]): util.List[Transformation[_]] = {
  ...
  // 生成逻辑执行计划
  val relNodes = modifyOperations.map(translateToRel)
  // 优化逻辑执行计划
  val optimizedRelNodes = optimize(relNodes)
  // 生成物理执行计划
  val execNodes = translateToExecNodePlan(optimizedRelNodes)
  // 将物理执行计划转换成 Transformation
  translateToPlan(execNodes)
}
```

对于 SQL，sqlQuery()方法的第一步就是将查询语句转换成 Operation 对象：

```
List<Operation> operations = parser.parse(query);
```

这样的转换使 Table API 和 SQL 在计算逻辑的封装上达成了一致。

## 3.4 总结

本章介绍了 DataStream API、DataSet API、Table API 和 SQL 的底层实现原理等。其中着重分析了 DataStream API 封装计算逻辑的数据结构。

在 DataStream API 中用 Function 接口定义业务逻辑，用算子 StreamOperator 封装函数。当任务被执行时，通过调用算子的方法进而执行到函数中的业务逻辑。在业务代码中定义业务逻辑的对象是数据流 DataStream 等类的实例，在对数据流进行转换操作时，这些转换操作被抽象成了 Transformation，Transformation 会在接下来的执行流程中直接被用于转换成执行图。

DataSet API 的整体设计思路与 DataStream API 的一致，只不过封装计算逻辑的类有所不同。这种区别主要来自对数据流和对数据集处理思路的不同。Table API 和 SQL 则是用 Operation 表示转换操作，最终 Operation 也会被转换成核心 API 这一层的抽象，进而生成执行图。

# 第 4 章
# Flink 的执行图

前文详细介绍了提供给开发人员的 API 的应用和实现原理等,主要涵盖从构造执行环境到调用执行环境的 execute() 方法的内容。对于 DataStream API,所有的计算逻辑都被封装在了 Transformation 对象中;对于 DataSet API,所有的计算逻辑最后都被封装在 DataSink 对象中;在 Table API 和 SQL 中,计算逻辑根据执行环境的不同,最终会被转换成 Transformation 或 DataSet 对象。

当计算逻辑被封装好以后,接下来需要做的事情就是生成执行图。对于 DataStream API,首先会将 Transformation 转换成 StreamGraph(位于 Flink-streaming-java 模块的 org.apache.Flink.streaming.api.graph 包中);对于 DataSet API,首先会将 DataSink 中封装的计算逻辑转换成 Plan(位于 Flink-core 模块的 org.apache.Flink.api.common 包中)。StreamGraph 和 Plan 都实现了接口 Pipeline(位于 Flink-core 模块的 org.apache.Flink.api.dag 包中)。调用 PipelineExecutor 接口中定义的 execute() 方法将 Pipeline 传入,随后会根据 Pipeline 的具体类型,使用不同的 FlinkPipelineTranslator 将 Pipeline 转换成 JobGraph(位于 Flink-runtime 模块的 org.apache.Flink.runtime.jobgraph 包中)。到了 JobGraph 这一层,DataStream 和 DataSet 就实现了统一。在 JobManager 中,JobGraph 会被转换成 ExecutionGraph(位于 Flink-runtime 模块的 org.apache.Flink.runtime.executiongraph 包中),最终 ExecutionGraph 中的节点会再次被转换成其他可以部署的对象,由调度器部署到 TaskManager 执行。

本书将 StreamGraph、JobGraph 和 ExecutionGraph 定义为执行图的不同层次。也有一些参考资料将 ExecutionGraph 这一层级称为"执行图",而将 StreamGraph 与 JobGraph 分别称为"流图"和"作业图"。这里特此说明,以避免读者在阅读不同资料时产生歧义。

图 4-1 展示了执行图的层级关系。

希望在学习完本章后,读者能够了解:

- StreamGraph/Plan、JobGraph 和 ExecutionGraph 的结构;

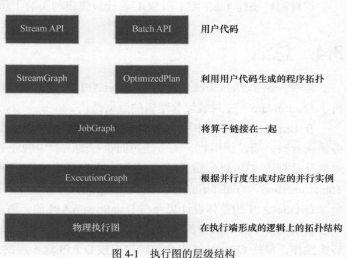

图 4-1 执行图的层级结构

- 执行图之间的关系和转换逻辑。

## 4.1 StreamGraph 的生成

在 StreamExecutionEnvironment 的 execute()方法中，会调用 getStreamGraph()方法以生成 StreamGraph：

```
public StreamGraph getStreamGraph(String jobName, boolean clearTransformations) {
    StreamGraph streamGraph = getStreamGraphGenerator().setJobName(jobName).generate();
    ...
    return streamGraph;
}
```

其中，调用 getStreamGraphGenerator()方法会返回一个 StreamGraphGenerator 对象：

```
private StreamGraphGenerator getStreamGraphGenerator() {
    if (transformations.size() <= 0) {
        throw new IllegalStateException("No operators defined in streaming topology. Cannot execute.");
    }
    return new StreamGraphGenerator(transformations, config, checkpointCfg)
        .setStateBackend(defaultStateBackend)
        .setChaining(isChainingEnabled)
        .setUserArtifacts(cacheFile)
        .setTimeCharacteristic(timeCharacteristic)
        .setDefaultBufferTimeout(bufferTimeout);
}
```

上述代码中，通过 StreamGraphGenerator 的构造方法生成了一个实例对象，并且给该对象设置了多个属性。

### 4.1.1 StreamGraphGenerator 分析

下面是 StreamGraphGenerator 中的一些重要字段。
- transformations：Transformation 类型，这个字段的值指向执行环境中的 transformations 字段。
- chaining：用于设置是否开启链接。如果开启，则可能实现把多个算子链接在一起，这样就可以把它们部署到一个任务中执行。
- scheduleMode：ScheduleMode 类型，表示调度策略。
- jobName：表示作业的名称。
- streamGraph：StreamGraph 类型。
- alreadyTransformed：Map<Transformation<?>, Collection<Integer>>类型，记录已经完成转换的 Transformation 对象。

StreamGraphGenerator 的构造方法和各种 set 方法的作用就是在对这些字段赋值，其中很重要的是将执行环境中的 transformations 字段赋值给 StreamGraphGenerator 的 transformations 字段。

通过上面的代码可知调用 generate()方法会生成一个完整的 StreamGraph 对象并返回。该方法的实现代码如下：

```java
public StreamGraph generate() {
    // 初始化 StreamGraph
    streamGraph = new StreamGraph(executionConfig, checkpointConfig, savepointRestoreSettings);
    streamGraph.setStateBackend(stateBackend);
    streamGraph.setChaining(chaining);
    streamGraph.setScheduleMode(scheduleMode);
    streamGraph.setUserArtifacts(userArtifacts);
    streamGraph.setTimeCharacteristic(timeCharacteristic);
    streamGraph.setJobName(jobName);
    streamGraph.setBlockingConnectionsBetweenChains(blockingConnectionsBetweenChains);
    alreadyTransformed = new HashMap<>();
    // 完成所有 Transformation 对象的转换,填充 StreamGraph
    for (Transformation<?> transformation: transformations) {
        transform(transformation);
    }
    final StreamGraph builtStreamGraph = streamGraph;
    // 清空状态
    alreadyTransformed.clear();
    alreadyTransformed = null;
    streamGraph = null;
    // 返回构建好的完整的 StreamGraph 对象
    return builtStreamGraph;
}
```

显而易见,主要的逻辑都在 **transform()** 方法中实现。

```java
private Collection<Integer> transform(Transformation<?> transform) {
    // (1) 如果 alreadyTransformed 字段中已经包含该 Transformation 对象,则直接取出并返回
    if (alreadyTransformed.containsKey(transform)) {
        return alreadyTransformed.get(transform);
    }
    ...
    // (2) 按照 Transformation 对象的具体实现类对其进行转换
    Collection<Integer> transformedIds;
    if (transform instanceof OneInputTransformation<?, ?>) {
        transformedIds = transformOneInputTransform((OneInputTransformation<?, ?>) transform);
    } else if (transform instanceof TwoInputTransformation<?, ?, ?>) {
        transformedIds = transformTwoInputTransform((TwoInputTransformation<?, ?, ?>) transform);
    } else if (transform instanceof SourceTransformation<?>) {
        transformedIds = transformSource((SourceTransformation<?>) transform);
    } else if (transform instanceof SinkTransformation<?>) {
        transformedIds = transformSink((SinkTransformation<?>) transform);
    } else if (transform instanceof UnionTransformation<?>) {
        transformedIds = transformUnion((UnionTransformation<?>) transform);
    } else if (transform instanceof SplitTransformation<?>) {
        transformedIds = transformSplit((SplitTransformation<?>) transform);
    } else if (transform instanceof SelectTransformation<?>) {
        transformedIds = transformSelect((SelectTransformation<?>) transform);
    } else if (transform instanceof FeedbackTransformation<?>) {
        transformedIds = transformFeedback((FeedbackTransformation<?>) transform);
    } else if (transform instanceof CoFeedbackTransformation<?>) {
        transformedIds = transformCoFeedback((CoFeedbackTransformation<?>) transform);
    } else if (transform instanceof PartitionTransformation<?>) {
        transformedIds = transformPartition((PartitionTransformation<?>) transform);
    } else if (transform instanceof SideOutputTransformation<?>) {
        transformedIds = transformSideOutput((SideOutputTransformation<?>) transform);
```

```
        } else {
            throw new IllegalStateException("Unknown transformation: " + transform);
        }
        // （3）将已经转换的 Transformation 对象添加到 alreadyTransformed 字段中
        if (!alreadyTransformed.containsKey(transform)) {
            alreadyTransformed.put(transform, transformedIds);
        }
        // （4）给 StreamGraph 对象设置一些属性值
        ...
        return transformedIds;
    }
```

代码实现的核心逻辑分为下面 4 步。

（1）如果 alreadyTransformed 字段中已经包含该 Transformation 对象，则直接取出并返回。

（2）按照 Transformation 对象的具体实现类对其进行转换。这里用 if-else 语句罗列了所有 Transformation 的实现类。这提示我们，如果想要自定义一种转换操作，则需要新实现一种 Transformation，并且在这里需要添加一个条件判断的分支，以便使自定义的 Transformation 完成转换。

（3）将已经转换的 Transformation 对象添加到 alreadyTransformed 字段中。

（4）给 StreamGraph 对象设置一些属性值。主要是把 Transformation 对象中的属性值（如 name、uid 等值）赋值给 StreamGraph 中的对应节点。

第 2 步调用的 transform×××()方法完成了对每种类型的 Transformation 的转换。其中有一些转换需要调用 determineSlotSharingGroup()方法，因此在分析 transform×××()方法前，先介绍 StreamGraphGenerator 中的 determineSlotSharingGroup()方法：

```
    private String determineSlotSharingGroup(String specifiedGroup, Collection<Integer> inputIds) {
        if (specifiedGroup != null) { // 如果指定了 SlotSharingGroup，则返回指定的值
            return specifiedGroup;
        } else { // 如果没有指定，则按照如下规则设置值
            String inputGroup = null;
            for (int id: inputIds) { // 遍历输入节点的 id
                String inputGroupCandidate = streamGraph.getSlotSharingGroup(id); // 获取输
// 入节点的 SlotSharingGroup 值如果之前没有赋过值，则将输入节点的值赋给当前节点；如果已经赋过值，且该值
// 跟输入节点的值不同，则返回默认值
                if (inputGroup == null) {
                    inputGroup = inputGroupCandidate;
                } else if (!inputGroup.equals(inputGroupCandidate)) {
                    return DEFAULT_SLOT_SHARING_GROUP;
                }
            }
            return inputGroup == null ? DEFAULT_SLOT_SHARING_GROUP : inputGroup;
        }
    }
```

通过调用这个方法，可以确定该算子的 SlotSharingGroup。SlotSharingGroup 在判断算子是否可以进行链接时会被用到。

接下来介绍几种常用的 Transformation 转换。首先是 transformSource()方法：

```
    private <T> Collection<Integer> transformSource(SourceTransformation<T> source) {
        // 确定 SlotSharingGroup
```

```
            String slotSharingGroup = determineSlotSharingGroup(source.getSlotSharingGroup(),
Collections.emptyList());
        // 添加 source 算子
        streamGraph.addSource(source.getId(),
            slotSharingGroup,
            source.getCoLocationGroupKey(),
            source.getOperatorFactory(),
            null,
            source.getOutputType(),
            "Source: " + source.getName());
        // 其他属性的设置
        ...
        // 返回 Transformation 对象的 id
        return Collections.singleton(source.getId());
    }
```

因为 source 任务是一个具体的任务，所以需要确定 SlotSharingGroup。接着，将 source 算子添加到 StreamGraph 中。然后设置一些属性，如并行度等。最后返回 Transformation 对象的 id。

再来看 transformOneInputTransform()方法：

```
    private <IN, OUT> Collection<Integer> transformOneInputTransform(OneInputTransformation
<IN, OUT> transform) {
        // 对输入进行转换
        Collection<Integer> inputIds = transform(transform.getInput());
        ...
        // 确定 SlotSharingGroup
        String slotSharingGroup = determineSlotSharingGroup(transform.getSlotSharingGroup(),
inputIds);
        // 添加算子
        streamGraph.addOperator(transform.getId(),
            slotSharingGroup,
            transform.getCoLocationGroupKey(),
            transform.getOperatorFactory(),
            transform.getInputType(),
            transform.getOutputType(),
            transform.getName());
        // 其他属性的设置
        ...
        // 添加边
        for (Integer inputId: inputIds) {
            streamGraph.addEdge(inputId, transform.getId(), 0);
        }
        // 返回 Transformation 对象的 id
        return Collections.singleton(transform.getId());
    }
```

因为该 Transformation 有输入，所以首先是递归地对输入进行转换。添加算子时调用的是 addOperator()方法，与添加 source 算子时调用的 addSource()方法不同，但逻辑是类似的。因为中间的算子还涉及与其他算子的链接，所以还需要调用 addEdge()方法添加边。

TwoInputTransformation 和 SinkTransformation 的转换均与上述步骤类似。

从前文的分析已经了解到，虽然所有的转换操作都对应一个 Transformation，但并不是所有 Transformation 都会转换成一个具体的任务去执行。比如对于 Union 转换，它只表示两个数据流合并到一起，并没有对数据进行具体的计算；对于 KeyBy 转换，它只影响数据如何分发到下游。这些 Transformation 的转换就会有所区别。如 transformUnion() 方法：

```java
private <T> Collection<Integer> transformUnion(UnionTransformation<T> union) {
    // 获得输入的 Transformation 对象
    List<Transformation<T>> inputs = union.getInputs();
    List<Integer> resultIds = new ArrayList<>();
    // 依次对输入的 Transformation 进行转换
    for (Transformation<T> input: inputs) {
        resultIds.addAll(transform(input));
    }
    // 返回 Transformation 对象的 id 列表
    return resultIds;
}
```

该方法只是递归地调用了 transform() 方法来转换输入的 Transformation 对象，最后返回 id 列表。

对于 PartitionTransformation，会进行如下转换：

```java
private <T> Collection<Integer> transformPartition(PartitionTransformation<T> partition) {
    // 获取输入的 Transformation 对象
    Transformation<T> input = partition.getInput();
    List<Integer> resultIds = new ArrayList<>();
    // 转换输入的 Transformation
    Collection<Integer> transformedIds = transform(input);
    for (Integer transformedId: transformedIds) { // 添加虚拟节点
        int virtualId = Transformation.getNewNodeId();
        streamGraph.addVirtualPartitionNode(
            transformedId, virtualId, partition.getPartitioner(), partition.getShuffleMode());
        resultIds.add(virtualId);
    }
    // 返回 Transformation 对象的 id 列表
    return resultIds;
}
```

因为重分区操作并不是真正的任务，所以这里用虚拟节点 VirtualPartitionNode 来表示。这个节点包含重分区的一些信息、虚拟节点的 id 以及上游节点的 id。这样，之后就可以通过虚拟节点的 id 找到它的上一个真实节点的 id。

## 4.1.2　StreamGraph 分析

StreamGraphGenerator 构建了 StreamGraph，还在不少方法中调用了 StreamGraph 添加节点和边的方法。StreamGraph 实现了 Pipeline 接口。该接口没有定义任何方法，仅是为了表示 DataStream 中的 StreamGraph 和 DataSet 中的 Plan 都属于 Pipeline 类型。

现在来分析 StreamGraph 的字段和方法，其核心字段如下。

- chaining：用于设置是否开启链接。如果开启，则可能实现把多个算子链接在一起，这样就可以把它们部署到一个任务中执行。
- scheduleMode：ScheduleMode 类型，表示调度策略。这部分内容会在后文详细介绍。
- jobName：表示作业的名称。
- allVerticesInSameSlotSharingGroupByDefault：用于设置是否把所有节点放入同一个

SlotSharingGroup。默认值为 true。
- **streamNodes**：Map<Integer, StreamNode>类型，表示一个会被执行的任务节点。
- **sources**：Integer 类型，表示 source 任务的 id。
- **sinks**：Integer 类型，表示 sink 任务的 id。
- **virtualSelectNodes**：Map<Integer, Tuple2<Integer, <String>>>类型，表示虚拟的 select 节点。
- **virtualSideOutputNodes**：Map<Integer, Tuple2<Integer, OutputTag>>类型，表示虚拟的 side output 节点。
- **virtualPartitionNodes**：Map<Integer, Tuple3<Integer, StreamPartitioner<?>, Shufflemode>>类型，表示虚拟的 partition 节点。

上面的许多字段直接来自 StreamGraphGenerator。

StreamGraph 中重要的方法是关于节点和边的添加方法。添加节点时，有 addSource()方法、addOperator()方法和 addSink()方法。其中，addSource()和 addSink()方法如下：

```
public <IN, OUT> void addSource(Integer vertexID,
    @Nullable String slotSharingGroup,
    @Nullable String coLocationGroup,
    StreamOperatorFactory<OUT> operatorFactory,
    TypeInformation<IN> inTypeInfo,
    TypeInformation<OUT> outTypeInfo,
    String operatorName) {
    addOperator(vertexID, slotSharingGroup, coLocationGroup, operatorFactory, inTypeInfo,
outTypeInfo, operatorName);
    sources.add(vertexID);
}
public <IN, OUT> void addSink(Integer vertexID,
    @Nullable String slotSharingGroup,
    @Nullable String coLocationGroup,
    StreamOperatorFactory<OUT> operatorFactory,
    TypeInformation<IN> inTypeInfo,
    TypeInformation<OUT> outTypeInfo,
    String operatorName) {
    addOperator(vertexID, slotSharingGroup, coLocationGroup, operatorFactory, inTypeInfo,
outTypeInfo, operatorName);
    sinks.add(vertexID);
}
```

这两个方法直接调用了 addOperator()方法，调用后将节点 id 分别添加到了 sources 和 sinks 中。addOperator()方法如下：

```
public <IN, OUT> void addOperator(
    Integer vertexID,
    @Nullable String slotSharingGroup,
    @Nullable String coLocationGroup,
    StreamOperatorFactory<OUT> operatorFactory,
    TypeInformation<IN> inTypeInfo,
    TypeInformation<OUT> outTypeInfo,
    String operatorName) {
    // 添加节点
    if (operatorFactory.isStreamSource()) {
        addNode(vertexID, slotSharingGroup, coLocationGroup, SourceStreamTask.class,
operatorFactory, operatorName);
```

```
        } else {
            addNode(vertexID, slotSharingGroup, coLocationGroup, OneInputStreamTask.class,
operatorFactory, operatorName);
        }
        // 设置数据类型、序列化器等
        ...
    }
```

其核心在于调用了 addNode() 方法。

```
protected StreamNode addNode(Integer vertexID,
    @Nullable String slotSharingGroup,
    @Nullable String coLocationGroup,
    Class<? extends AbstractInvokable> vertexClass,
    StreamOperatorFactory<?> operatorFactory,
    String operatorName) {
    ...
    // 创建 StreamNode
    StreamNode vertex = new StreamNode(
        vertexID,
        slotSharingGroup,
        coLocationGroup,
        operatorFactory,
        operatorName,
        new ArrayList<OutputSelector<?>>(),
        vertexClass);
    // 将 StreamNode 节点和节点 id 放入 streamNodes 中进行维护
    streamNodes.put(vertexID, vertex);
    return vertex;
}
```

在 addNode() 方法中，根据传入的参数创建了 StreamNode 对象并将其放入 streamNodes 字段中维护起来。这里要注意 vertexClass 参数。在 addOperator() 方法中，会根据任务是 source 任务还是后面只有一个输入的任务来分别传入 SourceStreamTask.class 或 OneInputStreamTask.class（添加其他节点时还可能传入其他参数）。这个参数最终会影响到 TaskManager 的任务在哪个类下被执行。

添加边时会调用 addEdge() 方法：

```
public void addEdge(Integer upStreamVertexID, Integer downStreamVertexID, int typeNumber) {
    addEdgeInternal(upStreamVertexID,
        downStreamVertexID,
        typeNumber,
        null,
        new ArrayList<String>(),
        null,
        null);
}
private void addEdgeInternal(Integer upStreamVertexID,
    Integer downStreamVertexID,
    int typeNumber,
    StreamPartitioner<?> partitioner,
    List<String> outputNames,
    OutputTag outputTag,
    ShuffleMode shuffleMode) {
    // 在虚拟节点中找到上一个真实节点的 id，递归地调用 addEdgeInternal() 方法
    ...
```

```
        } else {
            StreamNode upstreamNode = getStreamNode(upStreamVertexID);
            StreamNode downstreamNode = getStreamNode(downStreamVertexID);
            // 设置 partitioner、shuffleMode 等属性
            ...
            // 创建 StreamEdge 对象
            StreamEdge edge = new StreamEdge(upstreamNode, downstreamNode, typeNumber,
outputNames, partitioner, outputTag, shuffleMode);
            // 给节点添加出入边
            getStreamNode(edge.getSourceId()).addOutEdge(edge);
            getStreamNode(edge.getTargetId()).addInEdge(edge);
        }
    }
```

由于在构建节点时维护了虚拟节点和真实节点之间的连接关系，因此在构建边的时候会根据这一关系将真实节点连接起来。得到上下游节点的 id 后，获取对应的 StreamNode 对象，为它们添加出入边。

其中 getStreamNode() 方法如下：

```
public StreamNode getStreamNode(Integer vertexID) {
    return streamNodes.get(vertexID);
}
```

添加虚拟节点，以 **addVirtualPartitionNode()** 方法为例：

```
public void addVirtualPartitionNode(
    Integer originalId,
    Integer virtualId,
    StreamPartitioner<?> partitioner,
    ShuffleMode shuffleMode) {
 ...
    virtualPartitionNodes.put(virtualId, new Tuple3<>(originalId, partitioner, shuffleMode));
}
```

因此，StreamGraph 的核心方法的主要作用就是构造 StreamNode 对象并将其放入 streamNodes 字段中进行维护，同时添加虚拟节点放入 virtualPartitionNodes 等字段中进行维护，并且构造 StreamEdge 对象为节点添加出入边。

### 4.1.3　StreamNode 和 StreamEdge

在上述分析中，出现了 StreamNode 和 StreamEdge 的概念，它们分别表示节点和边。它们与 StreamGraph 的关系如图 4-2 所示。

1. StreamNode

StreamNode 的核心字段如下。

- parallelism：表示并行度。
- maxParallelism：表示最大并行度。
- operatorName：表示算子名称。
- slotSharingGroup：表示共享槽组。
- statePartitioner1：KeySelector 类型，表示

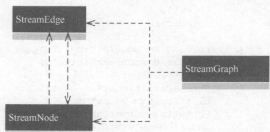

图 4-2　StreamNode、StreamEdge 与 StreamGraph 的关系

键的提取器。
- statePartitioner2：KeySelector 类型，表示键的提取器。如果有两个数据流汇聚到一起，则这个字段表示第二个数据流的 KeySelector 对象。
- operatorFactory：StreamOperatorFactory 类型。StreamOperatorFactory 是一个接口，表示一个工厂类，具有生成 StreamOperator 实例的功能。
- inEdges：StreamEdge 类型，表示这个 StreamNode 对象的入边。
- outEdges：StreamEdge 类型，表示这个 StreamNode 对象的出边。
- jobVertexClass：Class 类型，表示该任务的执行类。
- transformationUID：表示转换操作的 uid。

上述的许多字段是直接从 Transformation 对象传递来的。StreamNode 主要提供了添加入边和出边的方法：

```
public void addInEdge(StreamEdge inEdge) {
    if (inEdge.getTargetId() != getId()) {
        throw new IllegalArgumentException("Destination id doesn't match the StreamNode id");
    } else {
        inEdges.add(inEdge);
    }
}
public void addOutEdge(StreamEdge outEdge) {
    if (outEdge.getSourceId() != getId()) {
        throw new IllegalArgumentException("Source id doesn't match the StreamNode id");
    } else {
        outEdges.add(outEdge);
    }
}
```

2. StreamEdge

StreamEdge 表示连接两个节点的边，其核心字段如下。
- edgeId：表示边的 id。
- sourceId：表示输入的算子 id。
- targetId：表示输出的算子 id。
- outputPartitioner：StreamPartitioner 类型，表示分区器。
- sourceOperatorName：表示输入的算子名称。
- targetOperatorName：表示输出的算子名称。
- shuffleMode：ShuffleMode 类型，表示上下游算子的数据交换方式。

StreamEdge 中维护了上下游算子的一些关系，如输出关系等。这样，在 TaskManager 生成任务的实例时，就可以从边中获得分区器等信息。

## 4.2 Plan 的生成

StreamGraph 是针对 DataStream API 构造的执行图。在 DataSet API 中，这个执行图称为 Plan。

在 ExecutionEnvironment 的 execute()方法中，会调用 createProgramPlan()方法以生成 Plan：

```
public Plan createProgramPlan(String jobName, boolean clearSinks) {
   ...
   OperatorTranslation translator = new OperatorTranslation();
   Plan plan = translator.translateToPlan(this.sinks, jobName);
   ...
   return plan;
}
```

这里的 OperatorTranslation 就相当于生成 StreamGraph 时用到的 StreamGraphGenerator。

### 4.2.1　OperatorTranslation 分析

OperatorTranslation 类只有一个字段，即 translated，是 Map 类型，用来维护已经转换的 Operator 对象。这里的 Operator 是 org.apache.Flink.api.common.operators.Operator 类，并非继承自 DataSet 类的 org.apache.Flink.api.java.operators.Operator 类，前者正是由后者转换而来的。

转换时调用的方法为 translateToPlan()：

```
public Plan translateToPlan(List<DataSink<?>> sinks, String jobName) {
   List<GenericDataSinkBase<?>> planSinks = new ArrayList<>();
   for (DataSink<?> sink : sinks) {
      planSinks.add(translate(sink));
   }
   Plan p = new Plan(planSinks);
   p.setJobName(jobName);
   return p;
}
```

这个过程同样可以与 StreamGraph 的进行对比。递归地处理每一个 sink，调用 translate()方法对其进行转换。最终，将结果封装成 Plan 对象并返回。

translate()方法如下：

```
private <T> GenericDataSinkBase<T> translate(DataSink<T> sink) {
   // 递归地转换输入
   Operator<T> input = translate(sink.getDataSet());
   // 将自身转换成 Operator 对象
   GenericDataSinkBase<T> translatedSink = sink.translateToDataFlow(input);
   translatedSink.setResources(sink.getMinResources(), sink.getPreferredResources());
   return translatedSink;
}
```

对于 translateToDataFlow()方法，之前已经介绍过，它会把 DataSet 等相关类转换成 org.apache.Flink.api.common.operators.Operator 类。这里调用的 translate()方法是另一个重载方法：

```
private <T> Operator<T> translate(DataSet<T> dataSet) {
   ...
   // （1）如果 translated 字段中已经包含该 Operator 对象，则直接取出并返回
   Operator<?> previous = this.translated.get(dataSet);
   if (previous != null) {
      if (!(dataSet instanceof UnionOperator)) {
         Operator<T> typedPrevious = (Operator<T>) previous;
         return typedPrevious;
      }
   }
```

```java
        // （2）按照 dataSet 对象的不同类型对其进行转换
        Operator<T> dataFlowOp;
        if (dataSet instanceof DataSource) {
            DataSource<T> dataSource = (DataSource<T>) dataSet;
            dataFlowOp = dataSource.translateToDataFlow();
            dataFlowOp.setResources(dataSource.getMinResources(), dataSource.getPreferredResources());
        }
        else if (dataSet instanceof SingleInputOperator) {
            SingleInputOperator<?, ?, ?> singleInputOperator = (SingleInputOperator<?, ?, ?>) dataSet;
            dataFlowOp = translateSingleInputOperator(singleInputOperator);
            dataFlowOp.setResources(singleInputOperator.getMinResources(), singleInputOperator.getPreferredResources());
        }
        else if (dataSet instanceof TwoInputOperator) {
            TwoInputOperator<?, ?, ?, ?> twoInputOperator = (TwoInputOperator<?, ?, ?, ?>) dataSet;
            dataFlowOp = translateTwoInputOperator(twoInputOperator);
            dataFlowOp.setResources(twoInputOperator.getMinResources(), twoInputOperator.getPreferredResources());
        }
        else if (dataSet instanceof BulkIterationResultSet) {
            BulkIterationResultSet<?> bulkIterationResultSet = (BulkIterationResultSet<?>) dataSet;
            dataFlowOp = translateBulkIteration(bulkIterationResultSet);
            dataFlowOp.setResources(bulkIterationResultSet.getIterationHead().getMinResources(),
                bulkIterationResultSet.getIterationHead().getPreferredResources());
        }
        else if (dataSet instanceof DeltaIterationResultSet) {
            DeltaIterationResultSet<?, ?> deltaIterationResultSet = (DeltaIterationResultSet<?, ?>) dataSet;
            dataFlowOp = translateDeltaIteration(deltaIterationResultSet);
            dataFlowOp.setResources(deltaIterationResultSet.getIterationHead().getMinResources(),
                deltaIterationResultSet.getIterationHead().getPreferredResources());
        }
        else if (dataSet instanceof DeltaIteration.SolutionSetPlaceHolder || dataSet instanceof DeltaIteration.WorksetPlaceHolder) {
            throw new InvalidProgramException("A data set that is part of a delta iteration was used as a sink or action."
                + " Did you forget to close the iteration?");
        }
        else {
            throw new RuntimeException("Error while creating the data flow plan for the program: Unknown operator or data set type: " + dataSet);
        }
        // （3）将已转换的 dataSet 对象添加到 translated 字段中
        this.translated.put(dataSet, dataFlowOp);
        ...
        return dataFlowOp;
    }
```

代码实现的核心逻辑与 StreamGraph 转换过程的一致。如果 dataSet 对象属于 DataSource 类型，则调用其 translateToDataFlow()方法完成转换；如果属于 SingleInputOperator 类型，则调用 translateSingleInputOperator()方法，如下：

```java
    private <I, O> org.apache.Flink.api.common.operators.Operator<O> translateSingleInputOperator(SingleInputOperator<?, ?, ?> op) {
        SingleInputOperator<I, O, ?> typedOp = (SingleInputOperator<I, O, ?>) op;
```

```
        DataSet<I> typedInput = (DataSet<I>) op.getInput();
        Operator<I> input = translate(typedInput);
        org.apache.Flink.api.common.operators.Operator<O> dataFlowOp = typedOp.translate
ToDataFlow(input);
        ...
        return dataFlowOp;
    }
```

在这个方法中，仍然是调用 translateToDataFlow()方法完成转换，对于 dataSet 对象属于其他类型的情况也是一样的。

### 4.2.2　Plan 分析

Plan 类实现了 Pipeline 接口，同时还实现了 Visitable 接口。Visitable 接口定义了一个方法：

```
public interface Visitable<T extends Visitable<T>> {
    void accept(Visitor<T> visitor);
}
```

这里用到了设计模式中的访问者模式。accept()方法接收一个 Visitor 对象，该对象大致可以被理解为优化规则，它会访问每一个 Visitable 对象，对其进行一些修改。Visitable、Visitor 等的关系如图 4-3 所示。

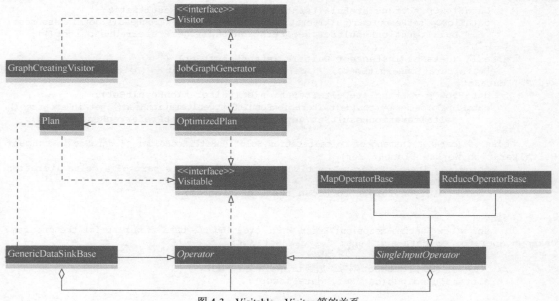

图 4-3　Visitable、Visitor 等的关系

Plan 的核心字段如下。

- sinks：GenericDataSinkBase<?>类型。
- jobName：表示作业名称。
- defaultParallelism：表示默认设置的并行度。
- jobId：JobID 类型，表示作业的唯一标识。

根据对 OperatorTranslation 的分析，可了解到转换生成的 org.apache.Flink.api.common.operators.

Operator 对象全部会被封装到 sinks 字段中。随后就可以在 accept()方法中对 sinks 进行优化了：

```
public void accept(Visitor<Operator<?>> visitor) {
    for (GenericDataSinkBase<?> sink : this.sinks) {
        sink.accept(visitor);
    }
}
```

## 4.3 从 StreamGraph 到 JobGraph

从 StreamGraph 到 JobGraph 的转换如图 4-4 所示。

StreamNode 与 JobVertex 对应，StreamEdge 与 JobEdge 对应。在 JobGraph 中，用 IntermediateDataSet 来表示 JobVertex 的输出。

从 StreamGraph 到 JobGraph 的转换调用的是 FlinkPipelineTranslationUtil 的静态方法 getJobGraph()：

```
public static JobGraph getJobGraph(
    Pipeline pipeline,
    Configuration optimizerConfiguration,
    int defaultParallelism) {
// 根据 pipeline 对象的具体类型返回对应的 pipelineTranslator
    FlinkPipelineTranslator pipelineTranslator = getPipelineTranslator(pipeline);
// 完成从 pipeline 到 JobGraph 的转换
    return pipelineTranslator.translateToJobGraph(pipeline,
        optimizerConfiguration,
        defaultParallelism);
}
```

上述代码首先调用了 getPipelineTranslator()方法，获取了对应的 translator。

```
private static FlinkPipelineTranslator getPipelineTranslator(Pipeline pipeline) {
    PlanTranslator planToJobGraphTransmogrifier = new PlanTranslator();
    if (planToJobGraphTransmogrifier.canTranslate(pipeline)) {
        return planToJobGraphTransmogrifier;
    }
    FlinkPipelineTranslator streamGraphTranslator = reflectStreamGraphTranslator();
    ...
    return streamGraphTranslator;
}
public boolean canTranslate(Pipeline pipeline) {
    return pipeline instanceof Plan;
}
```

如果 pipeline 是 StreamGraph 类型，则返回 StreamGraphTranslator；如果 pipeline 是 Plan 类型，则返回 PlanTranslator：

StreamGraphTranslator 的 translateToJobGraph()方法如下：

```
public JobGraph translateToJobGraph(
    Pipeline pipeline,
    Configuration optimizerConfiguration,
    int defaultParallelism) {
    StreamGraph streamGraph = (StreamGraph) pipeline;
    return streamGraph.getJobGraph(null);
}
```

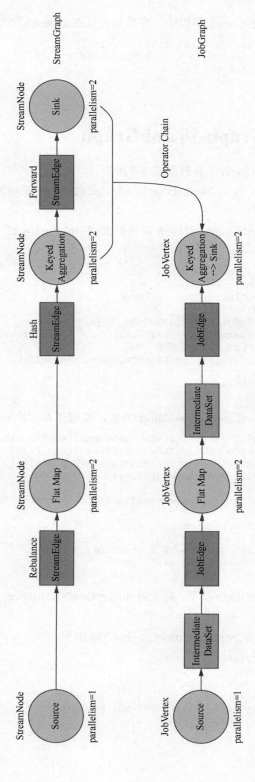

图 4-4 从 StreamGraph 到 JobGraph 的转换

其中调用的是 StreamGraph 的 getJobGraph()方法：

```
public JobGraph getJobGraph(@Nullable JobID jobID) {
    return StreamingJobGraphGenerator.createJobGraph(this, jobID);
}
public static JobGraph createJobGraph(StreamGraph streamGraph, @Nullable JobID jobID) {
    return new StreamingJobGraphGenerator(streamGraph, jobID).createJobGraph();
}
```

所有的转换逻辑发生在 StreamingJobGraphGenerator 中。

## 4.3.1 StreamingJobGraphGenerator 分析

StreamingJobGraphGenerator 是 JobGraph 的生成器，它的核心字段如下。
- streamGraph：StreamGraph 类型。
- jobVertices：Map<Integer, JobVertex>类型，用于维护 JobGraph 中的节点。
- jobGraph：JobGraph 类型。
- builtVertices：Collection<Integer>类型，用于维护已经构建的 StreamNode 的 id。
- physicalEdgesInOrder：StreamEdge 类型，表示物理边的集合，排除了链接内部的边。
- chainedConfigs：Map<Integer, Map<Integer, StreamConfig>> 类型，保存链接信息。
- vertexConfigs：Map<Integer, StreamConfig>类型，保存节点信息。
- chainedNames：Map<Integer, String>类型，表示链接的名称。

上述字段中，许多字段与链接有关。从 StreamGraph 到 JobGraph 的转换中，很重要的事情是将可以链接在一起的任务链接起来，这样它们可以作为同一个任务实现串行执行，从而可以达到节省资源、提高效率的目的。

调用 createJobGraph()方法可以生成 JobGraph：

```
private JobGraph createJobGraph() {
    // （1）验证
    preValidate();
    // （2）设置调度方式、计算哈希值等
    jobGraph.setScheduleMode(streamGraph.getScheduleMode());
    Map<Integer, byte[]> hashes = defaultStreamGraphHasher.traverseStreamGraphAndGenerateHashes(streamGraph);
    List<Map<Integer, byte[]>> legacyHashes = new ArrayList<>(legacyStreamGraphHashers.size());
    for (StreamGraphHasher hasher : legacyStreamGraphHashers) {
        legacyHashes.add(hasher.traverseStreamGraphAndGenerateHashes(streamGraph));
    }
    Map<Integer, List<Tuple2<byte[], byte[]>>> chainedOperatorHashes = new HashMap<>();
    // （3）将任务链接在一起，创建节点 JobVertex 和边 JobEdge
    setChaining(hashes, legacyHashes, chainedOperatorHashes);
    // （4）设置物理边
    setPhysicalEdges();
    // （5）设置共享槽组
    setSlotSharingAndCoLocation();
    // （6）设置内存占比
    setManagedMemoryFraction(
        Collections.unmodifiableMap(jobVertices),
```

```
        Collections.unmodifiableMap(vertexConfigs),
        Collections.unmodifiableMap(chainedConfigs),
        id -> streamGraph.getStreamNode(id).getMinResources(),
        id -> streamGraph.getStreamNode(id).getManagedMemoryWeight());
    // （7）配置检查点、保存点参数
    configureCheckpointing();
    jobGraph.setSavepointRestoreSettings(streamGraph.getSavepointRestoreSettings());
    // （8）设置分布式缓存文件
    JobGraphGenerator.addUserArtifactEntries(streamGraph.getUserArtifacts(), jobGraph);
    // （9）设置运行时的配置
    try {
        jobGraph.setExecutionConfig(streamGraph.getExecutionConfig());
    }
    catch (IOException e) {
        ...
    }
    // （10）返回完整的 JobGraph 对象
    return jobGraph;
}
```

上述代码的步骤如下。

（1）进行一些转换前的验证，比如检验配置是否有冲突等。

（2）设置调度方式、计算哈希值等。

（3）将任务链接在一起，创建节点 JobVertex 和边 JobEdge。这是整个过程中最重要的一步。

（4）设置物理边。

（5）设置共享槽组。

（6）设置内存占比。

（7）配置检查点、保存点参数。

（8）设置分布式缓存文件。

（9）设置运行时的配置。

（10）返回完整的 JobGraph 对象。

接下来对其中一些步骤进行详细介绍。

第 1 步是进行一些转换前的验证：

```
private void preValidate() {
    CheckpointConfig checkpointConfig = streamGraph.getCheckpointConfig();
    if (checkpointConfig.isCheckpointingEnabled()) {
        // 如果没有为迭代任务强制开启检查点机制，则会抛出异常
        if (streamGraph.isIterative() && !checkpointConfig.isForceCheckpointing()) {
            ...
        }
        ClassLoader classLoader = Thread.currentThread().getContextClassLoader();
        for (StreamNode node : streamGraph.getStreamNodes()) {
            StreamOperatorFactory operatorFactory = node.getOperatorFactory();
            if (operatorFactory != null) {
                Class<?> operatorClass = operatorFactory.getStreamOperatorClass(classLoader);
// 如果 StreamOperator 对象实现了 InputSelectable 接口，则抛出异常，因为目前该类型不支持开启检查点机制
                if (InputSelectable.class.isAssignableFrom(operatorClass)) {
                    ...
                }
```

```
            }
        }
    }
```

抛出异常的条件已在注释中给出。

第 2 步是设置调度方式、计算哈希值等。调度方式的值直接从 StreamGraph 中获得。计算哈希值的过程较为复杂,概括来说就是如果用户指定了 uid,则根据 uid 生成,否则根据拓扑结构生成。注意,这里生成了 3 个对象,一个是 hashes,一个是 legacyHashes,还有一个是 chainedOperatorHashes。hashes 的数据类型是映射（Map）,键为节点 id,值为哈希值。legacyHashes 的数据类型和含义与 hashes 的一样,只不过它的哈希值是由用户指定的 userHash 加上另一种算法计算出来的,这是为了显式地指定哈希值,以免版本变动导致哈希值不一致。chainedOperatorHashes 的数据类型为 Map<Integer, Tuple2<byte[], byte[]>>,其值是一个列表（List）,是将链接在一起的算子的哈希值放在同一个列表中进行维护。在 setChaining() 方法中就能看到它的值是如何被填充的。

第 3 步将任务链接在了一起:

```
private void setChaining(Map<Integer, byte[]> hashes, List<Map<Integer, byte[]>> legacyHashes, Map<Integer, List<Tuple2<byte[], byte[]>>> chainedOperatorHashes) {
    // 遍历 source 任务的 id
    for (Integer sourceNodeId : streamGraph.getSourceIDs()) {
        createChain(sourceNodeId, sourceNodeId, hashes, legacyHashes, 0, chainedOperatorHashes);
    }
}
```

在 createChain() 方法中,创建了 JobVertex 和 JobEdge。该方法涉及的代码较多,先来分析方法签名:

```
private List<StreamEdge> createChain(
    Integer startNodeId,
    Integer currentNodeId,
    Map<Integer, byte[]> hashes,
    List<Map<Integer, byte[]>> legacyHashes,
    int chainIndex,
    Map<Integer, List<Tuple2<byte[], byte[]>>> chainedOperatorHashes)
```

第 1 个参数为起始节点的 id,第 2 个参数为当前节点的 id,注意到上面代码中 setChaining() 方法调用 createChain() 方法时,这两个值都为 sourceNodeId,表示从 source 开始,并且当前正处于 source 节点。可以推测接下来会递归地调用这个方法,以 source 为起点,第 2 个参数会变为后续节点的 id。第 5 个参数为链接中节点的索引。其他几个参数为前面介绍过的哈希值的相关参数。

进入 createChain() 方法后,会先判断起始节点是否已经构建,如果没有构建则进入构建的逻辑,否则直接返回空列表。

```
// 物理出边
List<StreamEdge> transitiveOutEdges = new ArrayList<StreamEdge>();
// 存放可以被链接的出边的列表
List<StreamEdge> chainableOutputs = new ArrayList<StreamEdge>();
// 存放不可以被链接的出边的列表
List<StreamEdge> nonChainableOutputs = new ArrayList<StreamEdge>();
```

```java
        // 获得当前的 StreamNode
        StreamNode currentNode = streamGraph.getStreamNode(currentNodeId);
        // 遍历当前 StreamNode 的出边
        for (StreamEdge outEdge : currentNode.getOutEdges()) {
            if (isChainable(outEdge, streamGraph)) { // 根据出边,判断该节点与其下游节点是否可以被链接在一起
                chainableOutputs.add(outEdge);   // 如果可以被链接在一起,则把该出边添加到 chainableOutputs 中
            } else {
                nonChainableOutputs.add(outEdge); // 如果不可以被链接在一起,则把该出边添加到 nonChainable
                                                  // Outputs 中
            }
        }
        // 遍历可以被链接在一起的边,递归调用 createChain()方法,将结果全部添加到 transitiveOutEdges 中
        for (StreamEdge chainable : chainableOutputs) {
            transitiveOutEdges.addAll(
                    createChain(startNodeId, chainable.getTargetId(), hashes, legacyHashes,
chainIndex + 1, chainedOperatorHashes));
        }
        // 遍历不可以被链接在一起的边,将其直接添加到 transitiveOutEdges 中,并递归调用 createChain()方
        // 法为后面的节点构建链接
        for (StreamEdge nonChainable : nonChainableOutputs) {
            transitiveOutEdges.add(nonChainable);
            createChain(nonChainable.getTargetId(), nonChainable.getTargetId(), hashes, legacyHashes,
0, chainedOperatorHashes);
        }
        // 维护每个链接的哈希值
        List<Tuple2<byte[], byte[]>> operatorHashes =
                chainedOperatorHashes.computeIfAbsent(startNodeId, k -> new ArrayList<>());
        byte[] primaryHashBytes = hashes.get(currentNodeId);
        OperatorID currentOperatorId = new OperatorID(primaryHashBytes);
        for (Map<Integer, byte[]> legacyHash : legacyHashes) {
            operatorHashes.add(new Tuple2<>(primaryHashBytes, legacyHash.get(currentNodeId)));
        }
        // 配置链接的名称、资源等
        ...
        // 为每个节点创建 StreamConfig。如果当前节点是链接中的第一个节点,则创建一个 JobVertex,否则只需
        // 创建 StreamConfig
        StreamConfig config = currentNodeId.equals(startNodeId)
                ? createJobVertex(startNodeId, hashes, legacyHashes, chainedOperatorHashes)
                : new StreamConfig(new Configuration());
        // 设置 StreamConfig 中的属性
        setVertexConfig(currentNodeId, config, chainableOutputs, nonChainableOutputs);
        if (currentNodeId.equals(startNodeId)) {
            config.setChainStart();
            config.setChainIndex(0);
            config.setOperatorName(streamGraph.getStreamNode(currentNodeId).getOperatorName());
            config.setOutEdgesInOrder(transitiveOutEdges);
            config.setOutEdges(streamGraph.getStreamNode(currentNodeId).getOutEdges());
            for (StreamEdge edge : transitiveOutEdges) {
                // 遍历物理出边,构建 JobEdge
                connect(startNodeId, edge);
            }
            config.setTransitiveChainedTaskConfigs(chainedConfigs.get(startNodeId));
```

```
        } else {
            chainedConfigs.computeIfAbsent(startNodeId, k -> new HashMap<Integer, StreamConfig>());
            config.setChainIndex(chainIndex);
            StreamNode node = streamGraph.getStreamNode(currentNodeId);
            config.setOperatorName(node.getOperatorName());
            chainedConfigs.get(startNodeId).put(currentNodeId, config);
        }
        config.setOperatorID(currentOperatorId);
        if (chainableOutputs.isEmpty()) {
            config.setChainEnd();
        }
        return transitiveOutEdges;
```

因为用到了递归，所以整个方法比较复杂。关键在于理解一开始创建的 3 个列表的含义。chainableOutputs 和 nonChainableOutputs 分别表示可以被链接在一起的边和不可以被链接在一起的边。其中的元素都是 StreamEdge。是否可以被链接在一起是由 isChainable()方法决定的：

```
public static boolean isChainable(StreamEdge edge, StreamGraph streamGraph) {
    StreamNode upStreamVertex = streamGraph.getSourceVertex(edge);
    StreamNode downStreamVertex = streamGraph.getTargetVertex(edge);
    StreamOperatorFactory<?> headOperator = upStreamVertex.getOperatorFactory();
    StreamOperatorFactory<?> outOperator = downStreamVertex.getOperatorFactory();
    return downStreamVertex.getInEdges().size() == 1
            && outOperator != null
            && headOperator != null
            && upStreamVertex.isSameSlotSharingGroup(downStreamVertex)
            && outOperator.getChainingStrategy() == ChainingStrategy.ALWAYS
            && (headOperator.getChainingStrategy() == ChainingStrategy.HEAD ||
                headOperator.getChainingStrategy() == ChainingStrategy.ALWAYS)
            && (edge.getPartitioner() instanceof ForwardPartitioner)
            && edge.getShuffleMode() != ShuffleMode.BATCH
            && upStreamVertex.getParallelism() == downStreamVertex.getParallelism()
            && streaMGraph.isChainingEnabled();
}
```

总结起来，isChainable()方法中的判断条件如下。
- 该出边的上游算子是该出边的下游算子的唯一输入。
- 上下游的 StreamOperatorFactory 对象的值不能为 null。
- 上下游处于同一个 SlotSharingGroup。
- 下游的 ChainingStrategy 为 ALWAYS。
- 上游的 ChainingStrategy 为 ALWAYS 或 HEAD。
- 该出边的分区器为 ForwardPartitioner，即没有进行重分区等。
- 该出边的 ShuffleMode 不为 BATCH。
- 上下游的并行度相同。
- StreamGraph 的 chaining 属性的值为 true。

transitiveOutEdges 表示物理出边，即链接与链接之间的边，从 transitiveOutEdges.add (nonChainable)一行可以看出它添加的是不能被链接在一起的边。

由维护哈希值的一段代码可以看出 chainedOperatorHashes 按照链接来维护哈希值，其中的哈希值仍然是 hashes 和 legacyHashes 中的值，只不过换了一套封装。

接着，为每个 StreamNode 创建一个 StreamConfig，其中包含 StreamNode 的计算逻辑，如果 StreamNode 是这个链接的第一个节点，则调用 createJobVertex() 来创建；否则直接通过构造方法创建。同一个链接中的 StreamConfig 会在 chainedConfigs 字段中进行维护。另外，对于每一个链接的物理出边，会由 connect() 方法将其与起始节点连接在一起。

下面分别分析 createJobVertex() 方法和 connect() 方法。createJobVertex() 方法如下：

```
private StreamConfig createJobVertex(
    Integer streamNodeId,
    Map<Integer, byte[]> hashes,
    List<Map<Integer, byte[]>> legacyHashes,
    Map<Integer, List<Tuple2<byte[], byte[]>>> chainedOperatorHashes) {
    // 设置哈希值等
    ...
    if (chainedInputOutputFormats.containsKey(streamNodeId)) {
        ...
    } else {
        // 创建 JobVertex
        jobVertex = new JobVertex(
            chainedNames.get(streamNodeId),
            jobVertexId,
            legacyJobVertexIds,
            chainedOperatorVertexIds,
            userDefinedChainedOperatorVertexIds);
    }
    // 给 JobVertex 添加属性
    ...
    jobVertices.put(streamNodeId, jobVertex);
    builtVertices.add(streamNodeId);
    jobGraph.addVertex(jobVertex);
    return new StreamConfig(jobVertex.getConfiguration());
}
```

虽然方法涉及的代码较多，但大部分代码用于设置属性等。JobVertex 是通过构造方法直接创建的。其中的许多属性值直接来自 StreamNode。创建好以后，更新 jobVertices、builtVertices 等字段的值，并在 jobGraph 字段中添加该节点。

connect() 方法如下：

```
private void connect(Integer headOfChain, StreamEdge edge) {
    // 将这条边添加到物理出边中
    physicalEdgesInOrder.add(edge);
    Integer downStreamvertexID = edge.getTargetId();
    JobVertex headVertex = jobVertices.get(headOfChain);
    JobVertex downStreamVertex = jobVertices.get(downStreamvertexID);
    StreamConfig downStreamConfig = new StreamConfig(downStreamVertex.getConfiguration());
    downStreamConfig.setNumberOfInputs(downStreamConfig.getNumberOfInputs() + 1);
    StreamPartitioner<?> partitioner = edge.getPartitioner();
    ...
```

```
// 构造 JobEdge
JobEdge jobEdge;
if (partitioner instanceof ForwardPartitioner || partitioner instanceof RescalePartitioner)
                                                                                            {
    jobEdge = downStreamVertex.connectNewDataSetAsInput(
      headVertex,
      DistributionPattern.POINTWISE,
      resultPartitionType);
} else {
    jobEdge = downStreamVertex.connectNewDataSetAsInput(
        headVertex,
        DistributionPattern.ALL_TO_ALL,
        resultPartitionType);
}
...
}
```

上述代码的主要逻辑是通过边获取一个链接上下游的 JobVertex，然后构造一条边 JobEdge 将两个 JobVertex 连接起来。创建边时会根据 partitioner 的类型，调用 connectNewDataSetAsInput() 方法来创建，后文在分析 JobVertex 和 JobEdge 的结构时会对该方法进行介绍。

在 setVertexConfig() 方法中又会设置其他的属性，如计算逻辑、状态后端等。

至此分析完成 createJobGraph() 方法中的第 3 步。

下面分析第 4 步，设置物理边。

```
private void setPhysicalEdges() {
    Map<Integer, List<StreamEdge>> physicalInEdgesInOrder = new HashMap<Integer, List<StreamEdge>>();
    for (StreamEdge edge : physicalEdgesInOrder) {
        int target = edge.getTargetId();
        List<StreamEdge> inEdges = physicalInEdgesInOrder.computeIfAbsent(target, k -> new ArrayList<>());
        inEdges.add(edge);
    }
    for (Map.Entry<Integer, List<StreamEdge>> inEdges : physicalInEdgesInOrder.entrySet()) {
        int vertex = inEdges.getKey();
        List<StreamEdge> edgeList = inEdges.getValue();
        vertexConfigs.get(vertex).setInPhysicalEdges(edgeList);
    }
}
```

setPhysicalEdges() 方法用于给 StreamConfig 添加物理边。

剩下的第 5 步到第 9 步基本就是将 StreamGraph 中的属性按照一定的规则设置到 JobGraph 中。最后在第 10 步中返回 JobGraph 对象。

### 4.3.2 JobGraph 分析

StreamGraph 转换成 JobGraph 后，所有的信息都会封装到 JobGraph 中。JobGraph 的核心字段如下。
- taskVertices：Map<JobVertexID, JobVertex>类型，维护所有 JobGraph 中的节点 JobVertex。
- jobID：JobID 类型，表示作业的唯一标识。
- jobName：表示作业的名称。

- scheduleMode：ScheduleMode 类型，表示调度方式。

其他与配置有关的字段如内存的配置、保存点的配置等。还有几个字段，如 userJars、classpaths 等，通过设置这些字段，可以在 TaskManager 添加一些依赖。而 JobGraph 的方法也主要是针对这些字段的 get/set 方法。

### 4.3.3　JobVertex、JobEdge 和 IntermediateDataSet

JobVertex 表示一个链接后的任务节点，它的核心字段如下。

- id：JobVertexID 类型，表示 JobVertex 的唯一标识。
- operatorIDs：Array<OperatorID>类型，存放该节点中所有算子的 id。
- results：Array<IntermediateDataSet>类型，表示该节点的输出。
- inputs：Array<JobEdge>类型，表示该节点的输入。
- configuration：Configuration 类型，保存运行时需要被反序列化的信息，如自定义函数、序列化器等。
- invokableClassName：TaskManager 根据该类名创建执行任务的类的对象。
- name：表示节点名称。

其余还有一些表示共享槽组、资源等的字段。这里需要注意的是，节点的输入和输出的表示方式与之前的有所区别。这里用 JobEdge 表示节点的输入，用 IntermediateDataSet 表示节点的输出。在前文介绍过的 connectNewDataSetAsInput() 方法中创建了 IntermediateDataSet 对象和 JobEdge 对象：

```
public JobEdge connectNewDataSetAsInput(
    JobVertex input,
    DistributionPattern distPattern,
    ResultPartitionType partitionType) {
    IntermediateDataSet dataSet = input.createAndAddResultDataSet(partitionType);
    JobEdge edge = new JobEdge(dataSet, this, distPattern);
    this.inputs.add(edge);
    dataSet.addConsumer(edge);
    return edge;
}
public IntermediateDataSet createAndAddResultDataSet(ResultPartitionType partitionType) {
    return createAndAddResultDataSet(new IntermediateDataSetID(), partitionType);
}
public IntermediateDataSet createAndAddResultDataSet(
    IntermediateDataSetID id,
    ResultPartitionType partitionType) {
    IntermediateDataSet result = new IntermediateDataSet(id, partitionType, this);
    this.results.add(result);
    return result;
}
```

在 IntermediateDataSet 的构造方法中，传入了上游的 JobVertex 作为其 producer 字段：

```
public IntermediateDataSet(IntermediateDataSetID id, ResultPartitionType resultType,
JobVertex producer) {
    this.id = checkNotNull(id);
    this.producer = checkNotNull(producer);
```

```
    this.resultType = checkNotNull(resultType);
}
```

另外,创建出 JobEdge 后又会将它作为 IntermediateDataSet 的 consumers 字段:

```
public void addConsumer(JobEdge edge) {
    this.consumers.add(edge);
}
```

JobEdge 的核心字段如下。
- target:JobVertex 类型,表示这条边连接的下游节点。
- distributionPattern:DistributionPattern 类型,表示一个枚举类,包括 ALL_TO_ALL(每个下游任务都消费这条边的数据)、POINTWISE(一个或多个下游任务消费这条边的数据)。
- source:IntermediateDataSet 类型,表示数据源。

其构造方法如下:

```
public JobEdge(IntermediateDataSet source, JobVertex target, DistributionPattern 
distributionPattern) {
    ...
    this.target = target;
    this.distributionPattern = distributionPattern;
    this.source = source;
    this.sourceId = source.getId();
}
```

由此,上游节点、IntermediateDataSet、JobEdge 和下游节点就联系了起来。JobVertex、JobEdge、IntermediateDataSet 与 JobGraph 的关系如图 4-5 所示。

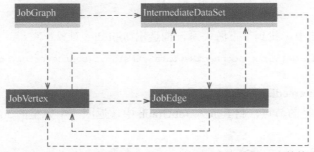

图 4-5  JobVertex、JobEdge、IntermediateDataSet 与 JobGraph 的关系

## 4.4 从 Plan 到 JobGraph

前文已经分析过,无论是 DataStream 还是 DataSet,都会调用 FlinkPipelineTranslationUtil 的静态方法 getJobGraph(),在该方法中,根据 pipeline 的不同类型返回不同的 pipelineTranslator。当 pipeline 是 Plan 类型时,由 PlanTranslator 进行处理。

```
public JobGraph translateToJobGraph(
        Pipeline pipeline,
        Configuration optimizerConfiguration,
        int defaultParallelism) {
```

```
        Plan plan = (Plan) pipeline;
        setDefaultParallelism(plan, defaultParallelism);
        return compilePlan(plan, optimizerConfiguration);
    }
    private JobGraph compilePlan(Plan plan, Configuration optimizerConfiguration) {
        Optimizer optimizer = new Optimizer(new DataStatistics(), optimizerConfiguration);
        OptimizedPlan optimizedPlan = optimizer.compile(plan);
        JobGraphGenerator jobGraphGenerator = new JobGraphGenerator(optimizerConfiguration);
        return jobGraphGenerator.compileJobGraph(optimizedPlan, plan.getJobId());
    }
```

在 compilePlan()方法中，首先用 Optimizer 对 Plan 进行优化，得到 OptimizedPlan 对象。然后构造一个 JobGraphGenerator，其与 DataStream 中的 StreamingJobGraphGenerator 相对应。通过调用 JobGraphGenerator 的 compileJobGraph()方法，返回一个完整的 JobGraph 对象。

JobGraphGenerator 是 JobGraph 的生成器，其逻辑十分复杂，其中核心的步骤是调用 optimizedPlan 的 accept()方法，并传入 JobGraphGenerator 自身。前文提到过，这个 accept()方法可以被理解为传入一个优化器对执行计划进行优化。传入 JobGraphGenerator 后，会在 preVisit()方法中根据节点的不同类型调用不同方法构造 JobVertex 对象：

```
final JobVertex vertex;
try {
    if (node instanceof SinkPlanNode) {
        vertex = createDataSinkVertex((SinkPlanNode) node);
    }
    else if (node instanceof SourcePlanNode) {
        vertex = createDataSourceVertex((SourcePlanNode) node);
    }
    ...
```

在 postVisit()方法中，找到上下游关系，构造出 JobEdge 对象：

```
JobEdge edge = targetVertex.connectNewDataSetAsInput(sourceVertex, distributionPattern,
resultType);
```

同时也会创建 IntermediateDataSet 对象。

最后，遍历创建好的节点，将其放入 JobGraph 中，返回一个完整的 JobGraph。

## 4.5 从 JobGraph 到 ExecutionGraph

虽然 JobGraph 中包含每个算子并行度的信息，但是每个任务仍然以一个对象来呈现。在创建 ExecutionGraph 的过程中，会按照并行度的大小创建任务的并行实例，同时也会创建 JobGraph 中的 IntermediateDataSet 和 JobEdge 在 ExecutionGraph 中对应的对象。

JobGraph 和 ExecutionGraph 的关系如图 4-6 所示。

JobGraph 被创建好以后，会被提交到 JobManager。在 JobManager 中需要创建调度器，在调度器的初始化过程中，会调用 ExecutionGraphBuilder 的静态方法 buildGraph()将 JobGraph 转换成 ExecutionGraph。ExecutionGraph 对象就是调度器中的一个字段。

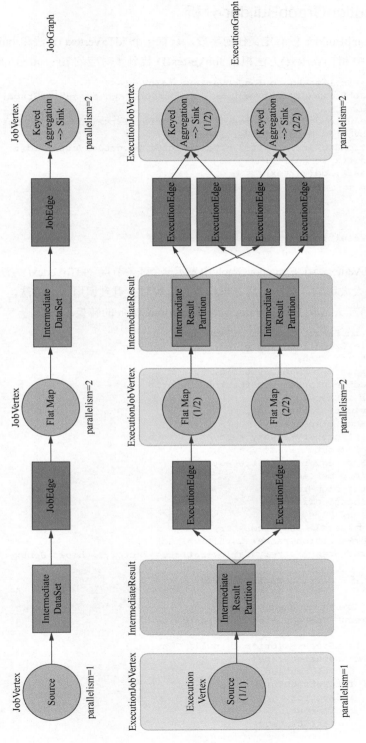

图 4-6 JobGraph 和 ExecutionGraph 的关系

## 4.5.1　ExecutionGraphBuilder 分析

ExecutionGraphBuilder 没有定义任何字段，只有一个 idToVertex()方法和 buildGraph()方法的两个重载方法。其中 idToVertex()方法根据 JobVertexID 找到了对应的 ExecutionJobVertex 对象。

```
private static List<ExecutionJobVertex> idToVertex(
    List<JobVertexID> jobVertices, ExecutionGraph executionGraph) throws Illegal
ArgumentException {
    List<ExecutionJobVertex> result = new ArrayList<>(jobVertices.size());
    for (JobVertexID id : jobVertices) {
        ExecutionJobVertex vertex = executionGraph.getJobVertex(id);
        if (vertex != null) {
          result.add(vertex);
        } else {
          ...
        }
    }
    return result;
}
```

ExecutionJobVertex 类在 ExecutionGraph 中表示一个任务节点。IdToVertex()方法对 ExecutionGraph 对象的生成没有太大影响，但在之后分析检查点流程时会看到调用这个方法。

buildGraph()方法完成了从 JobGraph 到 ExecutionGraph 的转换：

```
public static ExecutionGraph buildGraph(
   @Nullable ExecutionGraph prior,
   JobGraph jobGraph,
   Configuration jobManagerConfig,
   ScheduledExecutorService futureExecutor,
   Executor ioExecutor,
   SlotProvider slotProvider,
   ClassLoader classLoader,
   CheckpointRecoveryFactory recoveryFactory,
   Time rpcTimeout,
   RestartStrategy restartStrategy,
   MetricGroup metrics,
   BlobWriter blobWriter,
   Time allocationTimeout,
   Logger log,
   ShuffleMaster<?> shuffleMaster,
   JobMasterPartitionTracker partitionTracker,
   FailoverStrategy.Factory failoverStrategyFactory) throws JobExecutionException,
JobException {
    // 构造创建 ExecutionGraph 所需的参数
    ...
    // 初始化 ExecutionGraph
    final ExecutionGraph executionGraph;
    try {
        executionGraph = (prior != null) ? prior :
            new ExecutionGraph(
                jobInformation,
                futureExecutor,
                ioExecutor,
                rpcTimeout,
                restartStrategy,
```

```
                maxPriorAttemptsHistoryLength,
                failoverStrategyFactory,
                slotProvider,
                classLoader,
                blobWriter,
                allocationTimeout,
                partitionReleaseStrategyFactory,
                shuffleMaster,
                partitionTracker,
                jobGraph.getScheduleMode());
    } catch (IOException e) {
        ...
    }
    ...
    // 根据 JobGraph 中的节点创建 ExecutionGraph 中的节点等
    List<JobVertex> sortedTopology = jobGraph.getVerticesSortedTopologicallyFromSources();
    executionGraph.attachJobGraph(sortedTopology);
    // 设置检查点相关内容
    JobCheckpointingSettings snapshotSettings = jobGraph.getCheckpointingSettings();
    if (snapshotSettings != null) {
        ...
    }
    // 创建 ExecutionGraph 的度量 metric
    ...
    return executionGraph;
}
```

在 buildGraph()方法中，会先通过 ExecutionGraph 的构造方法实例化一个对象，但此时该对象中关于节点的属性值还为空。在 ExecutionGraph 的 attachJobGraph()方法中，根据 JobGraph 中的节点信息填充 ExecutionGraph 中的字段值。

### 4.5.2　ExecutionGraph 分析

ExecutionGraph 是协调分布式任务执行的关键的数据结构。它持有每一个任务的并行实例并维护它们的运行状态。它与其中一些关键组件的关系如图 4-7 所示。下文会依次对这些组件进行介绍。

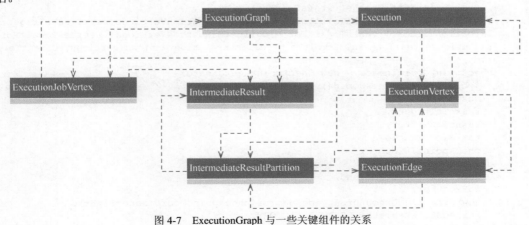

图 4-7　ExecutionGraph 与一些关键组件的关系

ExecutionGraph 的字段非常多，也较为复杂。其中许多字段的值直接来自 JobMaster 中的字段。这里先对其中一些核心字段进行简单的介绍，后文在必要时会再次提及，这些字段如下。

- jobInformation：JobInformation 类型，包含作业名称、作业 id 等信息。
- tasks：Map<JobVertexID, ExecutionJobVertex>类型，表示将 JobVertexID 与 ExecutionGraph 中的任务节点 ExecutionJobVertex 映射起来。
- verticesInCreationOrder：ExecutionJobVertex 类型，表示按照创建顺序维护的 ExecutionJobVertex 列表。
- intermediateResults： Map<IntermediateDataSetID, IntermediateResult>类型，维护所有的 IntermediateResult 对象。
- currentExecutions： Map<ExecutionAttemptID, Execution>类型，表示当前执行的任务。
- jobStatusListeners：JobStatusListener 类型，表示作业状态的监听器。
- failoverStrategy：FailoverStrategy 类型，表示作业失败后的策略。
- restartStrategy：RestartStrategy 类型，表示重启策略。
- slotProviderStrategy：SlotProviderStrategy 类型，用于给任务提供槽位的策略。
- scheduleMode：ScheduleMode 类型，表示调度方式。
- verticesFinished：表示已完成的节点个数。
- state：JobStatus 类型，表示作业的状态。
- checkpointCoordinator：CheckpointCoordinator 类型，表示检查点的协调器。

ExecutionGraph 提供的方法中，比较重要的如下。

```
public void enableCheckpointing(
    CheckpointCoordinatorConfiguration chkConfig,
    List<ExecutionJobVertex> verticesToTrigger,
    List<ExecutionJobVertex> verticesToWaitFor,
    List<ExecutionJobVertex> verticesToCommitTo,
    List<MasterTriggerRestoreHook<?>> masterHooks,
    CheckpointIDCounter checkpointIDCounter,
    CompletedCheckpointStore checkpointStore,
    StateBackend checkpointStateBackend,
    CheckpointStatsTracker statsTracker) {
ExecutionVertex[] tasksToTrigger = collectExecutionVertices(verticesToTrigger);
ExecutionVertex[] tasksToWaitFor = collectExecutionVertices(verticesToWaitFor);
ExecutionVertex[] tasksToCommitTo = collectExecutionVertices(verticesToCommitTo);
...
checkpointCoordinator = new CheckpointCoordinator(
    jobInformation.getJobId(),
    chkConfig,
    tasksToTrigger,
    tasksToWaitFor,
    tasksToCommitTo,
    checkpointIDCounter,
    checkpointStore,
    checkpointStateBackend,
    ioExecutor,
    new ScheduledExecutorServiceAdapter(checkpointCoordinatorTimer),
    SharedStateRegistry.DEFAULT_FACTORY,
```

```
        failureManager);
    ...
    if (chkConfig.getCheckpointInterval() != Long.MAX_VALUE) {
        registerJobStatusListener(checkpointCoordinator.createActivatorDeactivator());
    }
    ...
}
```

在这个方法中，实例化了 CheckpointCoordinator 对象，并且调用了其 createActivatorDeactivator() 方法生成了监听器并对其进行了注册。在作业状态发生改变的时候，该监听器就会触发一些与检查点有关的操作。CheckpointCoordinator 的作用会在第 10 章详细介绍。

在完善 ExecutionGraph 的过程中，关键的方法是 attachJobGraph()：

```
public void attachJobGraph(List<JobVertex> topologiallySorted) throws JobException {
    ...
    // 遍历每一个 JobVertex
    for (JobVertex jobVertex : topologiallySorted) {
        ...
        // 给每一个 JobVertex 对象创建一个 ExecutionJobVertex 对象
        ExecutionJobVertex ejv = new ExecutionJobVertex(
            this,
            jobVertex,
            1,
            maxPriorAttemptsHistoryLength,
            rpcTimeout,
            globalModVersion,
            createTimestamp);
        // 处理上游输出与下游输入之间的关系
        ejv.connectToPredecessors(this.intermediateResults);
        // 将构造的 ExecutionJobVertex 对象放入 tasks 字段进行维护
        ExecutionJobVertex previousTask = this.tasks.putIfAbsent(jobVertex.getID(), ejv);
        ...
        // 将构造的 ExecutionJobVertex 对象放入 verticesInCreationOrder 字段进行维护
        this.verticesInCreationOrder.add(ejv);
        ...
    }
    ...
}
```

在这个方法中，根据 JobGraph 中的 JobVertex 节点创建出 ExecutionGraph 中对应的 ExecutionJobVertex 节点，将其放入 ExecutionGraph 的字段进行维护，并且在 ExecutionGraph 中建立了上下游之间的关系。ExecutionJobVertex 的构造方法和 connectToPredecessors() 方法会在后文详细介绍。

ExecutionGraph 还提供了许多与任务状态改变有关的方法。当任务状态改变时会调用 updateState() 方法来更新状态：

```
public boolean updateState(TaskExecutionState state) {
    assertRunningInJobMasterMainThread();
    final Execution attempt = currentExecutions.get(state.getID());
    if (attempt != null) {
        try {
            // 调用 updateStateInternal() 方法来更新状态
            final boolean stateUpdated = updateStateInternal(state, attempt);
```

```java
                maybeReleasePartitions(attempt);
                return stateUpdated;
            }
            catch (Throwable t) {
                ...
            }
        }
        else {
            return false;
        }
    }
    private boolean updateStateInternal(final TaskExecutionState state, final Execution attempt) {
        Map<String, Accumulator<?, ?>> accumulators;
        switch (state.getExecutionState()) {
            case RUNNING:
                return attempt.switchToRunning();
            case FINISHED:
                accumulators = deserializeAccumulators(state);
                attempt.markFinished(accumulators, state.getIOMetrics());
                return true;
            case CANCELED:
                accumulators = deserializeAccumulators(state);
                attempt.completeCancelling(accumulators, state.getIOMetrics(), false);
                return true;
            case FAILED:
                accumulators = deserializeAccumulators(state);
                attempt.markFailed(state.getError(userClassLoader), accumulators, state.getIOMetrics(), !isLegacyScheduling());
                return true;
            default:
                attempt.fail(new Exception("TaskManager sent illegal state update: " + state.getExecutionState()));
                return false;
        }
    }
```

updateState()方法更新状态的逻辑主要在 updateStateInternal()方法中实现，该方法根据不同的状态进行不同的操作，主要是调用 Execution 实例对象的对应方法。

### 4.5.3　ExecutionJobVertex、ExecutionVertex 和 Execution 分析

在分析 ExecutionGraph 时可看到 JobVertex 转换成了 ExecutionJobVertex，因此，ExecutionJobVertex 的含义与 JobVertex 的类似，表示整个任务。ExecutionVertex 表示的则是每个任务的并行实例。每个 ExecutionJobVertex 都包含若干 ExecutionVertex。Execution 表示一个 ExecutionVertex 执行的抽象。每个 ExecutionVertex 对象可能会对应多个 Execution 对象，这是因为执行可能会失败，或是数据需要重新计算等。JobManager 与 TaskManager 之间都是用 Execution 的唯一标识 ExecutionAttemptID 来追踪任务的执行的。

在 ExecutionGraph 的 attachJobGraph()方法中，利用 ExecutionJobVertex 的构造方法创建了其实例。其构造方法主要是用于给其中的字段赋值。因此，在了解构造方法前，先来看看 ExecutionJobVertex

的主要字段。
- graph：ExecutionGraph 类型。
- jobVertex：JobVertex 类型。
- operatorIDs：OperatorID 类型，表示该节点中包含的所有算子 id。
- taskVertices：ExecutionVertex 数组类型，表示该节点中包含的所有任务的并行实例。
- producedDataSets：IntermediateResult 数组类型，表示该节点的所有输出。
- inputs：IntermediateResult 类型，表示该节点的输入，即上一个节点的输出。

剩余的字段中，主要是并行度、资源等从 JobVertex 中获取的值。下面介绍在构造方法中给字段赋值的过程：

```java
public ExecutionJobVertex(
    ExecutionGraph graph,
    JobVertex jobVertex,
    int defaultParallelism,
    int maxPriorAttemptsHistoryLength,
    Time timeout,
    long initialGlobalModVersion,
    long createTimestamp) throws JobException {
...
    this.graph = graph;
    this.jobVertex = jobVertex;
    int vertexParallelism = jobVertex.getParallelism();
    int numTaskVertices = vertexParallelism > 0 ? vertexParallelism : defaultParallelism;
...
    this.parallelism = numTaskVertices;
    this.resourceProfile = ResourceProfile.fromResourceSpec(jobVertex.getMinResources(), MemorySize.ZERO);
    this.taskVertices = new ExecutionVertex[numTaskVertices];
    this.operatorIDs = Collections.unmodifiableList(jobVertex.getOperatorIDs());
    this.userDefinedOperatorIds = Collections.unmodifiableList(jobVertex.getUserDefinedOperatorIDs());
    // 初始化 inputs
    this.inputs = new ArrayList<>(jobVertex.getInputs().size());
...
    this.producedDataSets = new IntermediateResult[jobVertex.getNumberOfProducedIntermediateDataSets()];
    // 构造 ExecutionJobVertex 的输出 IntermediateResult 对象，给 producedDataSets 字段赋值
    for (int i = 0; i < jobVertex.getProducedDataSets().size(); i++) {
        final IntermediateDataSet result = jobVertex.getProducedDataSets().get(i);
        this.producedDataSets[i] = new IntermediateResult(
            result.getId(),
            this,
            numTaskVertices,
            result.getResultType());
    }
    // 构造并行实例 ExecutionVertex 对象，给 taskVertices 字段赋值
    for (int i = 0; i < numTaskVertices; i++) {
        ExecutionVertex vertex = new ExecutionVertex(
            this,
            i,
            producedDataSets,
            timeout,
```

```
                initialGlobalModVersion,
                createTimestamp,
                maxPriorAttemptsHistoryLength);
            this.taskVertices[i] = vertex;
        }
        ...
    }
```

上面的代码中，先对并行度等字段赋值。然后通过 JobVertex 中的 IntermediateDataSet 对象构建 ExecutionJobVertex 中的 IntermediateResult 对象，并根据并行度创建 ExecutionVertex 对象。

ExecutionVertex 的字段如下。

- jobVertex：ExecutionJobVertex 类型，表示该并行实例属于哪个 ExecutionJobVertex 任务节点。
- resultPartitions：Map<IntermediateResultPartitionID, IntermediateResultPartition>类型，表示该并行实例的输出。
- inputEdges：ExecutionEdge[][]类型，表示该并行实例的输入。由于上游可能有多个输入，而每个输入又有多个并行实例，因此需要用二维数组来表示。
- subTaskIndex：表示任务并行实例的索引。
- executionVertexId：表示 ExecutionVertex 的唯一标识。
- currentExecution：Execution 类型，表示对该并行实例执行的一个抽象。

创建 ExecutionVertex 对象的代码如下：

```
public ExecutionVertex(
        ExecutionJobVertex jobVertex,
        int subTaskIndex,
        IntermediateResult[] producedDataSets,
        Time timeout,
        long initialGlobalModVersion,
        long createTimestamp,
        int maxPriorExecutionHistoryLength) {
    this.jobVertex = jobVertex;
    this.subTaskIndex = subTaskIndex;
    this.executionVertexId = new ExecutionVertexID(jobVertex.getJobVertexId(), subTaskIndex);
    this.taskNameWithSubtask = String.format("%s (%d/%d)",
        jobVertex.getJobVertex().getName(), subTaskIndex + 1, jobVertex.getParallelism());
    this.resultPartitions = new LinkedHashMap<>(producedDataSets.length, 1);
    // 构造 IntermediateResultPartition 对象
    for (IntermediateResult result : producedDataSets) {
        IntermediateResultPartition irp = new IntermediateResultPartition(result, this, subTaskIndex);
        result.setPartition(subTaskIndex, irp);
        resultPartitions.put(irp.getPartitionId(), irp);
    }
    // 初始化 inputEdges 数组
    this.inputEdges = new ExecutionEdge[jobVertex.getJobVertex().getInputs().size()][];
    ...
    // 构造 Execution 对象
    this.currentExecution = new Execution(
        getExecutionGraph().getFutureExecutor(),
        this,
        0,
        initialGlobalModVersion,
```

```
        createTimestamp,
        timeout);
    ...
    // 把该 Execution 对象注册到 ExecutionGraph 的 currentExecution 字段中
    getExecutionGraph().registerExecution(currentExecution);
    ...
}
```

在构造方法中，首先对索引、唯一 id 等字段赋值。然后，构造 IntermediateResultPartition 对象，将其维护在 resultPartitions 字段中，并通过 setPartition()方法给相应的 IntermediateResult 对象中的字段赋值。接着初始化 inputEdges 数组（并没有赋值），并构造 Execution 对象，然后将其注册到 ExecutionGraph 中。

其中，Execution 主要包含以下字段。
- vertex：ExecutionVertex 类型，表示它对应的任务并行实例。
- attemptId：ExecutionAttemptID 类型，表示该执行的唯一标识。
- taskRestore：JobManagerTaskRestore 类型，当检查点机制开启后，会用这个字段维护与恢复和状态有关的信息，后文会详细介绍。
- assignedAllocationID：表示当一个执行被部署成功后，会有一个与该部署对应的唯一标识。

在 ExecutionVertex 和 Execution 中有很多与部署有关的方法，它们会在任务生命周期的不同阶段被调用，也有一些是版本迭代中遗留下来的方法。这些与部署有关的方法往往最终会调用 Execution 的 deploy()方法：

```
public void deploy() throws JobException {
    // 设置槽
    final LogicalSlot slot = assignedResource;
    ...
    try {
        ...
        // 构造 TaskDeploymentDescriptor 对象
        final TaskDeploymentDescriptor deployment = TaskDeploymentDescriptorFactory
            .fromExecutionVertex(vertex, attemptNumber)
            .createDeploymentDescriptor(
                slot.getAllocationId(),
                slot.getPhysicalSlotNumber(),
                taskRestore,
                producedPartitions.values());
        taskRestore = null;
        final TaskManagerGateway taskManagerGateway = slot.getTaskManagerGateway();
        final ComponentMainThreadExecutor jobMasterMainThreadExecutor =
            vertex.getExecutionGraph().getJobMasterMainThreadExecutor();
        // 提交 TaskDeploymentDescriptor 对象到分配的槽对应的 TaskManager 中
        CompletableFuture.supplyAsync(() -> taskManagerGateway.submitTask(deployment, rpcTimeout), executor)
            .thenCompose(Function.identity())
            .whenCompleteAsync(…);
    }
    catch (Throwable t) {
        ...
    }
}
```

在 deploy() 方法中，先设置槽，然后利用 ExecutionVertex、Execution 等对象中的信息构造出一个 TaskDeploymentDescriptor 对象，这个对象被提交到 TaskManager 中，构造出一个 Task 对象去执行任务。

Execution 中还有一个重要方法是 triggerCheckpointHelper()：

```
private void triggerCheckpointHelper(long checkpointId, long timestamp, CheckpointOptions checkpointOptions, boolean advanceToEndOfEventTime) {
    ...
    final LogicalSlot slot = assignedResource;
    if (slot != null) {
        final TaskManagerGateway taskManagerGateway = slot.getTaskManagerGateway();
        // 触发 TaskManager 上 ExecutionAttemptID 为 attemptId 的任务的与检查点相关的操作
        taskManagerGateway.triggerCheckpoint(attemptId, getVertex().getJobId(), checkpointId, timestamp, checkpointOptions, advanceToEndOfEventTime);
    } else {
        ...
    }
}
```

若开启了检查点机制，则在 JobManager 会找出需要触发检查点操作的节点，即对应的 Execution 对象，然后调用 triggerCheckpointHelper() 方法，进而会在 TaskManager 找到对应的并行实例并触发相关的操作。

这些 Execution 中的行为在后文介绍调度逻辑、任务的生命周期等内容时还会提及。

### 4.5.4　IntermediateResult、IntermediateResultPartition 和 ExecutionEdge

在 JobGraph 中，用 IntermediateDataSet 表示上游节点的输出，用 JobEdge 表示下游节点的输入，由此建立上下游关系。在 ExecutionGraph 中，与 IntermediateDataSet 相对应的是 IntermediateResult。然而，ExecutionGraph 中需要体现出并行度，因此在 IntermediateResult 中还有 IntermediateResultPartition，用于表示每个并行实例的输出。因为每个并行实例的输出都由单独的对象来表示，所以下游的输入也需要一一对应，这就要用到 ExecutionEdge。

IntermediateResult 的字段主要如下。

- id：IntermediateDataSetID 类型，表示对应的 IntermediateDataSet 对象的唯一标识。
- producer：ExecutionJobVertex 类型，表示数据的生产者。
- partitions：IntermediateResultPartition 数组类型，表示并行实例的输出。
- resultType：ResultPartitionType 类型。

IntermediateResultPartition 的字段主要如下。

- totalResult：IntermediateResult 类型，表示该并行实例的输出属于哪个 IntermediateResult 对象。
- producer：ExecutionVertex 类型，表示该并行实例的输出的生产者。
- partitionNumber：表示并行实例的索引。
- partitionId：IntermediateResultPartitionID 类型，表示该 IntermediateResultPartition 对象的唯一标识。

- consumers：ExecutionEdge 类型，表示该输出的消费者。
- hasDataProduced：表示该分区是否已经生产了数据。

ExecutionEdge 的字段主要如下。
- source：IntermediateResultPartition 类型，表示数据源。
- target：ExecutionVertex 类型，表示数据被下游哪个并行实例消费。

在了解了以上概念的含义后，再回来看 ExecutionJobVertex 的 connectToPredecessors()方法是如何处理上下游关系的。

```
public void connectToPredecessors(Map<IntermediateDataSetID, IntermediateResult> intermediateDataSets) throws JobException {
    // 从 JobVertex 对象中获取它的 JobEdge 列表（它的输入）
    List<JobEdge> inputs = jobVertex.getInputs();
    // 遍历它的输入
    for (int num = 0; num < inputs.size(); num++) {
        JobEdge edge = inputs.get(num);
        // JobEdge 的 sourceId 字段表示它对应的上游的 IntermediateDataSet 对象的唯一标识
        // 根据该标识从 intermediateDataSets 字段中获取对应的 IntermediateResult 对象
        IntermediateResult ires = intermediateDataSets.get(edge.getSourceId());
        ...
        // 将该对象放入 ExecutionJobVertex 的 inputs 字段中维护
        this.inputs.add(ires);
        // 在 ExecutionVertex 对象中维护上下游关系
        int consumerIndex = ires.registerConsumer();
        for (int i = 0; i < parallelism; i++) {
            ExecutionVertex ev = taskVertices[i];
            ev.connectSource(num, ires, edge, consumerIndex);
        }
    }
}
```

在这个方法中，完成了对 ExecutionJobVertex 的 inputs 列表的赋值，并调用了 ExecutionVertex 的 connectSource()方法用于维护上下游关系。

```
public void connectSource(int inputNumber, IntermediateResult source, JobEdge edge, int consumerNumber) {
    final DistributionPattern pattern = edge.getDistributionPattern();
    final IntermediateResultPartition[] sourcePartitions = source.getPartitions();
    ExecutionEdge[] edges;
    // 根据 DistributionPattern 构造 ExecutionEdge 数组
    switch (pattern) {
      case POINTWISE:
          edges = connectPointwise(sourcePartitions, inputNumber);
          break;
      case ALL_TO_ALL:
          edges = connectAllToAll(sourcePartitions, inputNumber);
          break;
      default:
          throw new RuntimeException("Unrecognized distribution pattern.");
    }
    // inputNumber 表示输入的索引
    inputEdges[inputNumber] = edges;
    // 给 ExecutionEdge 对象中的 source 字段设置消费者
    for (ExecutionEdge ee : edges) {
```

```
            ee.getSource().addConsumer(ee, consumerNumber);
        }
    }
```

以 ALL_TO_ALL 方式为例来分析如何建立上下游关系：

```
private ExecutionEdge[] connectAllToAll(IntermediateResultPartition[] sourcePartitions,
int inputNumber) {
    // 根据上游 IntermediateResultPartition 的个数创建同样长度的 ExecutionEdge 数组
    ExecutionEdge[] edges = new ExecutionEdge[sourcePartitions.length];
    for (int i = 0; i < sourcePartitions.length; i++) {
        // 根据索引获取上游的 IntermediateResultPartition 对象
        IntermediateResultPartition irp = sourcePartitions[i];
        // 以该 IntermediateResultPartition 对象为输入，以并行实例为输出，创建一个 ExecutionEdge 对象
        edges[i] = new ExecutionEdge(irp, this, inputNumber);
    }
    return edges;
}
```

这种方式意味着，下游每个并行实例的 ExecutionEdge 对象的个数都与上游的并行度一致。这样，从上游节点到 IntermediateResultPartition，再到 ExecutionEdge，最后到下游节点的关系就建立起来了。

## 4.6 总结

本章梳理了 StreamGraph/Plan 的构造方式和 StreamGraph/Plan 转换成 JobGraph 并最终构造出 ExecutionGraph 的逻辑。在介绍转换逻辑时也分析了如何构造和关联每种图中涉及的上下游节点和连接它们的边。

# 第 5 章
# Flink 的运行时架构

Flink 的运行时架构由一个 Flink Master 和一个或多个任务管理器（TaskManager）组成。其中 Flink Master 包括 3 个独立的组件：资源管理器（ResourceManager）、派发器（Dispatcher）和作业管理器（JobManager）。Flink 的运行时架构如图 5-1 所示。

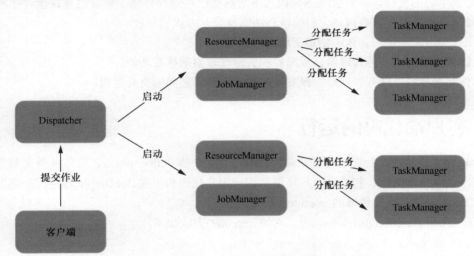

图 5-1　Flink 的运行时架构

很多时候可以将图 5-1 所示的架构简化成一个 JobManager 和一个或多个 TaskManager，这是典型的 Master-Worker 模式。其中，JobManager 主要用于协调分布式任务的执行，包括调度任务、协调容错机制等；TaskManager 主要用于负责任务的具体执行和任务间数据的传输。集群中至少有一个 JobManager 和一个 TaskManager。

这里需要注意的是，Flink 在版本迭代过程中，其架构也在不断优化。1.10 版本的 Flink 源码工程中已经没有叫作 JobManager 或者 TaskManager 的类。有些文章或图书中会将某些运行在同一进程中的组件统称为作业管理器或任务管理器。本书将 Flink-runtime 模块下的 org.apache.Flink.runtime.jobmaster.JobMaster 类视为架构层面的作业管理器角色，将同一模块下的 org.apache.Flink.runtime.taskexecutor.TaskExecutor 类视为架构层面的任务管理器角色，依据如下：

```
public class JobMaster extends FencedRpcEndpoint<JobMasterId> implements JobMaster
Gateway, JobMasterService {
    public static final String JOB_MANAGER_NAME = "jobmanager";
```

```
public class TaskExecutor extends RpcEndpoint implements TaskExecutorGateway {
    public static final String TASK_MANAGER_NAME = "taskmanager";
```

此外，Flink 的运行时架构中还有两个组件——派发器和资源管理器，它们分别在提交作业和分配资源时发挥着重要作用。这两个组件在源码中分别对应 Dispatcher（位于 Flink-runtime 模块的 org.apache.Flink.runtime.dispatcher 包下）和 ResourceManager（位于 Flink-runtime 模块的 org.apache.Flink.runtime.resourcemanager 包下）的实现类。

```
public abstract class Dispatcher extends PermanentlyFencedRpcEndpoint<DispatcherId>
implements DispatcherGateway {
    public static final String DISPATCHER_NAME = "dispatcher";
public abstract class ResourceManager<WorkerType extends ResourceIDRetrievable>
        extends FencedRpcEndpoint<ResourceManagerId>
        implements ResourceManagerGateway, LeaderContender {
    public static final String RESOURCE_MANAGER_NAME = "resourcemanager";
```

考虑到学习的便捷，本章会在本地模式下对这些组件进行介绍，同时也建议读者在本地模式下进行调试，这并不会影响到对它们的核心功能及交互方式的理解。

希望在学习完本章后，读者能够了解：

- 作业管理器、任务管理器、派发器和资源管理器的核心功能；
- 作业管理器、任务管理器、派发器和资源管理器之间的交互逻辑。

## 5.1 客户端代码的运行

Flink 程序启动后，会先运行客户端的代码，直到生成 JobGraph 后，将作业提交到派发器。Flink 程序读取基本配置、初始化执行环境、运行用户代码和生成 JobGraph 等逻辑全部发生在这一阶段。这里直接从执行环境的 execute()方法开始分析。

对于 StreamExecutionEnvironment，它的 execute()方法如下：

```
public JobExecutionResult execute() throws Exception {
    return execute(DEFAULT_JOB_NAME);
}
public JobExecutionResult execute(String jobName) throws Exception {
    // 调用 getStreamGraph()方法获取 StreamGraph 对象
    return execute(getStreamGraph(jobName));
}
public JobExecutionResult execute(StreamGraph streamGraph) throws Exception {
    // 执行作业。返回 JobClient 对象
    final JobClient jobClient = executeAsync(streamGraph);
    try {
        // 获取并返回 JobExecutionResult 对象。触发监听器
        final JobExecutionResult jobExecutionResult;
        if (configuration.getBoolean(DeploymentOptions.ATTACHED)) {
            jobExecutionResult = jobClient.getJobExecutionResult(userClassloader).get();
        } else {
            jobExecutionResult = new DetachedJobExecutionResult(jobClient.getJobID());
        }
        jobListeners.forEach(jobListener -> jobListener.onJobExecuted(jobExecutionResult, null));
        return jobExecutionResult;
```

```
    } catch (Throwable t) {
      ...
    }
  }
```

以上是 StreamGraph 的生成过程。

在最后一个 execute()方法中，主要是获取并返回 JobExecutionResult 对象，并在这个过程中触发监听器。通过 JobExecutionResult 对象，就可以获取累加器等对象。

作业的执行交给了 executeAsync()方法来实现：

```
public JobClient executeAsync(StreamGraph streamGraph) throws Exception {
  // 获取 PipelineExecutorFactory 对象
  final PipelineExecutorFactory executorFactory =
    executorServiceLoader.getExecutorFactory(configuration);
  // 获取 PipelineExecutor 对象并调用 execute()方法执行作业
  CompletableFuture<? extends JobClient> jobClientFuture = executorFactory
    .getExecutor(configuration)
    .execute(streamGraph, configuration);
  try {
    // 获取 JobClient 对象并返回。触发监听器
    JobClient jobClient = jobClientFuture.get();
    jobListeners.forEach(jobListener -> jobListener.onJobSubmitted(jobClient, null));
    return jobClient;
  } catch (Throwable t) {
    ...
  }
}
```

上述方法的核心逻辑是获取 PipelineExecutorFactory 对象，利用该对象获取 PipelineExecutor 对象并调用 execute()方法执行作业。若通过客户端启动，则 executorServiceLoader 字段的值是通过客户端设置并传递到执行环境中的；若在本地模式下启动则为默认的 DefaultExecutorServiceLoader 类对象。它的 getExecutorFactory()方法如下：

```
public PipelineExecutorFactory getExecutorFactory(final Configuration configuration) {
  final ServiceLoader<PipelineExecutorFactory> loader =
      ServiceLoader.load(PipelineExecutorFactory.class);
  final List<PipelineExecutorFactory> compatibleFactories = new ArrayList<>();
  final Iterator<PipelineExecutorFactory> factories = loader.iterator();
  while (factories.hasNext()) {
    try {
      final PipelineExecutorFactory factory = factories.next();
      if (factory != null && factory.isCompatibleWith(configuration)) {
        compatibleFactories.add(factory);
      }
    } catch (Throwable e) {
      ...
    }
  }
  ...
  return compatibleFactories.get(0);
}
```

该方法使用服务提供者接口（Service Provider Interface，SPI）机制发现所有的 PipelineExecutorFactory 实现类，再根据配置 configuration 选择出符合条件的 factory 对象。在本地模式下选择出的就是

LocalExecutorFactory 对象。LocalExecutorFactory 的 getExecutor()方法如下:

```java
public PipelineExecutor getExecutor(final Configuration configuration) {
    return LocalExecutor.create(configuration);
}
public static LocalExecutor create(Configuration configuration) {
    return new LocalExecutor(configuration, MiniCluster::new);
}
private LocalExecutor(Configuration configuration, Function<MiniClusterConfiguration,
MiniCluster> miniClusterFactory) {
    this.configuration = configuration;
    this.miniClusterFactory = miniClusterFactory;
}
```

因此,得到的执行器 PipelineExecutor 对象就是一个 LocalExecutor 对象。LocalExecutor 类的字段包括一个 Function 对象,如下:

```java
private final Function<MiniClusterConfiguration, MiniCluster> miniClusterFactory;
```

这个 Function 类就是 JDK 的 Function 类,表示一个函数,传入 MiniClusterConfiguration 对象,返回 MiniCluster 对象。从上面的 create()方法和 LocalExecutor 的构造方法得知,这个字段的值就是 MiniCluster 的构造方法。而 LocalExecutor 的 execute()方法如下:

```java
public CompletableFuture<? extends JobClient> execute(Pipeline pipeline, Configuration
configuration) throws Exception {
    ...
    // 获取 JobGraph 对象
    final JobGraph jobGraph = getJobGraph(pipeline, effectiveConfig);
    // 获取 PerJobMiniClusterFactory 对象,并提交 JobGraph 对象
    return PerJobMiniClusterFactory.createWithFactory(effectiveConfig, miniClusterFactory).
submitJob(jobGraph);
}
```

在 LocalExecutor 中将 StreamGraph 对象转换成了 JobGraph 对象,然后将 JobGraph 对象进行了提交。

而对于 ExecutionEnvironment,不仅整个流程与 StreamExecutionEnvironment 的一致,方法的实现也几乎一致:

```java
public JobExecutionResult execute() throws Exception {
    return execute(getDefaultName());
}
public JobExecutionResult execute(String jobName) throws Exception {
    final JobClient jobClient = executeAsync(jobName);
    try {
        if (configuration.getBoolean(DeploymentOptions.ATTACHED)) {
            lastJobExecutionResult = jobClient.getJobExecutionResult(userClassloader).get();
        } else {
            lastJobExecutionResult = new DetachedJobExecutionResult(jobClient.getJobID());
        }
        jobListeners.forEach(
            jobListener -> jobListener.onJobExecuted(lastJobExecutionResult, null));
    } catch (Throwable t) {
        ...
    }
    return lastJobExecutionResult;
```

```java
}
public JobClient executeAsync(String jobName) throws Exception {
    final Plan plan = createProgramPlan(jobName);
    final PipelineExecutorFactory executorFactory =
        executorServiceLoader.getExecutorFactory(configuration);
    CompletableFuture<? extends JobClient> jobClientFuture = executorFactory
        .getExecutor(configuration)
        .execute(plan, configuration);
    try {
        JobClient jobClient = jobClientFuture.get();
        jobListeners.forEach(jobListener -> jobListener.onJobSubmitted(jobClient, null));
        return jobClient;
    } catch (Throwable t) {
        ...
    }
}
```

稍有不同的是，Plan 对象是在 executeAsync()方法中生成的，最后同样进入了 LocalExecutor 的 execute()方法。

在 LocalExecutor 的 execute()方法中，在调用 createWithFactory()方法创建 PerJobMiniClusterFactory 对象后，调用其 submitJob()方法提交 JobGraph 对象：

```java
public CompletableFuture<? extends JobClient> submitJob(JobGraph jobGraph) throws Exception {
    MiniClusterConfiguration miniClusterConfig = getMiniClusterConfig(jobGraph.getMaximumParallelism());
    MiniCluster miniCluster = miniClusterFactory.apply(miniClusterConfig);
    // 实例化 MiniCluster 对象
    miniCluster.start(); // 启动 MiniCluster 中的各组件
    return miniCluster
        .submitJob(jobGraph) // 提交 JobGraph 对象
        .thenApply(result -> new PerJobMiniClusterJobClient(result.getJobID(), miniCluster))
        .whenComplete((ignored, throwable) -> {
            if (throwable != null) {
                shutDownCluster(miniCluster);
            }
        });
}
```

在上述方法中，首先获取配置，实例化 MiniCluster 对象。这个对象抽象了一个最小集群环境，一些重要的组件会在其 start()方法中启动。随后调用它的 submitJob()方法继续提交 JobGraph 对象。

下面先来看看 MiniCluster 类的一些重要字段。

- miniClusterConfiguration：MiniClusterConfiguration 类型，表示 MiniCluster 的配置。
- taskManagers：<TaskExecutor>类型，表示集群中的 TaskExecutor 列表。
- commonRpcService：RpcService 类型，用来连接 Rpc 端点，返回 RpcGateway 对象，用于远程调用。
- haServices：HighAvailabilityServices 类型，用于提供各个组件的高可用服务。
- heartbeatServices：HeartbeatServices 类型，用于提供心跳服务。
- resourceManagerLeaderRetriever：LeaderRetrievalService 类型，可以用于添加一个监听器，在资源管理器的 Leader 被选举出来后触发。

- dispatcherLeaderRetriever：LeaderRetrievalService 类型，可以用于添加一个监听器，在派发器的 Leader 被选举出来后触发。
- dispatcherGatewayRetriever：RpcGatewayRetriever<DispatcherId, DispatcherGateway>类型，用于设置和获取派发器组件的 RpcGateway 对象。
- resourceManagerGatewayRetriever：RpcGatewayRetriever<ResourceManagerId, ResourceManagerGateway>类型，用于设置和获取资源管理器组件的 RpcGateway 对象。

调用 start()方法会启动派发器、资源管理器、TaskManager 等组件。

```
public void start() throws Exception {
    synchronized (lock) {
    ...
      try {
        ...
        // 启动 TaskManager
        startTaskManagers();
        ...
        // 创建并启动资源管理器和派发器
        setupDispatcherResourceManagerComponents(configuration, dispatcherResource
ManagerComponentRpcServiceFactory, metricQueryServiceRetriever);
        // 获取资源管理器和派发器的 LeaderRetrievalService 对象
        resourceManagerLeaderRetriever = haServices.getResourceManagerLeaderRetriever();
        dispatcherLeaderRetriever = haServices.getDispatcherLeaderRetriever();
        ...
        // 创建派发器和资源管理器的 Leader 被选举成功后需要触发的监听器
        dispatcherGatewayRetriever = new RpcGatewayRetriever<>(
            commonRpcService,
            DispatcherGateway.class,
            DispatcherId::fromUuid,
            20,
            Time.milliseconds(20L));
        resourceManagerGatewayRetriever = new RpcGatewayRetriever<>(
            commonRpcService,
            ResourceManagerGateway.class,
            ResourceManagerId::fromUuid,
            20,
            Time.milliseconds(20L));
        ...
        // 触发监听器
        resourceManagerLeaderRetriever.start(resourceManagerGatewayRetriever);
        dispatcherLeaderRetriever.start(dispatcherGatewayRetriever);
        ...
      }
      catch (Exception e) {
        ...
      }
      ...
    }
}
```

在上述方法的整个过程中，涉及 TaskManager 的创建和启动以及派发器和资源管理器的创建、启动、选举、触发监听器等操作。下面会在各个组件中讨论这些过程。

除了 start()方法，MiniCluster 还有一个重要方法，即 submitJob()：

```java
public CompletableFuture<JobSubmissionResult> submitJob(JobGraph jobGraph) {
    // 获取 DispatcherGateway
    final CompletableFuture<DispatcherGateway> dispatcherGatewayFuture = getDispatcherGatewayFuture();
    ...
    final CompletableFuture<Acknowledge> acknowledgeCompletableFuture = jarUploadFuture
        .thenCombine(
            dispatcherGatewayFuture,
            // 提交 JobGraph 对象
            (Void ack, DispatcherGateway dispatcherGateway) -> dispatcherGateway.submitJob(jobGraph, rpcTimeout))
        .thenCompose(Function.identity());
    return acknowledgeCompletableFuture.thenApply(
        (Acknowledge ignored) -> new JobSubmissionResult(jobGraph.getJobID()));
}
```

该方法获取了 DispatcherGateway 对象，然后调用它的 submitJob()方法，通过远程过程调用（Remote Procedure Call，RPC）找到对应的派发器组件提交 JobGraph 对象。DispatcherGateway 对象就是在 start()方法中创建的。

至此，可以认为客户端代码结束，此后进入派发器组件的方法。

## 5.2 高可用相关组件

在实际场景中，各个组件一般需要高可用的服务，如 Leader 的选举等。在本地模式下，这些服务仍然存在，只不过对选举等过程进行了简化。在分析具体的组件之前，有必要先对几个反复出现的类进行介绍。

### 5.2.1 EmbeddedHaServices

EmbeddedHaServices 实现了 HighAvailability 接口。顾名思义，该接口用于提供组件的高可用服务。具体而言，HighAvailability 定义的方法包括以下几个：

```java
LeaderRetrievalService getResourceManagerLeaderRetriever();
LeaderRetrievalService getDispatcherLeaderRetriever();
LeaderRetrievalService getJobManagerLeaderRetriever(JobID jobID, String defaultJobManagerAddress);
LeaderElectionService getResourceManagerLeaderElectionService();
LeaderElectionService getDispatcherLeaderElectionService();
LeaderElectionService getJobManagerLeaderElectionService(JobID jobID);
```

该接口分别为资源管理器、派发器、JobManager 这几个需要高可用服务的组件提供了获取 LeaderRetrievalService 对象和 LeaderElectionService 对象的方法。LeaderElectionService 用于 Leader 的选举，LeaderRetrievalService 用于 Leader 的查找。

在本地模式下，该接口的实现类就是 EmbeddedHaServices。LeaderElectionService 和 LeaderRetrievalService 的实现类分别为 EmbeddedLeaderElectionService 和 EmbeddedLeaderRetrievalService。

下面是 EmbeddedHaServices 中的一些重要字段。

- executor：Executor 类型，就是 JUC 包下的 Executor 接口的实例对象，用于执行 Runnable

对象。

- resourceManagerLeaderService：EmbeddedLeaderService 类型，提供与资源管理器的 Leader 相关的操作。
- dispatcherLeaderService：EmbeddedLeaderService 类型，提供与派发器的 Leader 相关的操作。
- jobManagerLeaderServices：Hash <JobID, EmbeddedLeaderService>类型，每个作业都有一个 EmbeddedLeaderService 对象与之对应，该 Map 维护 JobID 与 EmbeddedLeaderService 对象的关系。

EmbeddedLeaderService 表示 Leader 的服务类，其中定义了内部类 EmbeddedLeaderElectionService 和 EmbeddedLeaderRetrievalService。

这些 EmbeddedLeaderService 类型的字段是通过调用 createEmbeddedLeaderService()方法来初始化的：

```
private EmbeddedLeaderService createEmbeddedLeaderService(Executor executor) {
    return new EmbeddedLeaderService(executor);
}
```

而 EmbeddedHaServices 对接口方法的实现如下：

```
public LeaderRetrievalService getResourceManagerLeaderRetriever() {
    return resourceManagerLeaderService.createLeaderRetrievalService();
}
public LeaderRetrievalService getDispatcherLeaderRetriever() {
    return dispatcherLeaderService.createLeaderRetrievalService();
}
public LeaderElectionService getResourceManagerLeaderElectionService() {
    return resourceManagerLeaderService.createLeaderElectionService();
}
public LeaderElectionService getDispatcherLeaderElectionService() {
    return dispatcherLeaderService.createLeaderElectionService();
}
public LeaderElectionService getJobManagerLeaderElectionService(JobID jobID) {
    synchronized (lock) {
        EmbeddedLeaderService service = getOrCreateJobManagerService(jobID);
        return service.createLeaderElectionService();
    }
}
public LeaderRetrievalService getJobManagerLeaderRetriever(JobID jobID, String defaultJobManagerAddress) {
    return getJobManagerLeaderRetriever(jobID);
}
```

其中，在获取 JobManager 相关服务对象时调用的方法如下：

```
private EmbeddedLeaderService getOrCreateJobManagerService(JobID jobID) {
    EmbeddedLeaderService service = jobManagerLeaderServices.get(jobID);
    if (service == null) {
        service = createEmbeddedLeaderService(executor);
        jobManagerLeaderServices.put(jobID, service);
    }
    return service;
}
private EmbeddedLeaderService createEmbeddedLeaderService(Executor executor) {
```

```
    return new EmbeddedLeaderService(executor);
}
```

总而言之，无论是 LeaderRetrievalService 对象还是 LeaderElectionService 对象，都可通过 EmbeddedLeaderService 的 createLeaderRetrievalService()方法或 createLeaderElectionService()方法获得。下面依次对相关的几个类进行介绍。

## 5.2.2 EmbeddedLeaderService

EmbeddedLeaderService 类的体量较大，主要是因为它包含多个内部类，几乎涵盖所有与 Leader 相关的操作逻辑。

首先对其中的几个重要字段进行介绍。

- allLeaderContenders：<EmbeddedLeaderElectionService>类型，维护所有参与 Leader 选举的竞争者对应的 EmbeddedLeaderElectionService 对象。
- listeners：Set 类型，当确认 Leader 时会触发监听器，这个集合会维护所有监听器。
- currentLeaderProposed：EmbeddedLeaderElectionService 类型，表示被提名但还没有被确认的 Leader 对应的 EmbeddedLeaderElectionService 对象。
- currentLeaderConfirmed：EmbeddedLeaderElectionService 类型，表示已被确认且监听器被触发的 Leader 对应的 EmbeddedLeaderElectionService 对象。
- currentLeaderSessionId：当前被提名的 Leader 的唯一标识。
- currentLeaderAddress：当前 Leader 的地址。

在 EmbeddedHaServices 中调用的 createLeaderElectionService()方法和 createLeaderRetrievalService()方法如下：

```
public LeaderElectionService createLeaderElectionService() {
    return new EmbeddedLeaderElectionService();
}
public LeaderRetrievalService createLeaderRetrievalService() {
    return new EmbeddedLeaderRetrievalService();
}
```

上述代码直接返回了 EmbeddedLeaderElectionService 的实例和 EmbeddedLeaderRetrievalService 的实例。这两个类是 EmbeddedLeaderService 的内部类。除了这两个内部类，EmbeddedLeaderService 还有 3 个实现了 Runnable 的内部类。EmbeddedLeaderService 中定义的方法基本是被这些内部类调用的。下面分别介绍这几个内部类。

### 1. EmbeddedLeaderElectionService

EmbeddedLeaderElectionService 类实现了 LeaderElectionService 接口，该接口定义了如下方法：

```
// 开始选举
void start(LeaderContender contender) throws Exception;
// 停止选举
void stop() throws Exception;
// 确认 Leader
void confirmLeadership(UUID leaderSessionID, String leaderAddress);
// 判断是否为 Leader
boolean hasLeadership(@Nonnull UUID leaderSessionId);
```

EmbeddedLeaderElectionService 类的字段如下。
- contender：LeaderContender 类型，所有需要参与 Leader 选举的组件都实现了 LeaderContender 接口，如 ResourceManager 类、DefaultDispatcherRunner 类和 JobManagerRunnerImpl 类等。
- isLeader：用于设置是否为 Leader。
- running：用于设置是否正在运行。

该类对接口方法的实现如下：

```java
public void start(LeaderContender contender) throws Exception {
    addContender(this, contender);
}
public void stop() throws Exception {
    removeContender(this);
}
public void confirmLeadership(UUID leaderSessionID, String leaderAddress) {
    confirmLeader(this, leaderSessionID, leaderAddress);
}
public boolean hasLeadership(@Nonnull UUID leaderSessionId) {
    return isLeader && leaderSessionId.equals(currentLeaderSessionId);
}
```

其中，开始选举时调用的是 EmbeddedLeaderService 的 addContender()方法：

```java
private void addContender(EmbeddedLeaderElectionService service, LeaderContender contender) {
    synchronized (lock) {
        try {
            // 将 EmbeddedLeaderElectionService 对象添加到 allLeaderContenders 中
            if (!allLeaderContenders.add(service)) {
                ...
            }
            // 更新 service 字段的值
            service.contender = contender;
            service.running = true;
            // 更新 Leader
            updateLeader().whenComplete((aVoid, throwable) -> {
                if (throwable != null) {
                    ...
                }
            });
        }
        catch (Throwable t) {
            ...
        }
    }
}
private CompletableFuture<Void> updateLeader() {
    if (currentLeaderConfirmed == null && currentLeaderProposed == null) {
        ...
        else {
            // 为当前竞争者生成唯一标识
            final UUID leaderSessionId = UUID.randomUUID();
            EmbeddedLeaderElectionService leaderService = allLeaderContenders.iterator().next();
            // 更新相关字段的值
            currentLeaderSessionId = leaderSessionId;
```

```
                currentLeaderProposed = leaderService;
                currentLeaderProposed.isLeader = true;
                // 将上面获取的 EmbeddedLeaderElectionService 中的 LeaderContender 对象选举为 Leader。
                // 选举过程异步执行
                return execute(new GrantLeadershipCall(leaderService.contender, leaderSessionId,
LOG));
            }
        } else {
          return CompletableFuture.completedFuture(null);
        }
    }
```

上述代码的核心逻辑已在注释中标识，概括起来就是将 LeaderContender 对象封装在 EmbeddedLeaderElectionService 对象中，并且将该 LeaderContender 对象选举为 Leader。GrantLeadershipCall 是 EmbeddedLeaderService 的内部类，实现了 Runnable 接口，其执行逻辑会在后文中介绍。

停止选举时调用 stop() 方法，主要是用于清空字段的值。

在确认 Leader 时，调用 confirmLeader() 方法：

```
    private void confirmLeader(
          final EmbeddedLeaderElectionService service,
          final UUID leaderSessionId,
          final String leaderAddress) {
       synchronized (lock) {
         ...
         try {
            if (service == currentLeaderProposed && currentLeaderSessionId.equals
(leaderSessionId)) {
                // 更新相关字段的值
                currentLeaderConfirmed = service;
                currentLeaderAddress = leaderAddress;
                currentLeaderProposed = null;
                // 通知所有监听器
                notifyAllListeners(leaderAddress, leaderSessionId);
            }
            else {
                ...
            }
         }
         catch (Throwable t) {
            ...
         }
       }
    }
    private CompletableFuture<Void> notifyAllListeners(String address, UUID leaderSessionId) {
       final List<CompletableFuture<Void>> notifyListenerFutures = new ArrayList<>(listeners.
size());
       for (EmbeddedLeaderRetrievalService listener : listeners) {
            // 通知所有监听器
            notifyListenerFutures.add(notifyListener(address, leaderSessionId, listener.listener));
       }
       return FutureUtils.waitForAll(notifyListenerFutures);
    }
    private CompletableFuture<Void> notifyListener(@Nullable String address, @Nullable
UUID leaderSessionId, LeaderRetrievalListener listener) {
```

```
// 异步调用监听器的相关方法
return CompletableFuture.runAsync(new NotifyOfLeaderCall(address, leaderSessionId,
listener, LOG), notificationExecutor);
    }
```

最终会异步执行 NotifyOfLeaderCall 的逻辑，调用 LeaderRetrievalListener 的相关方法对 Leader 的地址等信息进行通知。NotifyOfLeaderCall 也是 EmbeddedLeaderService 的内部类，实现了 Runnable 接口，其执行逻辑会在后文中介绍。

2. EmbeddedLeaderRetrievalService

EmbeddedLeaderRetrievalService 实现了 LeaderRetrievalService 接口。该接口定义了两个方法：

```
// 为 LeaderRetrievalService 添加监听器
void start(LeaderRetrievalListener listener) throws Exception;
// 停止 LeaderRetrievalService
void stop() throws Exception;
```

EmbeddedLeaderRetrievalService 的字段如下。

- listener：LeaderRetrievalListener 类型，表示选举成功时要触发的监听器。
- running：用于设置是否正在运行。

它对接口方法的实现如下：

```
public void start(LeaderRetrievalListener listener) throws Exception {
    addListener(this, listener);
}
public void stop() throws Exception {
    removeListener(this);
}
```

这里的 addListener() 方法与前面的 addContender() 方法可以类比。addContender() 方法是将 Contender 对象封装到 EmbeddedLeaderElectionService 对象中，并异步触发选举操作；addListener() 方法是将监听器封装到 EmbeddedLeaderRetrievalService 对象中，并异步触发监听器相关操作。

```
private void addListener(EmbeddedLeaderRetrievalService service, LeaderRetrievalListener
listener) {
    synchronized (lock) {
        try {
            // 添加监听器
            if (!listeners.add(service)) {
                ...
            }
            // 更新 service 字段的值
            service.listener = listener;
            service.running = true;
            // 如果已经确认 Leader，则立刻触发监听器
            if (currentLeaderConfirmed != null) {
                notifyListener(currentLeaderAddress, currentLeaderSessionId, listener);
            }
        }
        catch (Throwable t) {
            ...
        }
    }
}
```

触发监听器时调用的是 notifyListener()方法，这在前面已经介绍过。可以发现，在 Leader 被确认后，会触发所有的监听器，而在添加监听器时，如果发现此时 Leader 已经被确认，则会立刻触发该监听器。

removeListener()方法与 removeContender()方法类似，都可用于清空一些字段值。

3. GrantLeadershipCall

在 EmbeddedLeaderElectionService 中，添加竞争者时最终会异步地执行 GrantLeadershipCall。这个 Runnable 接口的实现类的作用是赋予竞争者 Leader 的地位。它的字段如下。

- contender：LeaderContender 类型，表示竞争者。
- leaderSessionId：表示 Leader 的唯一标识。

其 run()方法如下：

```
public void run() {
   try {
      contender.grantLeadership(leaderSessionId);
   }
   catch (Throwable t) {
      ...
   }
}
```

上述代码调用了 contender 的 grantLeadership()方法。每个 LeaderContender 的实现类都自行实现了该方法。

4. NotifyOfLeaderCall

在 EmbeddedLeaderRetrievalService 中，添加监听器后最终会异步地执行 NotifyOfLeaderCall。这个 Runnable 接口的实现类的作用是通知 Leader 的地址等信息。它的字段如下。

- address：表示 Leader 的地址。
- leaderSessionId：表示 Leader 的唯一标识。
- listener：LeaderRetrievalListener 类型，表示监听器。

其 run()方法如下：

```
public void run() {
   try {
      listener.notifyLeaderAddress(address, leaderSessionId);
   }
   catch (Throwable t) {
      ...
   }
}
```

上述代码调用了 listener 的 notifyLeaderAddress()方法。每个 LeaderRetrievalListener 的实现类都自行实现了该方法。

5. RevokeLeadershipCall

在 EmbeddedLeaderService 类中还有一个内部类，即 RevokeLeadershipCall，它也实现了 Runnable 接口。EmbeddedLeaderService 类中有一个 revokeLeadership()方法，表示废除 Leader，在该方法中会异步执行 RevokeLeadershipCall。RevokeLeadershipCall 的字段只有 LeaderContender 类

型的 contender 字段。其 run()方法如下：

```
public void run() {
    contender.revokeLeadership();
}
```

上述代码调用了 contender 的 revokeLeadership()方法。可见，在各个实现类中，run()方法主要是用来执行 close()或 clear()等相关方法的。

## 5.3 派发器的初始化与启动

派发器获取 JobGraph 后会创建 JobManager 来执行作业。它的方法主要跟提交作业、获取作业状态和获取作业结果有关。

下面是派发器的一些重要字段。

- jobGraphWriter：JobGraphWriter 类型，用于存储或者移除 JobGraph 对象。
- runningJobsRegistry：RunningJobsRegistry 类型，用于设置和维护各个作业的状态。
- highAvailabilityServices：HighAvailabilityServices 类型，用于提供高可用服务。
- resourceManagerGatewayRetriever：GatewayRetriever<ResourceManagerGateway>类型，用于获取资源管理器组件的 RpcGateway 对象。
- jobManagerSharedServices：JobManagerSharedServices 类型，封装 JobManager 之间可以共享的服务。
- heartbeatServices：HeartbeatServices 类型，用于提供心跳服务。
- jobManagerRunnerFutures：Map<JobID, CompletableFuture<JobManagerRunner>>类型，封装各个 JobID 与其对应的 JobManagerRunner 的 CompletableFuture 对象。
- recoveredJobs：Collection<JobGraph>类型，表示恢复的 JobGraph 集合。
- archivedExecutionGraphStore：ArchivedExecutionGraphStore 类型，表示归档的 ExecutionGraph 对象。
- jobManagerRunnerFactory：JobManagerRunnerFactory 类型，表示用于创建 JobManagerRunner 对象的工厂类。

派发器的构造方法主要完成了上述字段的赋值。

派发器的初始化要回到 MiniCluster 的 start()方法中调用的 setupDispatcherResourceManagerComponents()方法：

```
private void setupDispatcherResourceManagerComponents(Configuration configuration,
RpcServiceFactory dispatcherResourceManagerComponentRpcServiceFactory, MetricQueryServiceRetriever
metricQueryServiceRetriever) throws Exception {
    dispatcherResourceManagerComponents.addAll(createDispatcherResourceManagerComponents(
        configuration,
        dispatcherResourceManagerComponentRpcServiceFactory,
        haServices,
        blobServer,
        heartbeatServices,
        metricRegistry,
```

```
            metricQueryServiceRetriever,
            new ShutDownFatalErrorHandler()
        ));
        ...
    }
    protected Collection<? extends DispatcherResourceManagerComponent> createDispatcher
ResourceManagerComponents(
            Configuration configuration,
            RpcServiceFactory rpcServiceFactory,
            HighAvailabilityServices haServices,
            BlobServer blobServer,
            HeartbeatServices heartbeatServices,
            MetricRegistry metricRegistry,
            MetricQueryServiceRetriever metricQueryServiceRetriever,
            FatalErrorHandler fatalErrorHandler) throws Exception {
        DispatcherResourceManagerComponentFactory dispatcherResourceManagerComponentFactory =
createDispatcherResourceManagerComponentFactory();
        return Collections.singleton(
            dispatcherResourceManagerComponentFactory.create(
                configuration,
                ioExecutor,
                rpcServiceFactory.createRpcService(),
                haServices,
                blobServer,
                heartbeatServices,
                metricRegistry,
                new MemoryArchivedExecutionGraphStore(),
                metricQueryServiceRetriever,
                fatalErrorHandler));
    }
```

在 setupDispatcherResourceManagerComponents()方法中传入的 dispatcherResourceManagerComponentRpcServiceFactory 对象用于获取 RpcService 对象。

DispatcherResourceManagerComponentFactory 的 create()方法如下：

```
    public DispatcherResourceManagerComponent create(
            Configuration configuration,
            Executor ioExecutor,
            RpcService rpcService,
            HighAvailabilityServices highAvailabilityServices,
            BlobServer blobServer,
            HeartbeatServices heartbeatServices,
            MetricRegistry metricRegistry,
            ArchivedExecutionGraphStore archivedExecutionGraphStore,
            MetricQueryServiceRetriever metricQueryServiceRetriever,
            FatalErrorHandler fatalErrorHandler) throws Exception {
        LeaderRetrievalService dispatcherLeaderRetrievalService = null;
        ...
        DispatcherRunner dispatcherRunner = null;
        try {
            // （1）构造派发器的 LeaderRetrievalService 对象
            dispatcherLeaderRetrievalService = highAvailabilityServices.getDispatcherLeaderRetriever();
            // （2）构造派发器的 LeaderGatewayRetriever 对象（监听器）
            final LeaderGatewayRetriever<DispatcherGateway> dispatcherGatewayRetriever = new RpcGatewayRetriever<>(
```

```
            rpcService,
            DispatcherGateway.class,
            DispatcherId::fromUuid,
            10,
            Time.milliseconds(50L));
    ...
        // (3) 构造 DispatcherRunner 对象
        final PartialDispatcherServices partialDispatcherServices = new PartialDispatcherServices(
            configuration,
            highAvailabilityServices,
            resourceManagerGatewayRetriever,
            blobServer,
            heartbeatServices,
            () -> MetricUtils.instantiateJobManagerMetricGroup(metricRegistry, hostname),
            archivedExecutionGraphStore,
            fatalErrorHandler,
            historyServerArchivist,
            metricRegistry.getMetricQueryServiceGatewayRpcAddress());
        dispatcherRunner = dispatcherRunnerFactory.createDispatcherRunner(
            highAvailabilityServices.getDispatcherLeaderElectionService(),
            fatalErrorHandler,
            new HaServicesJobGraphStoreFactory(highAvailabilityServices),
            ioExecutor,
            rpcService,
            partialDispatcherServices);
    ...
        // (4) 添加派发器在 Leader 被选举成功后要触发的监听器并进行异步触发
        dispatcherLeaderRetrievalService.start(dispatcherGatewayRetriever);
        return new DispatcherResourceManagerComponent(
            dispatcherRunner,
            resourceManager,
            dispatcherLeaderRetrievalService,
            resourceManagerRetrievalService,
            webMonitorEndpoint);
    } catch (Exception exception) {
        ...
    }
}
```

create()方法实际上构造并启动了包括派发器、资源管理器在内的多个组件，但此处的部分代码只保留了与派发器有关的核心逻辑，主要包括下面 4 步。

（1）构造派发器的 LeaderRetrievalService 对象。

（2）构造派发器的 LeaderGatewayRetriever 对象。

（3）构造 DispatcherRunner 对象。

（4）添加派发器在 Leader 被选举成功后要触发的监听器并进行异步触发。

第 1 步得到的 dispatcherLeaderRetrievalService 对象属于 EmbeddedLeaderRetrievalService 类型；第 2 步得到的 dispatcherGatewayRetriever 对象是一个 LeaderRetrievalListener 实例；第 3 步得到的 DispatcherRunner 对象的实现类是 DefaultDispatcherRunner，它还实现了 LeaderContender 接口。这些类型在 5.2 节中已经介绍过。

由于第 4 步中添加了一个监听器，因此可以推测，在第 3 步中应该（异步地）进行了 Leader 的选举。

createDispatcherRunner()方法最终会调用 DefaultDispatcherRunner 的静态方法 create()：

```
public static DispatcherRunner create(
        LeaderElectionService leaderElectionService,
        FatalErrorHandler fatalErrorHandler,
        DispatcherLeaderProcessFactory dispatcherLeaderProcessFactory) throws Exception {
    final DefaultDispatcherRunner dispatcherRunner = new DefaultDispatcherRunner(
        leaderElectionService,
        fatalErrorHandler,
        dispatcherLeaderProcessFactory);
    return DispatcherRunnerLeaderElectionLifecycleManager.createFor(dispatcherRunner, leaderElectionService);
}
public static <T extends DispatcherRunner & LeaderContender> DispatcherRunner createFor(T dispatcherRunner, LeaderElectionService leaderElectionService) throws Exception {
    return new DispatcherRunnerLeaderElectionLifecycleManager<>(dispatcherRunner, leaderElectionService);
}
private DispatcherRunnerLeaderElectionLifecycleManager(T dispatcherRunner, LeaderElectionService leaderElectionService) throws Exception {
    this.dispatcherRunner = dispatcherRunner;
    this.leaderElectionService = leaderElectionService;
    // 开启选举
    leaderElectionService.start(dispatcherRunner);
}
```

注意，这里参与选举的是 DispatcherRunner，而不是派发器，因而还可以推测，DispatcherRunner 选举成功后会构造一个派发器。

根据之前的分析可知，start()方法最终会异步执行 GrantLeadershipCall，其 run()方法会执行 contender 的 grantLeadership()方法。此时的实现方法如下：

```
public void grantLeadership(UUID leaderSessionID) {
    runActionIfRunning(() -> startNewDispatcherLeaderProcess(leaderSessionID));
}
private void startNewDispatcherLeaderProcess(UUID leaderSessionID) {
    // 停止当前作为 Leader 的 DispatcherLeaderProcess 对象
    stopDispatcherLeaderProcess();
    // 创建新的作为 Leader 的 DispatcherLeaderProcess 对象
    dispatcherLeaderProcess = createNewDispatcherLeaderProcess(leaderSessionID);
    final DispatcherLeaderProcess newDispatcherLeaderProcess = dispatcherLeaderProcess;
    FutureUtils.assertNoException(
        // 启动派发器
        previousDispatcherLeaderProcessTerminationFuture.thenRun(newDispatcherLeaderProcess::start));
}
```

在 startNewDispatcherLeaderProcess()方法中，先停止了之前的 DispatcherLeaderProcess 对象，又创建了新的 DispatcherLeaderProcess 对象。这个 DispatcherLeaderProcess 类可以理解为对派发器的一层封装，控制了它的生命周期。在 createNewDispatcherLeaderProcess()方法中，创建了 DispatcherLeaderProcess 对象，并且定义了异步执行的 confirmLeadership()方法：

```
private DispatcherLeaderProcess createNewDispatcherLeaderProcess(UUID leaderSessionID) {
    final DispatcherLeaderProcess newDispatcherLeaderProcess = dispatcherLeaderProcessFactory.create(leaderSessionID);
    forwardShutDownFuture(newDispatcherLeaderProcess);
```

```java
            forwardConfirmLeaderSessionFuture(leaderSessionID, newDispatcherLeaderProcess);
            return newDispatcherLeaderProcess;
    }
    private void forwardConfirmLeaderSessionFuture(UUID leaderSessionID, DispatcherLeaderProcess newDispatcherLeaderProcess) {
        FutureUtils.assertNoException(
            newDispatcherLeaderProcess.getLeaderAddressFuture().thenAccept(
                leaderAddress -> {
                    if (leaderElectionService.hasLeadership(leaderSessionID)) {
                        // 确认 Leader 选举成功
                        leaderElectionService.confirmLeadership(leaderSessionID, leaderAddress);
                    }
                }));
    }
```

之前已经分析过该方法会触发监听器的操作。这里的操作主要就是连接 Rpc 服务，创建并设置 RpcGateway 的 CompletableFuture 对象。

DispatcherLeaderProcess 的 start()方法会创建 Dispatcher 对象并启动它：

```java
    public final void start() {
        runIfStateIs(
            State.CREATED,
            this::startInternal);
    }
    private void startInternal() {
        state = State.RUNNING;
        onStart();
    }
    protected void onStart() {
        startServices();
        onGoingRecoveryOperation = recoverJobsAsync()
            .thenAccept(this::createDispatcherIfRunning)
            .handle(this::onErrorIfRunning);
    }
    private void createDispatcherIfRunning(Collection<JobGraph> jobGraphs) {
        runIfStateIs(State.RUNNING, () -> createDispatcher(jobGraphs));
    }
    private void createDispatcher(Collection<JobGraph> jobGraphs) {
        final DispatcherGatewayService dispatcherService = dispatcherGatewayServiceFactory.create(
            DispatcherId.fromUuid(getLeaderSessionId()),
            jobGraphs,
            jobGraphStore);
        completeDispatcherSetup(dispatcherService);
    }
    public AbstractDispatcherLeaderProcess.DispatcherGatewayService create(
            DispatcherId fencingToken,
            Collection<JobGraph> recoveredJobs,
            JobGraphWriter jobGraphWriter) {
        final Dispatcher dispatcher;
        try {
            dispatcher = dispatcherFactory.createDispatcher(
                rpcService,
                fencingToken,
                recoveredJobs,
                PartialDispatcherServicesWithJobGraphStore.from(partialDispatcherServices, jobGraphWriter));
```

```
    } catch (Exception e) {
        ...
    }
    dispatcher.start();
    return DefaultDispatcherGatewayService.from(dispatcher);
}
```

在 createDispatcher()方法中就通过构造方法实例化了 Dispatcher 对象。最后返回的对象被封装成了 DispatcherLeaderProcess 的字段。Dispatcher 的 start()方法会触发对应的 Rpc 端点启动，最终调用 Dispatcher 的 onStart()方法：

```
public void onStart() throws Exception {
    try {
        startDispatcherServices();
    } catch (Exception e) {
        ...
    }
    startRecoveredJobs();
}
```

这个方法的主要作用就是注册指标、恢复作业等。

上面讲述了派发器的创建和启动流程，下面介绍一些派发器的重要方法。派发器重要的功能就是提交作业、创建 JobManager 来执行作业。提交作业的方法为 submitJob()：

```
public CompletableFuture<Acknowledge> submitJob(JobGraph jobGraph, Time timeout) {
    try {
        ...
        else {
            // 提交 JobGraph
            return internalSubmitJob(jobGraph);
        }
    } catch (FlinkException e) {
        ...
    }
}
private CompletableFuture<Acknowledge> internalSubmitJob(JobGraph jobGraph) {
    // 存储 JobGraph 对象并且执行作业
    final CompletableFuture<Acknowledge> persistAndRunFuture = waitForTerminatingJobManager(jobGraph.getJobID(), jobGraph, this::persistAndRunJob)
        .thenApply(ignored -> Acknowledge.get());
    ...
}
```

在介绍派发器的字段时，有一个字段是 jobGraphWriter，其类型为 JobGraphWriter，它的作用是存储或者移除 JobGraph 对象，在 persistAndRunJob()方法中，就是用这个对象对 JobGraph 进行存储。

```
private CompletableFuture<Void> persistAndRunJob(JobGraph jobGraph) throws Exception {
    // 存储 JobGraph
    jobGraphWriter.putJobGraph(jobGraph);
    // 执行作业
    final CompletableFuture<Void> runJobFuture = runJob(jobGraph);
    ...
}
```

至于 putJobGraph()方法对 JobGraph 进行了怎样的存储，由具体的实现类来决定。

runJob()方法如下：

```
private CompletableFuture<Void> runJob(JobGraph jobGraph) {
    // 创建 JobManagerRunner 的 CompletableFuture 对象
    final CompletableFuture<JobManagerRunner> jobManagerRunnerFuture = createJobManagerRunner(jobGraph);
    ...
    return jobManagerRunnerFuture
        .thenApply(FunctionUtils.uncheckedFunction(this::startJobManagerRunner)) // 启动 //JobManagerRunner
        ...
}
```

通过 runJob()方法启动 JobManagerRunner，JobGraph 对象就传递了下去。接下来是资源管理器的创建和启动流程。

## 5.4　资源管理器的初始化与启动

资源管理器主要负责 Flink 中的资源管理。它既与 JobManager 交互，又与 TaskManager 交互。本节主要介绍其初始化与启动。

下面是资源管理器的一些重要字段。

- resourceId：资源的唯一标识。
- jobManagerRegistrations：Map<JobID, JobManagerRegistration>类型，以每个作业的唯一标识 JobID 为键，维护当前所有注册的 JobManagerGateways。
- jmResourceIdRegistrations：Map<ResourceID, JobManagerRegistration>类型，以 ResourceID 为键，维护当前所有注册的 JobManagerGateways。
- taskExecutors：Map<ResourceID, WorkerRegistration<WorkerType>>类型，以 ResourceID 为键，维护当前所有注册的 TaskExecutor 对应的 worker 对象。这个 worker 对象在不同模式下有不同的含义，如在 YARN 模式下，这个 worker 表示一个 Container 的封装。
- taskExecutorGatewayFutures：Map<ResourceID, CompletableFuture<TaskExecutorGateway>>类型，以 ResourceID 为键，维护当前所有 TaskExecutorGateway 的 CompletableFuture 对象。
- highAvailabilityServices：HighAvailabilityServices 类型，用于提供高可用服务。
- heartbeatServices：HeartbeatServices 类型，用于提供心跳服务。
- slotManager：SlotManager 类型，负责槽的管理。
- leaderElectionService：LeaderElectionService 类型，用于 Leader 的选举。
- taskManagerHeartbeatManager：HeartbeatManager 类型，表示 TaskManager 的心跳管理器。
- jobManagerHeartbeatManager：HeartbeatManager 类型，表示 JobManager 的心跳管理器。

资源管理器的构造方法主要完成了上述字段的赋值。

资源管理器的初始化与派发器的一样，都是在 DispatcherResourceManagerComponentFactory 的 create()方法中完成。

```java
public DispatcherResourceManagerComponent create(
    Configuration configuration,
    Executor ioExecutor,
    RpcService rpcService,
    HighAvailabilityServices highAvailabilityServices,
    BlobServer blobServer,
    HeartbeatServices heartbeatServices,
    MetricRegistry metricRegistry,
    ArchivedExecutionGraphStore archivedExecutionGraphStore,
    MetricQueryServiceRetriever metricQueryServiceRetriever,
    FatalErrorHandler fatalErrorHandler) throws Exception {
  LeaderRetrievalService resourceManagerRetrievalService = null;
  ...
  ResourceManager<?> resourceManager = null;
  try {
        // （1）构造资源管理器的 LeaderRetrievalService 对象
        resourceManagerRetrievalService = highAvailabilityServices.getResourceManagerLeaderRetriever();
        // （2）构造资源管理器的 LeaderGatewayRetriever 对象（监听器）
        final LeaderGatewayRetriever<ResourceManagerGateway> resourceManagerGatewayRetriever =
        new RpcGatewayRetriever<>(
            rpcService,
            ResourceManagerGateway.class,
            ResourceManagerId::fromUuid,
            10,
            Time.milliseconds(50L));
        ...
        // （3）构造 ResourceManager 对象
        resourceManager = resourceManagerFactory.createResourceManager(
            configuration,
            ResourceID.generate(),
            rpcService,
            highAvailabilityServices,
            heartbeatServices,
            fatalErrorHandler,
            new ClusterInformation(hostname, blobServer.getPort()),
            webMonitorEndpoint.getRestBaseUrl(),
            resourceManagerMetricGroup);
        ...
        // （4）启动资源管理器
        resourceManager.start();
        // （5）添加资源管理器的 Leader 选举成功后要触发的监听器并进行异步触发
        resourceManagerRetrievalService.start(resourceManagerGatewayRetriever);
        return new DispatcherResourceManagerComponent(
            dispatcherRunner,
            resourceManager,
            dispatcherLeaderRetrievalService,
            resourceManagerRetrievalService,
            webMonitorEndpoint);
  } catch (Exception exception) {
        ...
  }
}
```

整个过程与派发器构造、选举、启动的过程类似。在第 3 步中实例化了一个 ResourceManager 对象。在它的 start() 方法中，会进行 Leader 的选举等操作。

start() 方法最终会调用 onStart() 方法：

```
public void onStart() throws Exception {
    try {
       startResourceManagerServices();
    } catch (Exception e) {
       ...
    }
}
private void startResourceManagerServices() throws Exception {
    try {
       // 获取资源管理器对应的 LeaderElectionService 对象
       leaderElectionService = highAvailabilityServices.getResourceManagerLeaderElectionService();
       ...
       // 开始选举 Leader
       leaderElectionService.start(this);
       ...
    } catch (Exception e) {
       ...
    }
}
```

选举过程之前已分析过，这里最终会调用资源管理器的 grantLeadership()方法：

```
public void grantLeadership(final UUID newLeaderSessionID) {
    final CompletableFuture<Boolean> acceptLeadershipFuture = clearStateFuture
       // 尝试接受 Leader 的地位
       .thenComposeAsync((ignored) -> tryAcceptLeadership(newLeaderSessionID),
getUnfencedMainThreadExecutor());
    final CompletableFuture<Void> confirmationFuture = acceptLeadershipFuture.thenAcceptAsync(
       (acceptLeadership) -> {
           if (acceptLeadership) {
              // 确认 Leader 的地位
              leaderElectionService.confirmLeadership(newLeaderSessionID, getAddress());
           }
       },
       getRpcService().getExecutor());
    ...
}
```

上述代码中，关键逻辑在 tryAcceptLeadership()方法和 confirmLeadership()方法中实现。

如果 tryAcceptLeadership()方法自身已经是 Leader，那么会调用 startServicesOnLeadership()方法开启与资源管理器相关的服务。

```
protected void startServicesOnLeadership() {
    // 开启心跳服务
    startHeartbeatServices();
    // 开启槽管理器
    slotManager.start(getFencingToken(), getMainThreadExecutor(), new ResourceActionsImpl());
}
```

confirmLeadership()方法中的逻辑与派发器中的类似，会触发监听器的操作，即连接 Rpc 服务，创建并设置 RpcGateway 的 CompletableFuture 对象。

资源管理器的方法大多与 JobManager 和 TaskManager 的注册、资源申请及汇报有关，这部分内容会在附录 A 提及。

## 5.5 TaskExecutor 的初始化与启动

在 Flink 的运行时架构中 TaskManager 用来执行具体的任务，并负责任务之间数据的传输。在源码工程中，这个组件对应的类就是 TaskExecutor。本节主要介绍其初始化过程与启动。

下面是 TaskExecutor 的一些重要字段。

- haServices：HighAvailabilityServices 类型，用于提供高可用服务。
- taskExecutorServices：TaskManagerServices 类型，它是 TaskExecutor 所需服务的总的封装。
- taskManagerConfiguration：TaskManagerConfiguration 类型，该 TaskExecutor 的配置。
- taskManagerLocation：TaskManagerLocation 类型，该 TaskExecutor 的连接信息。
- shuffleEnvironment：ShuffleEnvironment 类型，封装 TaskExecutor 的各个网络组件。
- jobManagerConnection：Map<ResourceID, JobManagerConnection>类型，以 ResourceID 为键，维护与该 TaskExecutor 连接的所有 JobMaster 的连接。
- taskSlotTable：TaskSlotTable 类型，任务槽的容器。
- jobManagerTable：JobManagerTable 类型，内部以 JobID 为键，维护与该 TaskExecutor 连接的所有 JobMaster 的连接。
- jobLeaderService：JobLeaderService 类型，提供与 JobMaster Leader 连接等服务。
- resourceManagerLeaderRetrieval：LeaderRetrievalService 类型，用于通知 TaskExecutor 当前资源管理器的地址并进行连接。
- jobManagerHeartbeatManager：HeartbeatManager 类型，表示 JobMaster 的心跳管理器。
- resourceManagerHeartbeatManager：HeartbeatManager 类型，表示资源管理器的心跳管理器。
- partitionTracker：TaskExecutorPartitionTracker 类型，表示分区的追踪器。
- backPressureSampleService：BackPressureSampleService 类型，提供反压（back pressure）采样服务。
- resourceManagerAddress：ResourceManagerAddress 类型，表示资源管理器的地址。
- establishedResourceManagerConnection：EstablishedResourceManagerConnection 类型，表示建立的资源管理器的连接，封装了 ResourceManagerGateway 对象。
- resourceManagerConnection：TaskExecutorToResourceManagerConnection 类型，表示 TaskExecutor 与资源管理器的连接，封装了 RpcService 等对象。

TaskExecutor 主要完成了上述字段的赋值。

在 MiniCluster 的 start()方法中，调用了 startTaskManagers()方法启动 TaskExecutor。

```
private void startTaskManagers() throws Exception {
    final int numTaskManagers = miniClusterConfiguration.getNumTaskManagers();
    for (int i = 0; i < numTaskManagers; i++) {
        startTaskExecutor();
    }
}
void startTaskExecutor() throws Exception {
    synchronized (lock) {
        final Configuration configuration = miniClusterConfiguration.getConfiguration();
```

```
        // 实例化 TaskExecutor
        final TaskExecutor taskExecutor = TaskManagerRunner.startTaskManager(
            configuration,
            new ResourceID(UUID.randomUUID().toString()),
            taskManagerRpcServiceFactory.createRpcService(),
            haServices,
            heartbeatServices,
            metricRegistry,
            blobCacheService,
            useLocalCommunication(),
            taskManagerTerminatingFatalErrorHandlerFactory.create(taskManagers.size()));
        // 启动与 TaskExecutor 相关的服务
        taskExecutor.start();
        taskManagers.add(taskExecutor);
    }
}
```

构造 TaskExecutor 对象时, 调用 TaskManagerRunner 的 startTaskManager()方法。在实际场景中也是通过这个方法实例化 TaskExecutor 对象的。

```
public static TaskExecutor startTaskManager(
    Configuration configuration,
    ResourceID resourceID,
    RpcService rpcService,
    HighAvailabilityServices highAvailabilityServices,
    HeartbeatServices heartbeatServices,
    MetricRegistry metricRegistry,
    BlobCacheService blobCacheService,
    boolean localCommunicationOnly,
    FatalErrorHandler fatalErrorHandler) throws Exception {
    InetAddress remoteAddress = InetAddress.getByName(rpcService.getAddress());
    final TaskExecutorResourceSpec taskExecutorResourceSpec = TaskExecutorResourceUtils.resourceSpecFromConfig(configuration);
    TaskManagerServicesConfiguration taskManagerServicesConfiguration =
        TaskManagerServicesConfiguration.fromConfiguration(
            configuration,
            resourceID,
            remoteAddress,
            localCommunicationOnly,
            taskExecutorResourceSpec);
    ...
    TaskManagerServices taskManagerServices = TaskManagerServices.fromConfiguration(
        taskManagerServicesConfiguration,
        taskManagerMetricGroup.f1,
        ioExecutor);
    TaskManagerConfiguration taskManagerConfiguration =
        TaskManagerConfiguration.fromConfiguration(configuration, taskExecutorResourceSpec);
    return new TaskExecutor(
        rpcService,
        taskManagerConfiguration,
        highAvailabilityServices,
        taskManagerServices,
        heartbeatServices,
        taskManagerMetricGroup.f0,
        metricQueryServiceAddress,
        blobCacheService,
```

```
        fatalErrorHandler,
        new TaskExecutorPartitionTrackerImpl(taskManagerServices.getShuffleEnvironment()),
        createBackPressureSampleService(configuration, rpcService.getScheduledExecutor()));
}
```

startTaskManager()方法主要会构造 TaskExecutor 的一些构造方法中所需的参数,然后通过其构造方法实例化一个 TaskExecutor 对象。

与其他组件类似,start()方法最终会调用 TaskExecutor 的 onStart()方法:

```
public void onStart() throws Exception {
    try {
        // 启动与 TaskExecutor 相关的服务
        startTaskExecutorServices();
    } catch (Exception e) {
        ...
    }
    ...
}
private void startTaskExecutorServices() throws Exception {
    try {
        // 连接资源管理器
        resourceManagerLeaderRetriever.start(new ResourceManagerLeaderListener());
        // 启动 TaskSlotTable
        taskSlotTable.start(new SlotActionsImpl(), getMainThreadExecutor());
        // 启动 JobLeaderService 服务
        jobLeaderService.start(getAddress(), getRpcService(), haServices, new JobLeaderListenerImpl());
        ...
    } catch (Exception e) {
        ...
    }
}
```

在连接资源管理器的步骤中,添加了监听器 ResourceManagerLeaderListener。当资源管理器的 Leader 被确认后,会调用该监听器的 notifyLeaderAddress()方法:

```
public void notifyLeaderAddress(final String leaderAddress, final UUID leaderSessionID) {
    runAsync(
        () -> notifyOfNewResourceManagerLeader(
            leaderAddress,
            ResourceManagerId.fromUuidOrNull(leaderSessionID)));
}
private void notifyOfNewResourceManagerLeader(String newLeaderAddress, ResourceManagerId newResourceManagerId) {
    resourceManagerAddress = createResourceManagerAddress(newLeaderAddress, newResourceManagerId);
    reconnectToResourceManager(new FlinkException(String.format("ResourceManager leader changed to new address %s", resourceManagerAddress)));
}
```

在 reconnectToResourceManager()方法中,通过 Rpc 服务与资源管理器建立了连接。

后面的启动 TaskSlotTable 和 JobLeaderService 服务的这两个步骤,基本上就是对 taskSlotTable 和 jobLeaderService 这两个对象中的字段进行初始化。

TaskExecutor 中还有许多与提供槽资源、提交任务、触发检查点有关的方法,后文会详细分析。

## 5.6　JobMaster 的初始化与启动

在 Flink 的运行时架构中 JobManager 主要负责作业的调度、资源的管理、检查点的定时执行等。在源码工程中，这个组件对应的类就是 JobMaster。

下面是 JobMaster 的一些重要字段。

- jobMasterConfiguration：JobMasterConfiguration 类型，表示 JobMaster 的配置。
- resourceId：表示资源的唯一标识。
- jobGraph：JobGraph 类型。
- highAvailabilityServices：HighAvailabilityServices 类型，用于提供高可用服务。
- heartbeatServices：HeartbeatServices 类型，用于提供心跳服务。
- slotPool：SlotPool 类型，表示槽资源池。
- scheduler：Scheduler 类型，实际上其可以更直观地表现为其父接口 SlotProvider 的功能，用于提供槽。
- schedulerNGFactory：SchedulerNGFactory 类型，用于创建调度器对象的工厂类。
- resourceManagerLeaderRetriever：LeaderRetrievalService 类型，用于通知 JobMaster 当前资源管理器的地址并进行连接。
- registeredTaskManagers：Map<ResourceID, Tuple2<TaskManagerLocation, TaskExecutorGateway>> 类型，以 ResourceID 为键，用于维护所有注册的 TaskExecutor。
- taskManagerHeartbeatManager：HeartbeatManager<AccumulatorReport, AllocatedSlotReport> 类型，表示 TaskExecutor 的心跳管理器。
- resourceManagerHeartbeatManager：HeartbeatManager<Void, Void>类型，表示资源管理器的心跳管理器。
- schedulerNG：SchedulerNG 类型，表示任务的调度器。
- jobStatusListener：JobManagerJobStatusListener 类型，表示作业状态的监听器。
- resourceManagerAddress：ResourceManagerAddress 类型，表示资源管理器的地址。
- resourceManagerConnection：ResourceManagerConnection 类型，表示 JobMaster 与资源管理器的连接，用于向资源管理器注册 JobMaster。
- establishedResourceManagerConnection：EstablishedResourceManagerConnection 类型，表示已建立的资源管理器的连接，封装了 ResourceManagerGateway 对象。

JobMaster 的构造方法主要完成了上述字段的赋值。

在 MiniCluster 的 start()方法中，完成了派发器、资源管理器、TaskManager 组件的初始化，但并没有涉及 JobManager。因为 JobManager 是与 JobGraph 一一对应的，所以 JobManager 的创建是在派发器提交 JobGraph 之后。

根据前文可知，在派发器提交作业后，会调用 runJob()方法，在该方法中创建 JobManagerRunner 的 CompletableFuture 对象。

```java
private CompletableFuture<Void> runJob(JobGraph jobGraph) {
    // 创建 JobManagerRunner 的 CompletableFuture 对象
    final CompletableFuture<JobManagerRunner> jobManagerRunnerFuture = createJobManagerRunner(jobGraph);
    ...
    return jobManagerRunnerFuture
        .thenApply(FunctionUtils.uncheckedFunction(this::startJobManagerRunner)) // 启动 JobManagerRunner
        ...
}
```

JobManager 组件与派发器组件类似的是，都有一个 Runner 实现类，其用于启动 JobManager 组件或派发器组件，且 Runner 实现类都实现了 LeaderContender 接口。在派发器中这个 Runner 实现类是 DefaultDispatcherRunner，在 JobManager 组件或派发器组件中这个实现类是 JobManagerRunnerImpl。在 createJobManagerRunner()方法中，构造一些参数，然后调用构造方法实例化 JobManagerRunnerImpl 对象，再将该对象作为参数，调用派发器中的 startJobManagerRunner() 方法，利用 JobManagerRunnerImpl 启动 JobManager 组件或派发器组件。

```java
private JobManagerRunner startJobManagerRunner(JobManagerRunner jobManagerRunner) throws Exception {
    ...
    // 启动 JobManager
    jobManagerRunner.start();
    return jobManagerRunner;
}
public void start() throws Exception {
    try {
        // 选举 Leader
        leaderElectionService.start(this);
    } catch (Exception e) {
        ...
    }
}
```

由此又进入熟悉的 Leader 选举环节。最后，调用 JobManagerRunnerImpl 的 grantLeadership() 方法：

```java
public void grantLeadership(final UUID leaderSessionID) {
    synchronized (lock) {
        ...
        leadershipOperation = leadershipOperation.thenCompose(
            (ignored) -> {
                synchronized (lock) {
                    // 启动 JobManager
                    return verifyJobSchedulingStatusAndStartJobManager(leaderSessionID);
                }
            });
        ...
    }
}
private CompletableFuture<Void> verifyJobSchedulingStatusAndStartJobManager(UUID leaderSessionId) {
    // 获取作业调度状态
```

```
        final CompletableFuture<JobSchedulingStatus> jobSchedulingStatusFuture =
getJobSchedulingStatus();
        return jobSchedulingStatusFuture.thenCompose(
            jobSchedulingStatus -> {
                if (jobSchedulingStatus == JobSchedulingStatus.DONE) {
                    return jobAlreadyDone();
                } else {
                    // 启动 JobMaster
                    return startJobMaster(leaderSessionId);
                }
            });
    }
```

当作业调度状态不为 DONE 时，启动 JobMaster。

```
    private CompletionStage<Void> startJobMaster(UUID leaderSessionId) {
        ...
        final CompletableFuture<Acknowledge> startFuture;
        try {
            // 启动 JobMaster
            startFuture = jobMasterService.start(new JobMasterId(leaderSessionId));
        } catch (Exception e) {
            ...
        }
        ...
    }
```

这个 jobMasterService 对象就是 JobMaster 对象，其 start()方法如下：

```
    public CompletableFuture<Acknowledge> start(final JobMasterId newJobMasterId) throws
Exception {
        ...
        // 开始作业的执行
        return callAsyncWithoutFencing(() -> startJobExecution(newJobMasterId), RpcUtils.
INF_TIMEOUT);
    }
    private Acknowledge startJobExecution(JobMasterId newJobMasterId) throws Exception {
        ...
        // 启动与 JobMaster 相关的服务
        startJobMasterServices();
        // 开始调度作业
        resetAndStartScheduler();
        return Acknowledge.get();
    }
```

其中，启动的与 JobMaster 相关的服务主要包括心跳服务、槽资源池服务、资源管理器的连接和触发监听器等。

```
    private void startJobMasterServices() throws Exception {
        startHeartbeatServices();
        slotPool.start(getFencingToken(), getAddress(), getMainThreadExecutor());
        scheduler.start(getMainThreadExecutor());
        reconnectToResourceManager(new FlinkException("Starting JobMaster component."));
        resourceManagerLeaderRetriever.start(new ResourceManagerLeaderListener());
    }
```

接着，在 resetAndStartScheduler()方法中，创建任务调度器，并且开始任务调度：

```
private void resetAndStartScheduler() throws Exception {
    ...
    } else {
        ...
        // 创建任务调度器
        final SchedulerNG newScheduler = createScheduler(newJobManagerJobMetricGroup);
        ...
    }
    // 开始任务调度
    schedulerAssignedFuture.thenRun(this::startScheduling);
}
```

## 5.7 总结

  Flink 的运行时架构主要包括派发器、资源管理器、作业管理器（JobMaster）和任务管理器（TaskExecutor）4 个组件。本章主要介绍了它们的初始化和启动流程。Flink 程序启动后，在客户端会运行用户代码，生成 JobGraph 并提交给派发器，派发器的主要作用是在获取 JobGraph 后创建 JobMaster。资源管理器主要负责资源管理，JobMaster 主要负责任务调度和检查点机制等，TaskExecutor 主要负责任务的执行、任务数据的传输等，后文会逐步对这些内容进行分析。

# 第 6 章

# 任务调度

针对每一个 JobGraph，派发器会创建一个作业管理器，然后将 JobGraph 封装到其中。作业管理器会生成调度器，在这个过程中 JobGraph 将会转换成 ExecutionGraph。ExecutionGraph 中的节点 ExecutionVertex 对应具体需要执行的任务并行实例，每一个 ExecutionVertex 对应一个 Execution 对象，由该对象负责将任务并行实例的信息进行另一层封装，最终部署到任务管理器端。

由于上下游任务存在依赖关系，因此在任务的调度和部署过程中就会有许多信息需要维护，这里要注意的一个关键问题是：对于作业中的任意一个任务，应该何时部署？

每一个作业管理器都有一个调度器（Scheduler），它负责大部分调度和部署的逻辑。根据作业模式（流处理或批处理）的不同，调度器还会利用不同的调度策略对任务进行调度。

实际上在调度过程中还涉及诸多其他操作，如槽资源的申请等。本章将重点放在"任务应该何时部署"这一问题上，槽资源的申请会在附录 A 进行介绍。

希望在学习完本章后，读者能够了解：

- 执行图中的各个对象在调度层面又被转换成什么数据结构；
- 任务的上下游关系如何影响每一个任务并行实例的部署时机；
- 如何选择调度策略。

## 6.1 调度器

在 Flink 中，调度器的顶层接口是 SchedulerNG（位于 Flink-runtime 模块的 org.apache.Flink.runtime.scheduler 包中）。该接口中与任务调度有关的方法主要是以下几个：

```
public interface SchedulerNG {
    void startScheduling();
    boolean updateTaskExecutionState(TaskExecutionState taskExecutionState);
    void scheduleOrUpdateConsumers(ResultPartitionID partitionID);
}
```

当构建好 ExecutionGraph 以后，就可以开始进行任务调度，这时会调用 startScheduling() 方法。部署完成后，这些任务并行实例会在任务管理器端运行。任务在其生命周期的不同阶段会有不同的状态，当需要更新状态时，就会调用 updateTaskExecutionState() 方法。

上述代码中最后一个与调度有关的方法是 scheduleOrUpdateConsumers()。顾名思义，该方法可用于调度或更新下游消费者（对于下游任务，其消费的是上游的输出数据）。

了解了调度器的主要职责后，下面对 Flink 中目前的实现类 DefaultScheduler 进行分析。

## 6.1.1 调度器的基本构成与初始化

Flink 中的调度器都继承自抽象类 SchedulerBase。下面是其中的一些重要字段。
- jobGraph：JobGraph 类型。
- executionGraph：ExecutionGraph 类型。
- schedulingTopology：SchedulingTopology<?, ?>类型，表示调度层面的拓扑结构。
- jobMasterConfiguration：Configuration 类型，表示作业管理器的配置。
- slotProvider：SlotProvider 类型，用于提供槽。
- restartStrategy：RestartStrategy 类型，表示重启策略。

SchedulerBase 的构造方法除了对字段进行初始化赋值，另外一件重要的事情就是构造 ExecutionGraph 对象。

Flink 的旧版本中实现过一个调度器，其已被废弃。当前用的是 DefaultScheduler。相比 SchedulerBase，DefaultScheduler 中定义的字段包括以下方面。
- executionSlotAllocator：ExecutionSlotAllocator 类型，用于槽的分配。
- schedulingStrategy：SchedulingStrategy 类型，表示调度策略。
- executionVertexOperations：ExecutionVertexOperations 类型，定义针对 ExecutionVertex 的一些操作，如部署、取消等。
- verticesWaitingForRestart：<ExecutionVertexID>类型，表示等待重启的任务。

DefaultScheduler 的构造方法除了调用父类（SchedulerBase）的构造方法，就是对 DefaultScheduler 自身的字段进行初始化赋值。

调度器对象的生成调用了 SchedulerNGFactory 的 createInstance()方法，而 SchedulerNGFactory 对象的生成要追溯到派发器的 createJobManagerRunner()方法中：

```
private CompletableFuture<JobManagerRunner> createJobManagerRunner(JobGraph jobGraph) {
    final RpcService rpcService = getRpcService();
    return CompletableFuture.supplyAsync(
        CheckedSupplier.unchecked(() ->
            jobManagerRunnerFactory.createJobManagerRunner(
                jobGraph,
                configuration,
                rpcService,
                highAvailabilityServices,
                heartbeatServices,
                jobManagerSharedServices,
                new DefaultJobManagerJobMetricGroupFactory(jobManagerMetricGroup),
                fatalErrorHandler)),
        rpcService.getExecutor());
}
public JobManagerRunner createJobManagerRunner(
        JobGraph jobGraph,
        Configuration configuration,
        RpcService rpcService,
        HighAvailabilityServices highAvailabilityServices,
```

```
                HeartbeatServices heartbeatServices,
                JobManagerSharedServices jobManagerServices,
                JobManagerJobMetricGroupFactory jobManagerJobMetricGroupFactory,
                FatalErrorHandler fatalErrorHandler) throws Exception {
        ...
        // 根据配置创建 SchedulerNGFactory 对象
        final SchedulerNGFactory schedulerNGFactory = SchedulerNGFactoryFactory.
createSchedulerNGFactory(configuration, jobManagerServices.getRestartStrategyFactory());
        ...
        return new JobManagerRunnerImpl(
            jobGraph,
            jobMasterFactory,
            highAvailabilityServices,
            jobManagerServices.getLibraryCacheManager(),
            jobManagerServices.getScheduledExecutorService(),
            fatalErrorHandler);
    }
    public static SchedulerNGFactory createSchedulerNGFactory(
            final Configuration configuration,
            final RestartStrategyFactory restartStrategyFactory) {
        // 根据 SCHEDULER 配置情况决定实例化哪种类型的调度器工厂类
        final String schedulerName = configuration.getString(JobManagerOptions.SCHEDULER);
        switch (schedulerName) {
            case SCHEDULER_TYPE_LEGACY:
                return new LegacySchedulerFactory(restartStrategyFactory);
            case SCHEDULER_TYPE_NG:
                return new DefaultSchedulerFactory();
            default:
                ...
        }
    }
```

这个 SCHEDULER 配置的定义如下:

```
    public static final ConfigOption<String> SCHEDULER =
        key("jobmanager.scheduler")
            .stringType()
            .defaultValue("ng")
            .withDescription(Description.builder()
                .text("Determines which scheduler implementation is used to schedule tasks. 
Accepted values are:")
                .list(
                    text("'legacy': legacy scheduler"),
                    text("'ng': new generation scheduler"))
                .build());
```

即配置 jobmanager.scheduler 就可以决定这里的工厂类的类型,其默认值为 "ng",表示 "new generation"。对应工厂类的 createInstance()方法如下:

```
    public SchedulerNG createInstance(
            final Logger log,
            final JobGraph jobGraph,
            final BackPressureStatsTracker backPressureStatsTracker,
            final Executor ioExecutor,
            final Configuration jobMasterConfiguration,
            final SlotProvider slotProvider,
            final ScheduledExecutorService futureExecutor,
```

```
        final ClassLoader userCodeLoader,
        final CheckpointRecoveryFactory checkpointRecoveryFactory,
        final Time rpcTimeout,
        final BlobWriter blobWriter,
        final JobManagerJobMetricGroup jobManagerJobMetricGroup,
        final Time slotRequestTimeout,
        final ShuffleMaster<?> shuffleMaster,
        final JobMasterPartitionTracker partitionTracker) throws Exception {
    // 根据 JobGraph 中的 ScheduleMode 决定返回哪种类型的调度策略工厂类
    final SchedulingStrategyFactory schedulingStrategyFactory = createSchedulingStrategyFactory
(jobGraph.getScheduleMode());
    ...
    // 根据 JobGraph 中的 ScheduleMode 决定返回哪种类型的 SlotProvider 策略
    final SlotProviderStrategy slotProviderStrategy = SlotProviderStrategy.from(
        jobGraph.getScheduleMode(),
        slotProvider,
        slotRequestTimeout);
    return new DefaultScheduler(
        log,
        jobGraph,
        backPressureStatsTracker,
        ioExecutor,
        jobMasterConfiguration,
        slotProvider,
        futureExecutor,
        new ScheduledExecutorServiceAdapter(futureExecutor),
        userCodeLoader,
        checkpointRecoveryFactory,
        rpcTimeout,
        blobWriter,
        jobManagerJobMetricGroup,
        slotRequestTimeout,
        shuffleMaster,
        partitionTracker,
        schedulingStrategyFactory,
        FailoverStrategyFactoryLoader.loadFailoverStrategyFactory(jobMasterConfiguration),
        restartBackoffTimeStrategy,
        new DefaultExecutionVertexOperations(),
        new ExecutionVertexVersioner(),
        new DefaultExecutionSlotAllocatorFactory(slotProviderStrategy));
}
```

在该方法中，会根据 ScheduleMode 选择实例化哪种调度策略工厂类和 SlotProvider 策略，然后将之作为参数传入 DefaultScheduler 的构造方法以生成调度器。

ScheduleMode 表示作业的调度模式，共有 EAGER、LAZY_FROM_SOURCES 和 LAZY_FROM_SOURCES_WITH_BATCH_SLOT_REQUEST 这 3 种模式。

流处理使用的是 EAGER 模式，表示一次性调度所有任务。批处理则使用 LAZY_FROM_SOURCES 或 LAZY_FROM_SOURCES_WITH_BATCH_SLOT_REQUEST 模式，两者均表示从 source 任务开始调度，等待上游任务结束后再调度下游任务，区别在于，如果作业中不包含 pipelined shuffles，那么使用 LAZY_FROM_SOURCES_WITH_BATCH_SLOT_REQUEST 模式可以占用更少的槽。

## 6.1.2　构造 ExecutionGraph

调度器本身依赖 ExecutionGraph 对象。在调度器初始化过程中就生成了 ExecutionGraph。调用的方法是 createAndRestoreExecutionGraph()方法：

```
private ExecutionGraph createAndRestoreExecutionGraph(
    JobManagerJobMetricGroup currentJobManagerJobMetricGroup,
    ShuffleMaster<?> shuffleMaster,
    JobMasterPartitionTracker partitionTracker) throws Exception {
    // 构造 ExecutionGraph 对象
    ExecutionGraph newExecutionGraph = createExecutionGraph(currentJobManagerJobMetricGroup, shuffleMaster, partitionTracker);
    // 从 ExecutionGraph 对象中获取 CheckpointCoordinator 对象
    final CheckpointCoordinator checkpointCoordinator = newExecutionGraph.getCheckpointCoordinator();
    // 如果 CheckpointCoordinator 对象的值不为 null，那么尝试从检查点或保存点恢复
    if (checkpointCoordinator != null) {
        // 尝试从最近的检查点恢复
        if (!checkpointCoordinator.restoreLatestCheckpointedState(
            new HashSet<>(newExecutionGraph.getAllVertices().values()),
            false,
            false)) {
            // 尝试从保存点恢复
            tryRestoreExecutionGraphFromSavepoint(newExecutionGraph, jobGraph.getSavepointRestoreSettings());
        }
    }
    return newExecutionGraph;
}
```

其中构造 ExecutionGraph 的主要逻辑都在 createExecutionGraph()方法中，在该方法中完成了 JobGraph 到 ExecutionGraph 的转换。如果可以从最近的检查点或保存点恢复，则还会更新 ExecutionGraph。

createExecutionGraph()方法如下：

```
private ExecutionGraph createExecutionGraph(
    JobManagerJobMetricGroup currentJobManagerJobMetricGroup,
    ShuffleMaster<?> shuffleMaster,
    final JobMasterPartitionTracker partitionTracker) throws JobExecutionException, JobException {
    ...
    return ExecutionGraphBuilder.buildGraph(
        null,
        jobGraph,
        jobMasterConfiguration,
        futureExecutor,
        ioExecutor,
        slotProvider,
        userCodeLoader,
        checkpointRecoveryFactory,
        rpcTimeout,
        restartStrategy,
        currentJobManagerJobMetricGroup,
        blobWriter,
```

```
            slotRequestTimeout,
            log,
            shuffleMaster,
            partitionTracker,
            failoverStrategy);
}
```

buildGraph()方法在第 4 章已经分析过,这里不再赘述。

## 6.2 调度拓扑

在 Flink 中,ExecutionGraph 表示执行图的最后一层,包含所有并行实例及其上下游关系。然而,在调度的过程中,调度器并不是直接调度 ExecutionGraph 中的对象,而是为 ExecutionGraph 构造对应的调度拓扑,即 Flink-runtime 模块下的 org.apache.Flink.runtime.scheduler.strategy.SchedulingTopology。

先回顾 ExecutionGraph 中的节点如何维护上下游关系。

ExecutionGraph 中用 ExecutionVertex 表示并行实例,它有如下字段。

- resultPartitions:表示其输出。
- inputEdges:表示其输入。

相关代码如下:

```
private final Map<IntermediateResultPartitionID, IntermediateResultPartition> resultPartitions;
private final ExecutionEdge[][] inputEdges;
```

IntermediateResultPartition 有如下字段。

- producer:表示该 IntermediateResultPartition 数据从何而来。
- consumers:表示下游数据被哪条边消费。

相关代码如下:

```
private final ExecutionVertex producer;
private List<List<ExecutionEdge>> consumers;
```

ExecutionEdge 有如下字段。

- source:表示它从哪个 IntermediateResultPartition 消费数据。
- target:表示它接收的数据被哪个 ExecutionVertex 处理。

相关代码如下:

```
private final IntermediateResultPartition source;
private final ExecutionVertex target;
```

根据这几个对象的相互依赖关系,可以清晰地了解到 ExecutionGraph 中上下游关系的维护方式。

在 SchedulingTopology 中,ExecutionVertex 对象对应 SchedulingExecutionVertex 对象,IntermediateResultPartition 对象对应 SchedulingResultPartition 对象。它们之间的依赖关系构建了 SchedulingTopology 中的上下游关系。

目前 SchedulingTopology 的实现类是 DefaultExecutionTopology,SchedulingExecutionVertex 的实现类是 DefaultExecutionVertex,SchedulingResultPartition 的实现类是 DefaultResultPartition。

下面是 DefaultExecutionTopology 中的一些重要字段。
- executionVerticesById：Map<ExecutionVertexID, DefaultExecutionVertex>类型，通过 ExecutionVertexID 找到该 ExecutionVertex 对象对应的 DefaultExecutionVertex 对象。
- executionVerticesList：<DefaultExecutionVertex>类型，维护所有的 ExecutionVertex 对应的 DefaultExecutionVertex 对象。
- resultPartitionsById：Map<IntermediateResultPartitionID, DefaultResultPartition>类型，通过 IntermediateResultPartitionID 找到该 IntermediateResultPartition 对应的 DefaultResultPartition 对象。

上下游关系则是通过 DefaultExecutionVertex 中的字段进行维护的：
- consumedResults：表示被消费的 DefaultResultPartition，即该节点的数据从哪个 DefaultResultPartition 而来。
- producedResults：表示生产的 DefaultResultPartition，即该节点的数据输出到哪个 DefaultResultPartition。

相关代码如下：

```
private final List<DefaultResultPartition> consumedResults;
private final List<DefaultResultPartition> producedResults;
```

DefaultResultPartition 有如下字段。
- producer：表示该 DefaultResultPartition 的数据由哪个节点生产。
- consumers：表示该 DefaultResultPartition 的数据由哪个节点消费。

相关代码如下：

```
private DefaultExecutionVertex producer;
private final List<DefaultExecutionVertex> consumers;
```

这个依赖关系是在 DefaultExecutionTopology 初始化时构建的，其构造方法会在 ExecutionGraph 的 attachJobGraph()方法中，在 ExecutionGraph 节点及上下游关系构建完成后调用。构造方法如下：

```
public DefaultExecutionTopology(ExecutionGraph graph) {
    ...
    this.executionVerticesById = new HashMap<>();
    this.executionVerticesList = new ArrayList<>(graph.getTotalNumberOfVertices());
    Map<IntermediateResultPartitionID, DefaultResultPartition> tmpResultPartitionsById = new HashMap<>();
    Map<ExecutionVertex, DefaultExecutionVertex> executionVertexMap = new HashMap<>();
    // 遍历所有 ExecutionVertex
    for (ExecutionVertex vertex : graph.getAllExecutionVertices()) {
        // 将该节点的所有 IntermediateResultPartition 映射成 DefaultResultPartition
        List<DefaultResultPartition> producedPartitions = generateProducedSchedulingResultPartition(vertex.getProducedPartitions());
        producedPartitions.forEach(partition -> tmpResultPartitionsById.put(partition.getId(), partition));
        // 将该 ExecutionVertex 映射成 DefaultExecutionVertex
        DefaultExecutionVertex schedulingVertex = generateSchedulingExecutionVertex(vertex, producedPartitions);
        this.executionVerticesById.put(schedulingVertex.getId(), schedulingVertex);
        this.executionVerticesList.add(schedulingVertex);
```

```
            executionVertexMap.put(vertex, schedulingVertex);
        }
        this.resultPartitionsById = tmpResultPartitionsById;
        // 建立 DefaultExecutionTopology 中的上下游关系
        connectVerticesToConsumedPartitions(executionVertexMap, tmpResultPartitionsById);
    }
```

该方法遍历所有的 ExecutionVertex，在遍历每一个对象的过程中，将该节点和它的 Intermediate ResultPartition 映射成 DefaultExecutionVertex 和 DefaultResultPartition 对象，并放入 List 或 Map 中进行维护。等所有的映射全部完成，便可构建出上下游关系。

ExecutionVertex 的 getProducedPartitions()方法用于获取其中的 resultPartitions 字段：

```
public Map<IntermediateResultPartitionID, IntermediateResultPartition> getProducedPartitions() {
    return resultPartitions;
}
```

generateProducedSchedulingResultPartition()方法如下：

```
private static List<DefaultResultPartition> generateProducedSchedulingResultPartition(
    Map<IntermediateResultPartitionID, IntermediateResultPartition> producedIntermediatePartitions) {
    List<DefaultResultPartition> producedSchedulingPartitions = new ArrayList<>
(producedIntermediatePartitions.size());
    producedIntermediatePartitions.values().forEach(
        irp -> producedSchedulingPartitions.add(
            new DefaultResultPartition(
                irp.getPartitionId(),
                irp.getIntermediateResult().getId(),
                irp.getResultType(),
                () -> irp.isConsumable()? ResultPartitionState.CONSUMABLE : ResultPartitionState.CREATED)));
    return producedSchedulingPartitions;
}
```

该方法将每一个 IntermediateResultPartition 对象映射成 DefaultResultPartition 对象，其获取一些 IntermediateResultPartition 对象中的字段作为参数，并调用 DefaultResultPartition 的构造方法：

```
DefaultResultPartition(
    IntermediateResultPartitionID partitionId,
    IntermediateDataSetID intermediateDataSetId,
    ResultPartitionType partitionType,
    Supplier<ResultPartitionState> resultPartitionStateSupplier) {
    this.resultPartitionId = checkNotNull(partitionId);
    this.intermediateDataSetId = checkNotNull(intermediateDataSetId);
    this.partitionType = checkNotNull(partitionType);
    this.resultPartitionStateSupplier = checkNotNull(resultPartitionStateSupplier);
    this.consumers = new ArrayList<>();
}
```

此时，consumers 字段被初始化成一个空的列表。

接着调用 generateSchedulingExecutionVertex()方法将 ExecutionVertex 对象映射成 DefaultExecutionVertex 对象：

```
private static DefaultExecutionVertex generateSchedulingExecutionVertex(
    ExecutionVertex vertex,
```

```
    List<DefaultResultPartition> producedPartitions) {
DefaultExecutionVertex schedulingVertex = new DefaultExecutionVertex(
    vertex.getID(),
    producedPartitions,
    () -> vertex.getExecutionState(),
    vertex.getInputDependencyConstraint());
producedPartitions.forEach(partition -> partition.setProducer(schedulingVertex));
return schedulingVertex;
}
```

在这个过程中，由于先前已经生成了 DefaultResultPartition 对象，因此在构造 DefaultExecutionVertex 时可以将其作为参数传入。构造方法如下：

```
DefaultExecutionVertex(
    ExecutionVertexID executionVertexId,
    List<DefaultResultPartition> producedPartitions,
    Supplier<ExecutionState> stateSupplier,
    InputDependencyConstraint constraint) {
    this.executionVertexId = checkNotNull(executionVertexId);
    this.consumedResults = new ArrayList<>();
    this.stateSupplier = checkNotNull(stateSupplier);
    this.producedResults = checkNotNull(producedPartitions);
    this.inputDependencyConstraint = checkNotNull(constraint);
}
```

既然有了 DefaultExecutionVertex 对象，那么先前的 DefaultResultPartition 对象的 producer 字段可以被赋值了。因此用 forEach() 方法调用每一个 DefaultResultPartition 对象的 setProducer() 方法。

至此，还有两个字段没有被赋值——一个是 DefaultExecutionVertex 中的 consumedResults，表示该节点的数据从何而来；另一个是 DefaultResultPartition 中的 consumers，表示该 DefaultResultPartition 的数据被下游哪个节点消费。这就需要调用 connectVerticesToConsumedPartitions() 方法：

```
private static void connectVerticesToConsumedPartitions(
    Map<ExecutionVertex, DefaultExecutionVertex> executionVertexMap,
    Map<IntermediateResultPartitionID, DefaultResultPartition> resultPartitions) {
    for (Map.Entry<ExecutionVertex, DefaultExecutionVertex> mapEntry : executionVertexMap.entrySet()) {
        final DefaultExecutionVertex schedulingVertex = mapEntry.getValue();
        final ExecutionVertex executionVertex = mapEntry.getKey();
        for (int index = 0; index < executionVertex.getNumberOfInputs(); index++) {
            for (ExecutionEdge edge : executionVertex.getInputEdges(index)) {
                DefaultResultPartition partition = resultPartitions.get(edge.getSource().getPartitionId());
                // 给 DefaultExecutionVertex 对象的 consumedResults 字段赋值
                schedulingVertex.addConsumedResult(partition);
                // 给 DefaultResultPartition 对象的 consumers 字段赋值
                partition.addConsumer(schedulingVertex);
            }
        }
    }
}
```

这个过程先获取了 ExecutionVertex 对象中的 ExecutionEdge 信息。根据本节之前的分析，可以知道 ExecutionEdge 是 ExecutionGraph 上下游关系中的重要一环，可以根据它找到上游的 IntermediateResultPartition（进而找到对应的 DefaultResultPartition）和下游的 ExecutionVertex（进

而找到对应的 DefaultExecutionVertex）。因而可以填充上面提到的两个字段。

SchedulingTopology 与其相关类的关系如图 6-1 所示。

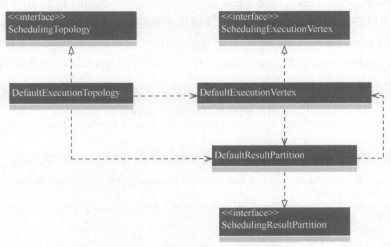

图 6-1　SchedulingTopology 与其相关类的关系

## 6.3　调度策略

在 DefaultScheduler 中有一个重要字段是 schedulingStrategy，其为 SchedulingStrategy 类型，表示调度策略，它封装了调度逻辑，当任务状态改变或者分区数据可以被消费时，会触发它的方法。另外，它还负责任务的重启。

SchedulingStrategy 接口的定义如下：

```
public interface SchedulingStrategy {
    // 开始调度
    void startScheduling();
    // 重启任务
    void restartTasks(Set<ExecutionVertexID> verticesToRestart);
    // 任务状态改变时触发的方法
    void onExecutionStateChange(ExecutionVertexID executionVertexId, ExecutionState executionState);
    // 分区数据可以被消费时触发的方法
    void onPartitionConsumable(ExecutionVertexID executionVertexId, ResultPartitionID resultPartitionId);
}
```

该接口有两个实现类——EagerSchedulingStrategy 和 LazyFromSourcesSchedulingStrategy。下面对这两种策略分别进行介绍。

### 6.3.1　EagerSchedulingStrategy

EagerSchedulingStrategy 用于调度流处理作业的任务，它会同时调度所有任务，它的字段如下。

- schedulerOperations：SchedulerOperations 类型，实际上就是 DefaultScheduler 对象。
- schedulingTopology：SchedulingTopology<?, ?>类型。
- deploymentOption：DeploymentOption 类型，其中封装布尔类型的字段，用于告知任务在输出数据时，分区是否在一旦有数据时就通知下游任务去调度和部署。在 EagerSchedulingStrategy 中其值默认为 false。

SchedulingStrategyFactory 的 createInstance()方法完成了 EagerSchedulingStrategy 对象的创建。

```
public SchedulingStrategy createInstance(
    SchedulerOperations schedulerOperations,
    SchedulingTopology<?, ?> schedulingTopology) {
    return new EagerSchedulingStrategy(schedulerOperations, schedulingTopology);
}
```

而 SchedulingStrategyFactory 对象的创建是在调度器的初始化过程中完成的。

```
final SchedulingStrategyFactory schedulingStrategyFactory = createSchedulingStrategyFactory(jobGraph.getScheduleMode());
    private SchedulingStrategyFactory createSchedulingStrategyFactory(final ScheduleMode scheduleMode) {
        switch (scheduleMode) {
            case EAGER:
                return new EagerSchedulingStrategy.Factory();
            case LAZY_FROM_SOURCES_WITH_BATCH_SLOT_REQUEST:
            case LAZY_FROM_SOURCES:
                return new LazyFromSourcesSchedulingStrategy.Factory();
            default:
                throw new IllegalStateException("Unsupported schedule mode " + scheduleMode);
        }
    }
```

在流处理作业中，scheduleMode 的值为 EAGER。这个值来源于 StreamGraphGenerator 中的默认值：

```
public static final ScheduleMode DEFAULT_SCHEDULE_MODE = ScheduleMode.EAGER;
private ScheduleMode scheduleMode = DEFAULT_SCHEDULE_MODE;
```

了解了初始化过程后，下面来看看调度策略如何实现 SchedulingStrategy 接口中的方法。

首先是 startScheduling()方法：

```
public void startScheduling() {
    allocateSlotsAndDeploy(SchedulingStrategyUtils.getAllVertexIdsFromTopology(schedulingTopology));
}
static Set<ExecutionVertexID> getAllVertexIdsFromTopology(final SchedulingTopology<?, ?> topology) {
    return IterableUtils.toStream(topology.getVertices())
        .map(SchedulingExecutionVertex::getId)
        .collect(Collectors.toSet());
}
private void allocateSlotsAndDeploy(final Set<ExecutionVertexID> verticesToDeploy) {
    final List<ExecutionVertexDeploymentOption> executionVertexDeploymentOptions =
        SchedulingStrategyUtils.createExecutionVertexDeploymentOptionsInTopologicalOrder(
            schedulingTopology,
            verticesToDeploy,
```

```
                id -> deploymentOption);
        schedulerOperations.allocateSlotsAndDeploy(executionVertexDeploymentOptions);
    }
    static List<ExecutionVertexDeploymentOption> createExecutionVertexDeploymentOptionsIn
TopologicalOrder(
            final SchedulingTopology<?, ?> topology,
            final Set<ExecutionVertexID> verticesToDeploy,
            final Function<ExecutionVertexID, DeploymentOption> deploymentOptionRetriever) {
        return IterableUtils.toStream(topology.getVertices())
            .map(SchedulingExecutionVertex::getId)
            .filter(verticesToDeploy::contains)
            .map(executionVertexID -> new ExecutionVertexDeploymentOption(
                executionVertexID,
                deploymentOptionRetriever.apply(executionVertexID)))
            .collect(Collectors.toList());
    }
```

前面已经介绍过 SchedulingTopology 的结构和作用。这里先从中获取所有的 ExecutionVertexID。在 allocateSlotsAndDeploy() 方法中，为每一个 ExecutionVertexID 映射了一个 ExecutionVertexDeploymentOption 对象。然后调用 schedulerOperations 的 allocateSlotsAndDeploy() 方法继续接下来的调度。调用 allocateSlotsAndDeploy() 方法是 DefaultScheduler 中的行为，这会在后文介绍。

总之，可以了解到，startScheduling() 方法获取所有的节点 id，并封装成 ExecutionVertexDeploymentOption 对象去调度和部署。

EagerSchedulingStrategy 的 restartTasks() 方法的逻辑与之类似：

```
public void restartTasks(Set<ExecutionVertexID> verticesToRestart) {
    allocateSlotsAndDeploy(verticesToRestart);
}
```

该方法传入的参数就是需要调度和部署的 ExecutionVertexID 的集合，直接调用上面介绍的 allocateSlotsAndDeploy() 方法继续实现下面的流程。

接口中剩余的两个方法在 EagerSchedulingStrategy 中都没有实现：

```
public void onExecutionStateChange(ExecutionVertexID executionVertexId, ExecutionState executionState) {
}
public void onPartitionConsumable(ExecutionVertexID executionVertexId, ResultPartitionID resultPartitionId) {
}
```

这是因为所有任务都在一开始就一次性部署了，所以不需要在任务状态改变或分区数据可以被消费时再判断是否要调度和部署下游任务。

## 6.3.2 LazyFromSourcesSchedulingStrategy

LazyFromSourcesSchedulingStrategy 用于调度批处理作业的任务，当输入数据准备好时调度消费端的节点。它的字段如下。

- schedulerOperations：SchedulerOperations 类型，实际上就是 DefaultScheduler 对象。
- schedulingTopology：SchedulingTopology\<?, ?>类型。

- deploymentOptions：Map<ExecutionVertexID, DeploymentOption>类型，以 ExecutionVertexID 为键，每个节点对应一个 DeploymentOption 对象。DeploymentOption 中封装了一个布尔类型的字段，用于告知任务在输出数据时，分区是否在一旦有数据时就通知下游任务去调度和部署。在 LazyFromSourcesSchedulingStrategy 中，它的值由分区的 ResultPartitionType 类型决定。
- inputConstraintChecker：InputDependencyConstraintChecker 类型，用于检查是否可以调度和部署下游任务。

LazyFromSourcesSchedulingStrategy 对象的创建流程与 EagerSchedulingStrategy 的类似。当 scheduleMode 的值不为 EAGER 时，可创建 LazyFromSourcesSchedulingStrategy 对象。

下面来看看它如何实现 SchedulingStrategy 接口中的方法。

首先看一下 startScheduling()方法：

```
public void startScheduling() {
    final DeploymentOption updateOption = new DeploymentOption(true);
    final DeploymentOption nonUpdateOption = new DeploymentOption(false);
    // 遍历每一个节点
    for (SchedulingExecutionVertex<?, ?> schedulingVertex : schedulingTopology.getVertices()) {
        DeploymentOption option = nonUpdateOption;
        // 遍历每个节点的所有分区
        for (SchedulingResultPartition<?, ?> srp : schedulingVertex.getProducedResults()) {
            // 当分区的ResultPartitionType的isPipelined字段的值为true时,将option设置为updateOption
            if (srp.getResultType().isPipelined()) {
                option = updateOption;
            }
            // 将分区对象添加到 inputConstraintChecker 中
            inputConstraintChecker.addSchedulingResultPartition(srp);
        }
        // 将节点与其 option 信息封装到 deploymentOptions 字段中
        deploymentOptions.put(schedulingVertex.getId(), option);
    }
    // 分配槽并部署节点
    allocateSlotsAndDeployExecutionVertices(schedulingTopology.getVertices());
}
```

startScheduling()方法的主要逻辑是遍历节点和对应的分区，对 inputConstraintChecker 和 deploymentOptions 执行一些初始化操作，最后调用 allocateSlotsAndDeployExecutionVertices()方法调度和部署 SchedulingTopology 中的所有节点。

```
private void allocateSlotsAndDeployExecutionVertices(
        final Iterable<? extends SchedulingExecutionVertex<?, ?>> vertices) {
    final Set<ExecutionVertexID> verticesToDeploy = IterableUtils.toStream(vertices)
        // 过滤出所有符合条件的节点
        .filter(IS_IN_CREATED_EXECUTION_STATE.and(isInputConstraintSatisfied()))
        .map(SchedulingExecutionVertex::getId)
        .collect(Collectors.toSet());
    // 将所有符合条件的节点封装成 ExecutionVertexDeploymentOption 对象
    final List<ExecutionVertexDeploymentOption> vertexDeploymentOptions =
        SchedulingStrategyUtils.createExecutionVertexDeploymentOptionsInTopologicalOrder(
            schedulingTopology,
            verticesToDeploy,
```

```
        deploymentOptions::get);
    // 调用 DefaultScheduler 的 allocateSlotsAndDeploy()方法去部署这些
//ExecutionVertexDeploymentOption 对象
    schedulerOperations.allocateSlotsAndDeploy(vertexDeploymentOptions);
}
```

在该方法中，首先过滤出符合条件的节点，然后将它们封装成 ExecutionVertexDeploymentOption 对象，调用 DefaultScheduler 的 allocateSlotsAndDeploy()方法去部署。其中的过滤条件有两个：

（1）节点的 ExecutionState 的值为 CREATED。

```
private static final Predicate<SchedulingExecutionVertex<?, ?>> IS_IN_CREATED_EXECUTION_
STATE = schedulingExecutionVertex -> CREATED == schedulingExecutionVertex.getState();
```

（2）根据节点的 InputDependencyConstraint 类型，判断它是否可以部署消费上游的数据。

```
private Predicate<SchedulingExecutionVertex<?, ?>> isInputConstraintSatisfied() {
    return inputConstraintChecker::check;
}
public boolean check(final SchedulingExecutionVertex<?, ?> schedulingExecutionVertex) {
    if (Iterables.isEmpty(schedulingExecutionVertex.getConsumedResults())) {
        return true;
    }
    final InputDependencyConstraint inputConstraint = schedulingExecutionVertex.
getInputDependencyConstraint();
    switch (inputConstraint) {
        case ANY:
            return checkAny(schedulingExecutionVertex);
        case ALL:
            return checkAll(schedulingExecutionVertex);
        default:
            throw new IllegalStateException("Unknown InputDependencyConstraint " +
inputConstraint);
    }
}
```

InputDependencyConstraint 是一个枚举类，其中的枚举值 ANY 表示上游任意一个分区可被消费时就可以调度下游的消费者，ALL 表示上游所有分区都可被消费时才能调度下游的消费者。显然，第一次调度的时候只会调度 source 任务，因为这些任务没有上游任务，没有任何依赖。

过滤完以后会调用 createExecutionVertexDeploymentOptionsInTopologicalOrder()方法构造 ExecutionVertexDeploymentOption 对象，该方法在介绍 EagerSchedulingStrategy 时已经介绍过。

最后，仍然调用 DefaultScheduler 的 allocateSlotsAndDeploy()方法完成部署。

实现的 restartTasks()方法如下：

```
public void restartTasks(Set<ExecutionVertexID> verticesToRestart) {
    verticesToRestart
        .stream()
        .map(schedulingTopology::getVertexOrThrow)
        .flatMap(vertex -> IterableUtils.toStream(vertex.getProducedResults()))
        .forEach(inputConstraintChecker::resetSchedulingResultPartition);
    allocateSlotsAndDeployExecutionVertices(
        SchedulingStrategyUtils.getVerticesFromIds(schedulingTopology, verticesToRestart));
}
```

其核心逻辑其实与 EagerSchedulingStrategy 的类似，都是部署传入的这些任务节点。只不过在 LazyFromSourcesSchedulingStrategy 策略中，需要更新 inputConstraintChecker 中的一些值，并且还需要通过 allocateSlotsAndDeployExecutionVertices()方法中的流程再去部署过滤后的节点。

由于在 LazyFromSourcesSchedulingStrategy 策略中，不会一开始就部署所有的任务，因此需要在上游任务状态更新时或者在上游数据可被消费时判断是否可以部署下游任务，对于这两种情况，分别通过 onExecutionStateChange()方法和 onPartitionConsumable()方法得以做了实现。

```
public void onExecutionStateChange(ExecutionVertexID executionVertexId, ExecutionState 
executionState) {
    if (!FINISHED.equals(executionState)) {
        return;
    }
    final Set<SchedulingExecutionVertex<?, ?>> verticesToSchedule = IterableUtils
        // 获取该节点的所有分区
        .toStream(schedulingTopology.getVertexOrThrow(executionVertexId).getProducedResults())
        // 过滤出所有 ResultPartitionType 的 isPipelined 字段的值为 false 的分区
        .filter(partition -> partition.getResultType().isBlocking())
        // 在 inputConstraintChecker 中对每个分区标记任务完成
        .flatMap(partition -> inputConstraintChecker.markSchedulingResultPartitionFinished(partition).stream())
        // 获取分区的消费者
        .flatMap(partition -> IterableUtils.toStream(partition.getConsumers()))
        .collect(Collectors.toSet());
    allocateSlotsAndDeployExecutionVertices(verticesToSchedule);
}
```

在这个过程中，最复杂的部分是关于 inputConstraintChecker 的逻辑，下文会详细分析。

```
public void onPartitionConsumable(ExecutionVertexID executionVertexId, ResultPartitionID 
resultPartitionId) {
    final SchedulingResultPartition<?, ?> resultPartition = schedulingTopology
        .getResultPartitionOrThrow(resultPartitionId.getPartitionId());
    if (!resultPartition.getResultType().isPipelined()) {
        return;
    }
    ...
    // 部署分区的消费者
    allocateSlotsAndDeployExecutionVertices(resultPartition.getConsumers());
}
```

在 onPartitionConsumable()方法中主要是判断分区的 ResultPartitionType，如果 isPipelined 的值为 false，则直接返回。最后调用 allocateSlotsAndDeployExecutionVertices()方法部署消费者。

### 6.3.3　InputDependencyConstraintChecker

在 LazyFromSourcesSchedulingStrategy 策略中依赖 InputDependencyConstraintChecker 对象。它主要用于检查任务是否可以部署。要理解它检查的逻辑，还得从它的底层依赖的数据结构说起。

在 LazyFromSourcesSchedulingStrategy 策略中，底层有一个数据结构叫作 SchedulingIntermediateDataSet，可以认为它在调度层面对应一个 IntermediateDataSet。SchedulingIntermediateDataSet 中有以下两个字段。

- partitions：SchedulingResultPartition 类型，表示该 IntermediateDataSet 中的所有子分区。

- producingPartitionIds：IntermediateResultPartitionID 类型，表示该 IntermediateDataSet 中的所有子分区的唯一标识。

这里的基本逻辑是，每一个子任务完成后，对应的子分区的 id 就会从 producingPartitionIds 中移除。当所有 id 都被移除时，就表示下游可以部署了，这时会将 partitions 字段返回，从中获取 consumers 进行部署。

下面是 SchedulingIntermediateDataSet 提供的方法。

以下方法会在子任务完成时被调用，用于移除对应的 id，并判断集合是否为空：

```
boolean markPartitionFinished(IntermediateResultPartitionID partitionId) {
    producingPartitionIds.remove(partitionId);
    return producingPartitionIds.isEmpty();
}
```

以下方法会在重启任务时被调用，用于重新将 id 添加到集合中：

```
void resetPartition(IntermediateResultPartitionID partitionId) {
    producingPartitionIds.add(partitionId);
}
```

以下方法用于判断是否所有子分区对应的 id 都被移除：

```
boolean allPartitionsFinished() {
    return producingPartitionIds.isEmpty();
}
```

以下方法会在初始化 SchedulingIntermediateDataSet 时被调用，用于将所有的 partition 对象和 id 都添加到集合中：

```
void addSchedulingResultPartition(SchedulingResultPartition<?, ?> partition) {
    partitions.add(partition);
    producingPartitionIds.add(partition.getId());
}
```

以下方法会在 id 集合为空时被调用，返回所有的分区对象用于下游任务的部署：

```
List<SchedulingResultPartition<?, ?>> getSchedulingResultPartitions() {
    return Collections.unmodifiableList(partitions);
}
```

InputDependencyConstraintChecker 不会直接操作 SchedulingIntermediateDataSet，而是通过 SchedulingIntermediateDataSetManager 来操作。SchedulingIntermediateDataSetManager 只有以下一个字段。

intermediateDataSets：Map<IntermediateDataSetID, SchedulingIntermediateDataSet> 类型，以 IntermediateDataSetID 为键，维护所有的 SchedulingIntermediateDataSet 对象。

SchedulingIntermediateDataSetManager 的方法主要是对 SchedulingIntermediateDataSet 的方法进行封装。最后，SchedulingIntermediateDataSetManager 对象作为 InputDependencyConstraintChecker 的字段，对任务、分区信息进行检查。

根据对 LazyFromSourcesSchedulingStrategy 的分析可知，下面几处会用到 InputDependencyConstraintChecker。

（1）在 startScheduling() 方法中。

```java
public void startScheduling() {
    final DeploymentOption updateOption = new DeploymentOption(true);
    final DeploymentOption nonUpdateOption = new DeploymentOption(false);
    for (SchedulingExecutionVertex<?, ?> schedulingVertex : schedulingTopology.getVertices()) {
        DeploymentOption option = nonUpdateOption;
        for (SchedulingResultPartition<?, ?> srp : schedulingVertex.getProducedResults()) {
            if (srp.getResultType().isPipelined()) {
                option = updateOption;
            }
            // 将子分区添加到 inputConstraintChecker
            inputConstraintChecker.addSchedulingResultPartition(srp);
        }
        deploymentOptions.put(schedulingVertex.getId(), option);
    }
    allocateSlotsAndDeployExecutionVertices(schedulingTopology.getVertices());
}
```

添加逻辑如下：

```java
void addSchedulingResultPartition(SchedulingResultPartition<?, ?> srp) {
    intermediateDataSetManager.addSchedulingResultPartition(srp);
}
void addSchedulingResultPartition(SchedulingResultPartition<?, ?> srp) {
    SchedulingIntermediateDataSet sid = getOrCreateSchedulingIntermediateDataSetIfAbsent(srp.getResultId());
    sid.addSchedulingResultPartition(srp);
}
private SchedulingIntermediateDataSet getOrCreateSchedulingIntermediateDataSetIfAbsent(
        final IntermediateDataSetID intermediateDataSetId) {
    return getSchedulingIntermediateDataSetInternal(intermediateDataSetId, true);
}
private SchedulingIntermediateDataSet getSchedulingIntermediateDataSetInternal(
        final IntermediateDataSetID intermediateDataSetId,
        boolean createIfAbsent) {
    return intermediateDataSets.computeIfAbsent(
        intermediateDataSetId,
        (key) -> {
            if (createIfAbsent) {
                return new SchedulingIntermediateDataSet();
            } else {
                throw new IllegalArgumentException("can not find data set for " + intermediateDataSetId);
            }
        });
}
```

由此将所有分区对象和 id 都放入 SchedulingIntermediateDataSet 的集合中。

（2）在 restartTasks() 方法中。

```java
public void restartTasks(Set<ExecutionVertexID> verticesToRestart) {
    verticesToRestart
        .stream()
        .map(schedulingTopology::getVertexOrThrow)
        .flatMap(vertex -> IterableUtils.toStream(vertex.getProducedResults()))
        // 将所有需要重启的任务的子分区都添加到 inputConstraintChecker
```

```
        .forEach(inputConstraintChecker::resetSchedulingResultPartition);
    allocateSlotsAndDeployExecutionVertices(
        SchedulingStrategyUtils.getVerticesFromIds(schedulingTopology, verticesToRestart));
}
```

其中，resetSchedulingResultPartition()方法如下：

```
void resetSchedulingResultPartition(SchedulingResultPartition<?, ?> srp) {
    intermediateDataSetManager.resetSchedulingResultPartition(srp);
}
void resetSchedulingResultPartition(SchedulingResultPartition<?, ?> srp) {
    SchedulingIntermediateDataSet sid = getSchedulingIntermediateDataSet(srp.getResultId());
    sid.resetPartition(srp.getId());
}
private SchedulingIntermediateDataSet getSchedulingIntermediateDataSet(
        final IntermediateDataSetID intermediateDataSetId) {
    return getSchedulingIntermediateDataSetInternal(intermediateDataSetId, false);
}
```

在任务状态更新时，会利用 InputConstraintChecker 标记任务完成：

```
public void onExecutionStateChange(ExecutionVertexID executionVertexId, ExecutionState executionState) {
    if (!FINISHED.equals(executionState)) {
        return;
    }
    final Set<SchedulingExecutionVertex<?, ?>> verticesToSchedule = IterableUtils
        .toStream(schedulingTopology.getVertexOrThrow(executionVertexId).getProducedResults())
        .filter(partition -> partition.getResultType().isBlocking())
        // 标记任务完成
        .flatMap(partition -> inputConstraintChecker.markSchedulingResultPartitionFinished(partition).stream())
        .flatMap(partition -> IterableUtils.toStream(partition.getConsumers()))
        .collect(Collectors.toSet());
    allocateSlotsAndDeployExecutionVertices(verticesToSchedule);
}
```

其中，markSchedulingResultPartitionFinished()方法如下：

```
List<SchedulingResultPartition<?, ?>> markSchedulingResultPartitionFinished(SchedulingResultPartition<?, ?> srp) {
    return intermediateDataSetManager.markSchedulingResultPartitionFinished(srp);
}
List<SchedulingResultPartition<?, ?>> markSchedulingResultPartitionFinished(SchedulingResultPartition<?, ?> srp) {
    SchedulingIntermediateDataSet intermediateDataSet = getSchedulingIntermediateDataSet(srp.getResultId());
    if (intermediateDataSet.markPartitionFinished(srp.getId())) {
        // 如果所有子任务都已完成，则返回所有子分区对象
        return intermediateDataSet.getSchedulingResultPartitions();
    }
    return Collections.emptyList();
}
```

（3）在部署任务之前，会进行如下检查：

```
private void allocateSlotsAndDeployExecutionVertices(
        final Iterable<? extends SchedulingExecutionVertex<?, ?>> vertices) {
    final Set<ExecutionVertexID> verticesToDeploy = IterableUtils.toStream(vertices)
```

```
            // 检查输入是否已经满足条件
            .filter(IS_IN_CREATED_EXECUTION_STATE.and(isInputConstraintSatisfied()))
            .map(SchedulingExecutionVertex::getId)
            .collect(Collectors.toSet());
        final List<ExecutionVertexDeploymentOption> vertexDeploymentOptions =
            SchedulingStrategyUtils.createExecutionVertexDeploymentOptionsInTopologicalOrder(
                schedulingTopology,
                verticesToDeploy,
                deploymentOptions::get);
        schedulerOperations.allocateSlotsAndDeploy(vertexDeploymentOptions);
    }
    public boolean check(final SchedulingExecutionVertex<?, ?> schedulingExecutionVertex) {
        if (Iterables.isEmpty(schedulingExecutionVertex.getConsumedResults())) {
            return true;
        }
        final InputDependencyConstraint inputConstraint = schedulingExecutionVertex.getInputDependencyConstraint();
        switch (inputConstraint) {
            case ANY:
                return checkAny(schedulingExecutionVertex);
            case ALL:
                return checkAll(schedulingExecutionVertex);
            default:
                throw new IllegalStateException("Unknown InputDependencyConstraint " + inputConstraint);
        }
    }
```

其中，checkAny()方法与checkAll()方法分别如下：

```
    private boolean checkAny(final SchedulingExecutionVertex<?, ?> schedulingExecutionVertex) {
        for (SchedulingResultPartition<?, ?> consumedResultPartition : schedulingExecutionVertex.getConsumedResults()) {
            if (partitionConsumable(consumedResultPartition)) {
                return true;
            }
        }
        return false;
    }

    private boolean checkAll(final SchedulingExecutionVertex<?, ?> schedulingExecutionVertex) {
        for (SchedulingResultPartition<?, ?> consumedResultPartition : schedulingExecutionVertex.getConsumedResults()) {
            if (!partitionConsumable(consumedResultPartition)) {
                return false;
            }
        }
        return true;
    }
```

checkAny()方法在上游任意一个子分区可被消费时，就返回true；而checkAll()方法只有在上游所有子分区都可被消费时，才返回true。

partitionConsumable()方法如下：

```
    private boolean partitionConsumable(SchedulingResultPartition<?, ?> partition) {
```

```
        if (BLOCKING.equals(partition.getResultType())) { //判断子分区的ResultPartitionType是否为
                                                          //BLOCKING
            return intermediateDataSetManager.allPartitionsFinished(partition);//该IntermediateDataSet
                                                                              //中所有子分区都已完成
        } else {
            final ResultPartitionState state = partition.getState();
            return ResultPartitionState.CONSUMABLE.equals(state);
        }
    }
```

从上面的分析可以看出，**InputConstraintChecker** 在任务状态更新和部署下游任务前做了两次检查，一次是检查该子任务对应的 IntermediateDataSet 中的所有子分区是否都已完成，另一次是检查将要部署的下游任务所要消费的上游的子分区是否都可以被消费。乍一看两者好像是一回事，其实不然，因为下游任务可能有多个上游任务作为它的输入。可能一个任务的所有子分区已全部完成，而另一个任务的则没有。

## 6.4 调度过程的实现

了解了调度层面涉及的数据结构和调度策略后，下面来分析调度过程的具体实现。

### 6.4.1 开始调度

在 JobMaster 的启动过程中，会调用 resetAndStartScheduler()方法，以开始调度。

```
private void resetAndStartScheduler() throws Exception {
    ...
    schedulerAssignedFuture.thenRun(this::startScheduling);
}
private void startScheduling() {
    jobStatusListener = new JobManagerJobStatusListener();
    // 注册作业状态的监听器
    schedulerNG.registerJobStatusListener(jobStatusListener);
    // 开始调度
    schedulerNG.startScheduling();
}
```

从上述代码可观察到，这里涉及作业状态的监听。在 Flink 中，作业状态由 JobStatus 来表示，作业状态变化如图 6-2 所示。

状态变化的过程分散在作业生命周期的不同阶段，读者可以在学习任务调度的过程中明白这些状态的转换，这里不再赘述。

监听器会被注册到 ExecutionGraph 中：

```
public void registerJobStatusListener(final JobStatusListener jobStatusListener) {
    executionGraph.registerJobStatusListener(jobStatusListener);
}
public void registerJobStatusListener(JobStatusListener listener) {
    if (listener != null) {
        jobStatusListeners.add(listener);
    }
}
```

开始调度时调用调度器的 startScheduling() 方法：

```
public final void startScheduling() {
    ...
    startSchedulingInternal();
}
protected void startSchedulingInternal() {
    // 调度前的准备
    prepareExecutionGraphForNgScheduling();
    // 利用调度策略开始调度
    schedulingStrategy.startScheduling();
}
```

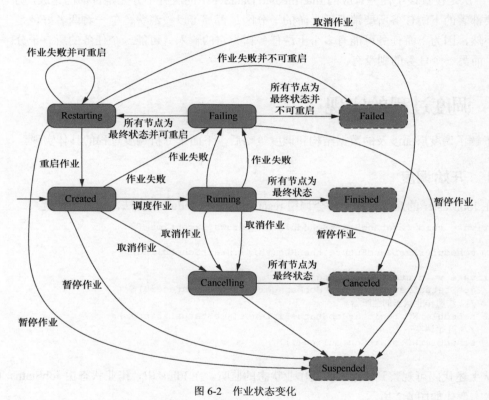

图 6-2 作业状态变化

其中，prepareExecutionGraphForNgScheduling() 方法如下：

```
protected final void prepareExecutionGraphForNgScheduling() {
    executionGraph.enableNgScheduling(new UpdateSchedulerNgOnInternalFailuresListener(this, jobGraph.getJobID()));
    executionGraph.transitionToRunning();
}
public void enableNgScheduling(final InternalFailuresListener internalTaskFailuresListener) {
    this.internalTaskFailuresListener = internalTaskFailuresListener;
    this.legacyScheduling = false;
}
public void transitionToRunning() {
    if (!transitionState(JobStatus.CREATED, JobStatus.RUNNING)) {
```

```
        ...
    }
}
```

该方法对 ExecutionGraph 字段进行初始化赋值,并且改变作业的状态。改变作业状态的方法如下:

```
public boolean transitionState(JobStatus current, JobStatus newState) {
    return transitionState(current, newState, null);
}
private boolean transitionState(JobStatus current, JobStatus newState, Throwable error) {
    ...
    if (state == current) {
        state = newState;
        // 通知作业状态改变
        notifyJobStatusChange(newState, error);
        return true;
    }
    else {
        return false;
    }
}
```

在 notifyJobStatusChange() 方法中,会触发监听器的操作:

```
private void notifyJobStatusChange(JobStatus newState, Throwable error) {
    if (jobStatusListeners.size() > 0) {
        ...
        for (JobStatusListener listener : jobStatusListeners) {
            try {
                listener.jobStatusChanges(getJobID(), newState, timestamp, serializedError);
            } catch (Throwable t) {
                ...
            }
        }
    }
}
```

由此可以知道,触发监听器的操作发生在改变作业状态的方法 transitionState() 中。

最后,调用调度策略的 startScheduling() 方法开始调度。

## 6.4.2 更新任务状态

在一个任务的生命周期中,会涉及不同状态的转换。在 Flink 中,任务状态由 ExecutionState 来表示。任务状态变化如图 6-3 所示。

当任务状态更新时,会调用调度器的 updateTaskExecutionState() 方法。

```
public final boolean updateTaskExecutionState(final TaskExecutionState taskExecutionState) {
    // 根据 ExecutionAttemptID 获取对应的 ExecutionVertexID
    final Optional<ExecutionVertexID> executionVertexId = getExecutionVertexId
(taskExecutionState.getID());
    // 更新任务状态
    boolean updateSuccess = executionGraph.updateState(taskExecutionState);
    if (updateSuccess) {
        if (isNotifiable(executionVertexId.get(), taskExecutionState)) {
```

```
            // 当任务状态为 FINISHED 或 FAILED 时调用下面的方法进行更新
            updateTaskExecutionStateInternal(executionVertexId.get(), taskExecutionState);
        }
        return true;
    } else {
        return false;
    }
}
```

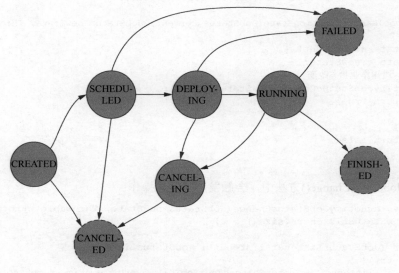

图 6-3　任务状态变化

前文介绍过,在执行层面每个 Execution 对象都有一个唯一标识 ExecutionAttemptID。在任务管理器与作业管理器通信时,都是用该标识来表示任务并行实例的,表示任务状态的 TaskExecutionState 类中封装了该字段。因此,在 updateTaskExecutionState() 方法中,首先需要获取对应的 ExecutionVertexID。

```
protected Optional<ExecutionVertexID> getExecutionVertexId(final ExecutionAttemptID
executionAttemptId) {
    return Optional.ofNullable(executionGraph.getRegisteredExecutions().get(execution
AttemptId))
        .map(this::getExecutionVertexId);
}
```

该方法从 ExecutionGraph 中的 currentExecutions 字段获取 ExecutionVertexID。

接着,在 updateTaskExecutionState() 方法中,分别调用了 ExecutionGraph 的 updateState() 方法和调度器本身的 updateTaskExecutionStateInternal() 方法进行了更新。下面分别介绍这两个方法具体进行了什么更新。

```
public boolean updateState(TaskExecutionState state) {
    // 获取 Execution 对象
    final Execution attempt = currentExecutions.get(state.getID());
    if (attempt != null) {
        try {
            // 更新任务状态
            final boolean stateUpdated = updateStateInternal(state, attempt);
```

```
            ...
        }
```

在 updateState()方法中，第一步就是获取 Execution 对象。其中，updateStateInternal()方法用于更新任务状态：

```
    private boolean updateStateInternal(final TaskExecutionState state, final Execution attempt) {
        Map<String, Accumulator<?, ?>> accumulators;
        switch (state.getExecutionState()) {
            case RUNNING:
                return attempt.switchToRunning();
            case FINISHED:
                accumulators = deserializeAccumulators(state);
                attempt.markFinished(accumulators, state.getIOMetrics());
                return true;
            case CANCELED:
                accumulators = deserializeAccumulators(state);
                attempt.completeCancelling(accumulators, state.getIOMetrics(), false);
                return true;
            case FAILED:
                accumulators = deserializeAccumulators(state);
                attempt.markFailed(state.getError(userClassLoader), accumulators, state.getIOMetrics(), !isLegacyScheduling());
                return true;
            default:
                attempt.fail(new Exception("TaskManager sent illegal state update: " + state.getExecutionState()));
                return false;
        }
    }
```

这里根据任务状态的不同，调用了 Execution 对象的不同方法。后文会对 Execution 的各个方法进行具体的介绍。当任务状态为 FINISHED 时，会触发下游任务的调度：

```
    void markFinished(Map<String, Accumulator<?, ?>> userAccumulators, IOMetrics metrics) {
        while (true) {
            ExecutionState current = this.state;
            if (current == RUNNING || current == DEPLOYING) {
                if (transitionState(current, FINISHED)) {
                    try {
                        // 调度下游消费者
                        finishPartitionsAndScheduleOrUpdateConsumers();
                        ...
                    }
                    finally {
                        vertex.executionFinished(this);
                    }
                    return;
                }
            }
            ...
        }
    }
    private void finishPartitionsAndScheduleOrUpdateConsumers() {
```

```
        final List<IntermediateResultPartition> newlyFinishedResults = getVertex().
finishAllBlockingPartitions();
        if (newlyFinishedResults.isEmpty()) {
          return;
        }
        final HashSet<ExecutionVertex> consumerDeduplicator = new HashSet<>();
        for (IntermediateResultPartition finishedPartition : newlyFinishedResults) {
            final IntermediateResultPartition[] allPartitionsOfNewlyFinishedResults =
                finishedPartition.getIntermediateResult().getPartitions();
            for (IntermediateResultPartition partition : allPartitionsOfNewlyFinishedResults) {
              // 调度下游消费者
              scheduleOrUpdateConsumers(partition.getConsumers(), consumerDeduplicator);
            }
        }
    }
```

这里关键的方法是 scheduleOrUpdateConsumers()方法：

```
    private void scheduleOrUpdateConsumers(
        final List<List<ExecutionEdge>> allConsumers,
        final HashSet<ExecutionVertex> consumerDeduplicator) {
      ...
      for (ExecutionEdge edge : allConsumers.get(0)) {
        final ExecutionVertex consumerVertex = edge.getTarget();
        final Execution consumer = consumerVertex.getCurrentExecutionAttempt();
        final ExecutionState consumerState = consumer.getState();
        if (consumerState == CREATED) {
          // 判断是否是 LegacyScheduling
          if (isLegacyScheduling() && consumerDeduplicator.add(consumerVertex) &&
              (consumerVertex.getInputDependencyConstraint() == InputDependencyConstraint.ANY ||
              consumerVertex.checkInputDependencyConstraints())) {
            // 调度下游消费者
            scheduleConsumer(consumerVertex);
          }
        }
        ...
      }
    }
```

注意，在调用 scheduleConsumer()方法前，要先进行判断，即调用 isLegacyScheduling()方法：

```
    private boolean isLegacyScheduling() {
      return getVertex().isLegacyScheduling();
    }
    public boolean isLegacyScheduling() {
      return getExecutionGraph().isLegacyScheduling();
    }
    public boolean isLegacyScheduling() {
      return legacyScheduling;
    }
```

最终返回的是 ExecutionGraph 的 legacyScheduling 字段。该字段在之前介绍过的 startScheduling() 方法中赋过值 false，表示不用旧版本的调度器。因此，整个 updateState()方法都不会调度下游的任务。

回到 updateTaskExecutionState()方法，另一个用于更新的方法 updateTaskExecutionStateInternal() 如下：

```
protected void updateTaskExecutionStateInternal(final ExecutionVertexID executionVertexId,
final TaskExecutionState taskExecutionState) {
    // 调用调度策略的 onExecutionStateChange()方法
    schedulingStrategy.onExecutionStateChange(executionVertexId, taskExecutionState.
getExecutionState());
    ...
}
```

调度策略的 onExecutionStateChange()方法先前已经介绍过，可以知道对于 LazyFromSources SchedulingStrategy，该方法会调度下游的任务。

### 6.4.3 调度或更新消费者

对一些作业来说，当上游任务的数据可以被消费时，就可以调度下游任务了。这时会调用调度器的 scheduleOrUpdateConsumers()方法。

```
public final void scheduleOrUpdateConsumers(final ResultPartitionID partitionId) {
    try {
    // 调用 ExecutionGraph 的方法调度或更新下游消费者
        executionGraph.scheduleOrUpdateConsumers(partitionId);
    } catch (ExecutionGraphException e) {
        ...
    }
    final ExecutionVertexID producerVertexId = getExecutionVertexIdOrThrow(partitionId.
getProducerId());
    // 调用调度器的方法调度或更新下游消费者
    scheduleOrUpdateConsumersInternal(producerVertexId, partitionId);
}
```

ExecutionGraph 的 scheduleOrUpdateConsumers()方法与前面介绍的 updateState()方法类似，最终会判断是否用旧版本的调度器，因此该方法不会做任何调度。调度在 scheduleOrUpdateConsumersInternal()方法中实现：

```
protected void scheduleOrUpdateConsumersInternal(final ExecutionVertexID producerVertexId,
final ResultPartitionID partitionId) {
    schedulingStrategy.onPartitionConsumable(producerVertexId, partitionId);
}
```

这里利用了具体的调度策略的 onPartitionConsumable()方法，该方法在 6.3 节中已经介绍过。

## 6.5 任务的部署

根据前文可了解到，在开始调度和更新任务状态去部署下游任务时，都会调用 DefaultScheduler 的 allocateSlotsAndDeploy()方法。该方法包含分配槽和部署任务的逻辑。本节内容只关注部署任务部分。相关代码如下：

```
public void allocateSlotsAndDeploy(final List<ExecutionVertexDeploymentOption>
executionVertexDeploymentOptions) {
    validateDeploymentOptions(executionVertexDeploymentOptions);
    // 数据结构的转换
    final Map<ExecutionVertexID, ExecutionVertexDeploymentOption> deploymentOptionsByVertex =
```

```java
        groupDeploymentOptionsByVertexId(executionVertexDeploymentOptions);
    final List<ExecutionVertexID> verticesToDeploy = executionVertexDeploymentOptions.stream()
        .map(ExecutionVertexDeploymentOption::getExecutionVertexId)
        .collect(Collectors.toList());
    final Map<ExecutionVertexID, ExecutionVertexVersion> requiredVersionByVertex =
        executionVertexVersioner.recordVertexModifications(verticesToDeploy);
    // 改变 Execution 的状态
    transitionToScheduled(verticesToDeploy);
    // 分配槽
    final List<SlotExecutionVertexAssignment> slotExecutionVertexAssignments =
        allocateSlots(executionVertexDeploymentOptions);
    // 将上面的数据结构封装成 DeploymentHandle 对象
    final List<DeploymentHandle> deploymentHandles = createDeploymentHandles(
        requiredVersionByVertex,
        deploymentOptionsByVertex,
        slotExecutionVertexAssignments);
    // 判断任务是否独立部署,即调度策略是否为 LazyFromSourcesSchedulingStrategy
    if (isDeployIndividually()) {
        // 如果调度策略是 LazyFromSourcesSchedulingStrategy,则独立部署
        deployIndividually(deploymentHandles);
    } else {
        // 如果调度策略不是 LazyFromSourcesSchedulingStrategy,则等待所有任务的资源分配好后统一部署
        waitForAllSlotsAndDeploy(deploymentHandles);
    }
}
```

在上面的方法中出现了几种数据类型,如 ExecutionVertexDeploymentOption、ExecutionVertexVersion、SlotExecutionVertexAssignment 和 DeploymentHandle。前三者都表示对节点 id 的简单封装,DeploymentHandle 表示对前三者的封装。

(1) ExecutionVertexDeploymentOption:

```java
public class ExecutionVertexDeploymentOption {
    private final ExecutionVertexID executionVertexId;
    private final DeploymentOption deploymentOption;
```

(2) ExecutionVertexVersion:

```java
public class ExecutionVertexVersion {
    private final ExecutionVertexID executionVertexId;
    private final long version;
```

(3) SlotExecutionVertexAssignment:

```java
public class SlotExecutionVertexAssignment {
    private final ExecutionVertexID executionVertexId;
    private final CompletableFuture<LogicalSlot> logicalSlotFuture;
```

(4) DeploymentHandle:

```java
class DeploymentHandle {
    private final ExecutionVertexVersion requiredVertexVersion;
    private final ExecutionVertexDeploymentOption executionVertexDeploymentOption;
    private final SlotExecutionVertexAssignment slotExecutionVertexAssignment;
```

最后在部署的时候做了判断,并根据不同的调度策略进行不同的部署。如果调度策略是 LazyFromSourcesSchedulingStrategy,那么任务可以独立部署;如果调度策略不是 LazyFromSources

SchedulingStrategy,则在当前的实现中意味着是 EagerSchedulingStrategy,那么需要等待所有任务的资源分配好后统一部署。

独立部署方式的 deployIndividually()方法如下:

```
private void deployIndividually(final List<DeploymentHandle> deploymentHandles) {
    for (final DeploymentHandle deploymentHandle : deploymentHandles) {
        FutureUtils.assertNoException(
            deploymentHandle
                .getSlotExecutionVertexAssignment()
                .getLogicalSlotFuture()
                // 分配资源
                .handle(assignResourceOrHandleError(deploymentHandle))
                // 部署任务
                .handle(deployOrHandleError(deploymentHandle)));
    }
}
```

统一部署方式的 waitForAllSlotsAndDeploy()方法如下:

```
private void waitForAllSlotsAndDeploy(final List<DeploymentHandle> deploymentHandles) {
    FutureUtils.assertNoException(
        // 分配资源
        assignAllResources(deploymentHandles)
        // 部署任务
        .handle(deployAll(deploymentHandles)));
}

private BiFunction<Void, Throwable, Void> deployAll(final List<DeploymentHandle> deploymentHandles) {
    return (ignored, throwable) -> {
        propagateIfNonNull(throwable);
        for (final DeploymentHandle deploymentHandle : deploymentHandles) {
            final SlotExecutionVertexAssignment slotExecutionVertexAssignment = deploymentHandle.getSlotExecutionVertexAssignment();
            final CompletableFuture<LogicalSlot> slotAssigned = slotExecutionVertexAssignment.getLogicalSlotFuture();
            checkState(slotAssigned.isDone());
            FutureUtils.assertNoException(
                // 部署任务
                slotAssigned.handle(deployOrHandleError(deploymentHandle)));
        }
        return null;
    };
}
```

以上两种部署方式最终都会调用 deployOrHandleError()方法:

```
private BiFunction<Object, Throwable, Void> deployOrHandleError(final DeploymentHandle deploymentHandle) {
    final ExecutionVertexVersion requiredVertexVersion = deploymentHandle.getRequiredVertexVersion();
    final ExecutionVertexID executionVertexId = requiredVertexVersion.getExecutionVertexId();
    return (ignored, throwable) -> {
        ...
        if (throwable == null) {
            // 部署任务
            deployTaskSafe(executionVertexId);
```

```
        } else {
            // 处理部署失败
            handleTaskDeploymentFailure(executionVertexId, throwable);
        }
        return null;
    };
}
```

其中，handleTaskDeploymentFailure()方法最终会调用 Execution 标记部署失败的相应方法。deployTaskSafe()方法用于部署任务，经过几步调用后也会进入 Execution 的 deploy()方法完成部署。

## 6.6　Execution 对象在调度过程中的行为

在第 4 章介绍 ExecutionGraph 时，提到 Execution 表示其中的执行对象。每个 Execution 对象都对应一个 ExecutionVertex 对象，但一个 ExecutionVertex 对象可能对应多个 Execution 对象，因为可能会有执行失败的情况发生。在调度和部署的过程中，最终会调用 Execution 的 deploy()方法完成部署。本节将介绍 Execution 在任务状态改变时的行为。

从前文可了解到，在任务状态发生改变时，会调用 ExecutionGraph 的 updateStateInternal()方法：

```
private boolean updateStateInternal(final TaskExecutionState state, final Execution attempt) {
    Map<String, Accumulator<?, ?>> accumulators;
    switch (state.getExecutionState()) {
        case RUNNING:
            return attempt.switchToRunning();
        case FINISHED:
            accumulators = deserializeAccumulators(state);
            attempt.markFinished(accumulators, state.getIOMetrics());
            return true;
        case CANCELED:
            accumulators = deserializeAccumulators(state);
            attempt.completeCancelling(accumulators, state.getIOMetrics(), false);
            return true;
        case FAILED:
            accumulators = deserializeAccumulators(state);
            attempt.markFailed(state.getError(userClassLoader), accumulators, state.getIOMetrics(), !isLegacyScheduling());
            return true;
        default:
            attempt.fail(new Exception("TaskManager sent illegal state update: " + state.getExecutionState()));
            return false;
    }
}
```

这个方法根据任务状态的不同调用 Execution 对象的不同方法。下面依次进行分析。

### 1. RUNNING

当任务管理器端传回的任务状态为 RUNNING 时，会调用 switchToRunning()方法：

```
boolean switchToRunning() {
    // 将任务状态从 DEPLOYING 转换为 RUNNING
```

```
    if (transitionState(DEPLOYING, RUNNING)) {
        // 如果转换成功,则发送 PartitionInfo 对象到任务管理器端以构建新的输入通道并与上游建立连接
        sendPartitionInfos();
        return true;
    }
    else {  // 如果转换失败,则进入下面的逻辑
        ExecutionState currentState = this.state;
        if (currentState == FINISHED || currentState == CANCELED) {
            // 如果当前状态为 FINISHED 或者 CANCELED,则不执行任何操作

        }
        else if (currentState == CANCELING || currentState == FAILED) {
            // 如果当前状态为 CANCELING 或者 FAILED,则通知任务管理器端取消该任务
            sendCancelRpcCall(NUM_CANCEL_CALL_TRIES);

        }
        else {
            // 如果当前状态是其他状态,则先通知任务管理器端取消该任务,再调用 markFailed()方法标记任务失败
            sendCancelRpcCall(NUM_CANCEL_CALL_TRIES);
            markFailed(new Exception(message));
        }
        return false;
    }
}
```

其中,**transitionState()方法如下**:

```
private boolean transitionState(ExecutionState currentState, ExecutionState targetState) {
    return transitionState(currentState, targetState, null);
}
private boolean transitionState(ExecutionState currentState, ExecutionState targetState,
Throwable error) {
    ...
    if (STATE_UPDATER.compareAndSet(this, currentState, targetState)) {
        ...
        try {
            // 通知 ExecutionVertex 状态改变
            vertex.notifyStateTransition(this, targetState, error);
        }
        catch (Throwable t) {
            ...
        }
        return true;
    } else {
        return false;
    }
}
```

如果更新状态成功,则会调用 ExecutionVertex 的 **notifyStateTransition()方法**:

```
void notifyStateTransition(Execution execution, ExecutionState newState, Throwable error) {
    if (currentExecution == execution) {
        getExecutionGraph().notifyExecutionChange(execution, newState, error);
    }
}
void notifyExecutionChange(
    final Execution execution,
```

```
        final ExecutionState newExecutionState,
        final Throwable error) {
    if (!isLegacyScheduling()) {
        return;
    }
    ...
}
```

在 notifyExecutionChange()方法中，调用了 isLegacyScheduling()方法，这意味着如果不是旧版本的调度器，则直接返回。因此可以认为，在调用 transitionState()的过程中，除了改变 Execution 的 state 字段的值，基本没有执行其他的操作。

再来看 sendPartitionInfos()方法：

```
private void sendPartitionInfos() {
    if (!partitionInfos.isEmpty()) {
        sendUpdatePartitionInfoRpcCall(new ArrayList<>(partitionInfos));
        partitionInfos.clear();
    }
}
```

该方法会将 partitionInfos 的值发送到任务管理器端，根据情况构建新的输入通道与上游建立连接。

如果更新状态失败，则 Execution 对象的当前状态发生改变，它会根据当前状态的不同执行不同的操作。其中，sendCancelRpcCall()方法用于通知任务管理器端取消该任务。

```
private void sendCancelRpcCall(int numberRetries) {
    final LogicalSlot slot = assignedResource;
    if (slot != null) {
        ...
        CompletableFuture<Acknowledge> cancelResultFuture = FutureUtils.retry(
            () -> taskManagerGateway.cancelTask(attemptId, rpcTimeout), // 通知任务管理
// 器端取消该任务
            numberRetries,
            jobMasterMainThreadExecutor);
        ...
    }
}
```

而 markFailed()方法则是 Execution 中用于标记任务失败的方法，该方法如下：

```
void markFailed(Throwable t) {
    processFail(t, true);
}
private void processFail(Throwable t, boolean isCallback) {
    processFail(t, isCallback, null, null, true, false);
}
private void processFail(Throwable t, boolean isCallback, Map<String, Accumulator<?, ?>> userAccumulators, IOMetrics metrics, boolean releasePartitions, boolean fromSchedulerNg) {
    while (true) {
        ExecutionState current = this.state;
        // 当前状态为 FAILED、CANCELED 或 FINISHED 时，直接返回
        if (current == FAILED) {
            return;
        }
```

```
            if (current == CANCELED || current == FINISHED) {
                return;
            }
            if (current == CANCELING) {
                // 将状态转换为 CANCELED，或者将所有任务重启
                completeCancelling(userAccumulators, metrics, true);
                return;
            }
            ...
             else if (transitionState(current, FAILED, t)) {
                ...
                // 在 ExecutionGraph 中撤销该 Execution
                vertex.getExecutionGraph().deregisterExecution(this);
                ...
                return;
            }
        }
    }
```

其中，completeCancelling()方法如下：

```
    void completeCancelling(Map<String, Accumulator<?, ?>> userAccumulators, IOMetrics metrics, boolean releasePartitions) {
        while (true) {
            ExecutionState current = this.state;
            if (current == CANCELED) {
                return;
            }
            else if (current == CANCELING || current == RUNNING || current == DEPLOYING) {
                if (transitionState(current, CANCELED)) {
                    // 释放资源，在 ExecutionGraph 中撤销该 Execution
                    finishCancellation(releasePartitions);
                    return;
                }
            }
            else {
                if (current != FAILED) {
                    // 重启所有任务
                    vertex.getExecutionGraph().failGlobal(new Exception(message));
                }
                return;
            }
        }
    }
    private void finishCancellation(boolean releasePartitions) {
        releaseAssignedResource(new FlinkException("Execution " + this + " was cancelled."));
        vertex.getExecutionGraph().deregisterExecution(this);
        handlePartitionCleanup(releasePartitions, releasePartitions);
    }
```

可以看到，这里又出现了 deregisterExecution()方法，该方法如下：

```
    void deregisterExecution(Execution exec) {
        Execution contained = currentExecutions.remove(exec.getAttemptId());
        if (contained != null && contained != exec) {
            // 重启所有任务
            failGlobal(new Exception("De-registering execution " + exec + " failed. Found for same ID execution " + contained));
```

            }
        }

该方法从 ExecutionGraph 的 currentExecutions 字段中移除该 Execution 对象。如果遇到异常状况，则重启所有任务。

2. FINISHED

当任务管理器端传回的状态为 FINISHED 时，会调用如下方法：

```
void markFinished(Map<String, Accumulator<?, ?
>> userAccumulators, IOMetrics metrics) {
    while (true) {
        ExecutionState current = this.state;
        if (current == RUNNING || current == DEPLOYING) {
            if (transitionState(current, FINISHED)) {
                try {
                    // 调度下游消费者
                    finishPartitionsAndScheduleOrUpdateConsumers();
                    ...
                    // 在 ExecutionGraph 中撤销该 Execution
                    vertex.getExecutionGraph().deregisterExecution(this);
                }
                finally {
                    // 告知 ExecutionGraph 该任务完成
                    vertex.executionFinished(this);
                }
                return;
            }
        }
        else if (current == CANCELING) {
            // 将状态转换为 CANCELED，或者将所有任务重启
            completeCancelling(userAccumulators, metrics, true);
            return;
        }
        else if (current == CANCELED || current == FAILED) {
            return;
        }
        else {
            // 标记任务失败
            markFailed(new Exception("Vertex received FINISHED message while being in state " + state));
            return;
        }
    }
}
```

上面代码中的许多方法已经介绍过，这里介绍 finishPartitionsAndScheduleOrUpdateConsumers() 方法和 executionFinished()方法。

finishPartitionsAndScheduleOrUpdateConsumers()方法如下：

```
private void finishPartitionsAndScheduleOrUpdateConsumers() {
    final List<IntermediateResultPartition> newlyFinishedResults = getVertex().finishAllBlockingPartitions();
    if (newlyFinishedResults.isEmpty()) {
        return;
```

```java
            }
            final HashSet<ExecutionVertex> consumerDeduplicator = new HashSet<>();
            for (IntermediateResultPartition finishedPartition : newlyFinishedResults) {
                final IntermediateResultPartition[] allPartitionsOfNewlyFinishedResults =
                    finishedPartition.getIntermediateResult().getPartitions();
                for (IntermediateResultPartition partition : allPartitionsOfNewlyFinishedResults) {
                    // 调度或更新下游消费者
                    scheduleOrUpdateConsumers(partition.getConsumers(), consumerDeduplicator);
                }
            }
        }
        List<IntermediateResultPartition> finishAllBlockingPartitions() {
            List<IntermediateResultPartition> finishedBlockingPartitions = null;
            for (IntermediateResultPartition partition : resultPartitions.values()) {
                if (partition.getResultType().isBlocking() && partition.markFinished()) {
                    if (finishedBlockingPartitions == null) {
                        finishedBlockingPartitions = new LinkedList<IntermediateResultPartition>();
                    }
                    finishedBlockingPartitions.add(partition);
                }
            }
            if (finishedBlockingPartitions == null) {
                return Collections.emptyList();
            }
            else {
                return finishedBlockingPartitions;
            }
        }
```

finishAllBlockingPartitions()方法会根据分区是否已经完成以及分区的ResultPartitionType返回符合条件的分区，finishPartitionsAndScheduleOrUpdateConsumers()方法会根据这些返回的分区，找到下游消费者，并对其进行调度。

```java
        private void scheduleOrUpdateConsumers(
            final List<List<ExecutionEdge>> allConsumers,
            final HashSet<ExecutionVertex> consumerDeduplicator) {
            ...
            for (ExecutionEdge edge : allConsumers.get(0)) {
                final ExecutionVertex consumerVertex = edge.getTarget();
                final Execution consumer = consumerVertex.getCurrentExecutionAttempt();
                final ExecutionState consumerState = consumer.getState();
                if (consumerState == CREATED) {
                    ...
                }
                else if (consumerState == DEPLOYING || consumerState == RUNNING) {
                    final PartitionInfo partitionInfo = createPartitionInfo(edge);
                    if (consumerState == DEPLOYING) {
                        consumerVertex.cachePartitionInfo(partitionInfo);
                    } else {
                        consumer.sendUpdatePartitionInfoRpcCall(Collections.singleton(partitionInfo));
                    }
                }
            }
        }
```

对于scheduleOrUpdateConsumers()方法，先前介绍过，当状态为CREATED时，因为调度器为新版调度器，所以不会执行任何操作，但是如果当前状态为 DEPLOYING 或者 RUNNING，则会构建分区信息：

```
private static PartitionInfo createPartitionInfo(ExecutionEdge executionEdge) {
    IntermediateDataSetID intermediateDataSetID = executionEdge.getSource().getIntermediateResult().getId();
    ShuffleDescriptor shuffleDescriptor = getConsumedPartitionShuffleDescriptor(executionEdge, false);
    return new PartitionInfo(intermediateDataSetID, shuffleDescriptor);
}
```

如果当前状态为 DEPLOYING，则会将分区信息添加到 Execution 的 partitionInfos 字段中：

```
void cachePartitionInfo(PartitionInfo partitionInfo){
    getCurrentExecutionAttempt().cachePartitionInfo(partitionInfo);
}
public Execution getCurrentExecutionAttempt() {
    return currentExecution;
}
void cachePartitionInfo(PartitionInfo partitionInfo) {
    partitionInfos.add(partitionInfo);
}
```

如果当前状态不为 RUNNING，则直接将分区信息发送到任务管理器端，构建输入通道并向上游发送请求。

另外，再介绍一下 markFinished() 方法中调用的 ExecutionVertex 的 executionFinished() 方法：

```
void executionFinished(Execution execution) {
    getExecutionGraph().vertexFinished();
}
void vertexFinished() {
    // 使 verticesFinished 字段的值加 1
    final int numFinished = ++verticesFinished;
    // 如果已经完成的任务数等于总任务数，则进入 if 分支
    if (numFinished == numVerticesTotal) {
        if (state == JobStatus.RUNNING) {
            ...
            // 将作业的状态从 RUNNING 转换成 FINISHED
            if (transitionState(JobStatus.RUNNING, JobStatus.FINISHED)) {
                onTerminalState(JobStatus.FINISHED);
            }
        }
    }
}
```

在这个方法中，会给 verticesFinished 字段计数。当已经完成的任务数等于总任务数时，就会将作业状态设置为 FINISHED，并调用 onTerminalState() 方法，该方法主要用于关闭资源。

3. CANCELED

当任务管理器端传回的状态为 CANCELED 时，会调用 completeCancelling() 方法。

4. FAILED

当任务管理器端传回的状态为 FAILED 时，会调用 markFailed() 方法。

5. 其他状态

当任务管理器端传回的状态为除 RUNNING、FINISHED、CANCELED 和 FAILED 以外的状态时，会调用 Execution 的 fail()方法：

```
public void fail(Throwable t) {
    processFail(t, false);
}
private void processFail(Throwable t, boolean isCallback) {
    processFail(t, isCallback, null, null, true, false);
}
```

processFail()方法的逻辑在之前介绍 markFailed()方法时已经介绍过，只不过这里传入的参数值不同。总体而言，其逻辑与 markFailed()方法的逻辑类似。

## 6.7 总结

本章分析了作业调度的逻辑，可以了解到在调度层面有一个调度拓扑 SchedulingTopology 与 ExecutionGraph 对应，整个调度过程是对 SchedulingTopology 对象进行操作的。调度逻辑定义在调度器中，由具体的调度策略实现。调度过程中伴随着作业状态和任务状态的变化。在部署任务或任务状态发生变化时，具体行为最终都由 Execution 对象负责。

# 第 7 章

# 任务的生命周期

执行图中的任务被部署到任务管理器端后,执行图中的各个对象会被映射成任务管理器端的对象,在逻辑关系上会对应物理执行图,如图 7-1 所示。

在部署任务的过程中,Flink 会在作业管理器端将任务相关信息封装成描述符对象——各种 Descriptor 类。这些类均位于 Flink-runtime 模块中。任务管理器端会根据这些"描述符"实例化所需的对象,将任务进行初始化并执行。本章会以此为出发点,对任务的整个生命周期进行分析。

希望在学习完本章后,读者能够了解:

- 任务的提交流程;
- 任务的初始化流程;
- 任务的执行流程;
- AbstractInvokable 实现类的生命周期和重要方法的实现。

## 7.1 任务的提交

任务的提交在 Execution 的 deploy()方法中实现:

```
public void deploy() throws JobException {
  ...
    try {
      ...
      // 构造 TaskDeploymentDescriptor 对象
      final TaskDeploymentDescriptor deployment = TaskDeploymentDescriptorFactory
        .fromExecutionVertex(vertex, attemptNumber)
        .createDeploymentDescriptor(
          slot.getAllocationId(),
          slot.getPhysicalSlotNumber(),
          taskRestore,
          producedPartitions.values());
      final TaskManagerGateway taskManagerGateway = slot.getTaskManagerGateway();
      // 将任务提交到任务管理器端
      CompletableFuture.supplyAsync(() -> taskManagerGateway.submitTask(deployment, rpcTimeout), executor)
        .thenCompose(Function.identity())
        ...
    }
    catch (Throwable t) {
      ...
    }
}
```

# 7.1 任务的提交

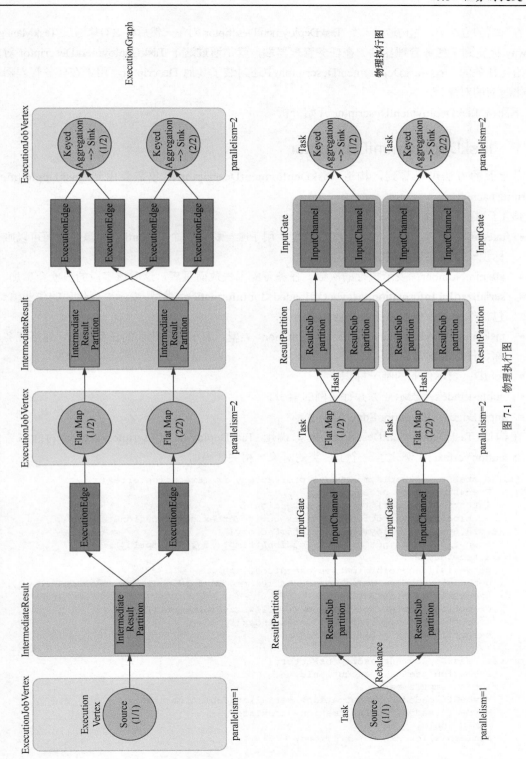

图 7-1 物理执行图

在部署的方法中，先构造了一个 TaskDeploymentDescriptor 对象，然后将该对象通过 TaskManager Gateway 提交到了任务管理器端。在任务管理器端，就是通过这个 TaskDeploymentDescriptor 对象来初始化任务的。在 TaskDeploymentDescriptor 中还封装了其他 Descriptor，用于在任务管理器端初始化对应的对象。

下面从 TaskDeploymentDescriptor 开始分析。

## 7.1.1　TaskDeploymentDescriptor

从上面的方法可以看到，构造 TaskDeploymentDescriptor 对象需要用到 TaskDeploymentDescriptorFactory。

该工厂类的字段如下。

- executionId：ExecutionAttemptID 类型，用于唯一标识一个 Execution 对象，进而可以唯一标识一个要部署的任务。
- attemptNumber：表示执行的次数。任务可能因失败而重启，这时该字段的值就会累加。
- serializedJobInformation：MaybeOffloaded<JobInformation>类型，表示序列化后的作业信息，包括作业 id、作业名称、作业配置等。
- taskInfo：MaybeOffloaded<TaskInformation>类型，表示序列化后的任务信息，包括任务名称、任务配置等。
- jobID：表示作业的唯一标识。
- subtaskIndex：表示任务并行实例的索引。
- inputEdges：ExecutionEdge[][]类型。

在构造 TaskDeploymentDescriptor 时，调用 TaskDeploymentDescriptorFactory 的静态方法 fromExecutionVertex()实例化了一个工厂类的对象。相关代码如下：

```
public static TaskDeploymentDescriptorFactory fromExecutionVertex(
        ExecutionVertex executionVertex,
        int attemptNumber) throws IOException {
    ExecutionGraph executionGraph = executionVertex.getExecutionGraph();
    return new TaskDeploymentDescriptorFactory(
        executionVertex.getCurrentExecutionAttempt().getAttemptId(),
        attemptNumber,
        getSerializedJobInformation(executionGraph),
        getSerializedTaskInformation(executionVertex.getJobVertex().getTaskInformationOrBlobKey()),
        executionGraph.getJobID(),
        executionGraph.getScheduleMode().allowLazyDeployment(),
        executionVertex.getParallelSubtaskIndex(),
        executionVertex.getAllInputEdges());
}
private TaskDeploymentDescriptorFactory(
        ExecutionAttemptID executionId,
        int attemptNumber,
        MaybeOffloaded<JobInformation> serializedJobInformation,
        MaybeOffloaded<TaskInformation> taskInfo,
        JobID jobID,
        boolean allowUnknownPartitions,
```

```
        int subtaskIndex,
        ExecutionEdge[][] inputEdges) {
    this.executionId = executionId;
    this.attemptNumber = attemptNumber;
    this.serializedJobInformation = serializedJobInformation;
    this.taskInfo = taskInfo;
    this.jobID = jobID;
    this.allowUnknownPartitions = allowUnknownPartitions;
    this.subtaskIndex = subtaskIndex;
    this.inputEdges = inputEdges;
}
```

该方法从 ExecutionGraph 中获取作业的信息，同时从 ExecutionVertex 中获取任务的信息。

实例化工厂类的对象后，调用 createDeploymentDescriptor() 方法构造一个 TaskDeploymentDescriptor 对象。该方法如下：

```
public TaskDeploymentDescriptor createDeploymentDescriptor(
        AllocationID allocationID,
        int targetSlotNumber,
        @Nullable JobManagerTaskRestore taskRestore,
        Collection<ResultPartitionDeploymentDescriptor> producedPartitions) {
    return new TaskDeploymentDescriptor(
        jobID,
        serializedJobInformation,
        taskInfo,
        executionId,
        allocationID,
        subtaskIndex,
        attemptNumber,
        targetSlotNumber,
        taskRestore,
        new ArrayList<>(producedPartitions),
        createInputGateDeploymentDescriptors());
}
```

上述代码调用了 TaskDeploymentDescriptor 的构造方法实例化了一个 TaskDeploymentDescriptor 对象。

下面对 TaskDeploymentDescriptor 对象进行分析。它的核心字段如下。

- serializedJobInformation：MaybeOffloaded<JobInformation>类型，表示序列化后的作业信息。
- serializedTaskInformation：MaybeOffloaded<TaskInformation>类型，表示序列化后的任务信息。
- jobId：表示作业的唯一标识。
- executionId：ExecutionAttemptID 类型，用于唯一标识一个 Execution 对象，进而可以唯一标识一个要部署的任务。
- subtaskIndex：表示任务并行实例的索引。
- attemptNumber：表示执行的次数。因为任务可能因失败而重启，这时该字段的值就会累加。
- producedPartitions：Collection<ResultPartitionDeploymentDescriptor>类型，用于在任务管理器端构造 ResultPartitionWriter。
- inputGates：Collection<InputGateDeploymentDescriptor\>类型，用于在任务管理器端构造输入网关（InputGate），负责数据的输入。

- taskRestore：JobManagerTaskRestore 类型，表示重启任务所需的信息。

TaskDeploymentDescriptor 的构造方法会对这些字段进行赋值。

其中，resultPartitionDeploymentDescriptors 的值来自 Execution 中的 producedPartitions 字段，inputGates 的值来自 createInputGateDeploymentDescriptors()方法。下面分别对这两个类及它们的构造过程进行分析。

## 7.1.2　ResultPartitionDeploymentDescriptor

ResultPartitionDeploymentDescriptor 对象用于在任务管理器端构造 ResultPartitionWriter。构造 ResultPartitionWriter 在部署任务的过程中实现，下面以 deployIndividually()方法为例进行介绍：

```
private void deployIndividually(final List<DeploymentHandle> deploymentHandles) {
    for (final DeploymentHandle deploymentHandle : deploymentHandles) {
        FutureUtils.assertNoException(
            deploymentHandle
                .getSlotExecutionVertexAssignment()
                .getLogicalSlotFuture()
                .handle(assignResourceOrHandleError(deploymentHandle))
                .handle(deployOrHandleError(deploymentHandle)));
    }
}
```

其中，**assignResourceOrHandleError()方法如下：**

```
private BiFunction<LogicalSlot, Throwable, Void> assignResourceOrHandleError(final DeploymentHandle deploymentHandle) {
    ...
    return (logicalSlot, throwable) -> {
        ...
        if (throwable == null) {
            final ExecutionVertex executionVertex = getExecutionVertex(executionVertexId);
            final boolean sendScheduleOrUpdateConsumerMessage = deploymentHandle.getDeploymentOption().sendScheduleOrUpdateConsumerMessage();
            executionVertex
                .getCurrentExecutionAttempt()
                // 注册 producedPartitions
                .registerProducedPartitions(logicalSlot.getTaskManagerLocation(), sendScheduleOrUpdateConsumerMessage);
            executionVertex.tryAssignResource(logicalSlot);
        } else {
            handleTaskDeploymentFailure(executionVertexId, maybeWrapWithNoResourceAvailableException(throwable));
        }
        return null;
    };
}
```

当槽分配好以后，会调用 registerProducedPartitions()方法构造 ResultPartitionDeploymentDescriptor 对象。

```
public CompletableFuture<Execution> registerProducedPartitions(
    TaskManagerLocation location,
    boolean sendScheduleOrUpdateConsumersMessage) {
```

```java
    return FutureUtils.thenApplyAsyncIfNotDone(
        // 注册 producedPartitions
        registerProducedPartitions(vertex, location, attemptId, sendScheduleOrUpdateConsumersMessage),
        vertex.getExecutionGraph().getJobMasterMainThreadExecutor(),
        producedPartitionsCache -> {
            // 给 producedPartitions 字段赋值
            producedPartitions = producedPartitionsCache;
            startTrackingPartitions(location.getResourceID(), producedPartitionsCache.values());
            return this;
        });
}
static CompletableFuture<Map<IntermediateResultPartitionID, ResultPartitionDeploymentDescriptor>> registerProducedPartitions(
        ExecutionVertex vertex,
        TaskManagerLocation location,
        ExecutionAttemptID attemptId,
        boolean sendScheduleOrUpdateConsumersMessage) {
    // 构造 ProducerDescriptor 对象
    ProducerDescriptor producerDescriptor = ProducerDescriptor.create(location, attemptId);
    Collection<IntermediateResultPartition> partitions = vertex.getProducedPartitions().values();
    Collection<CompletableFuture<ResultPartitionDeploymentDescriptor>> partitionRegistrations =
        new ArrayList<>(partitions.size());
    for (IntermediateResultPartition partition : partitions) {
        // 构造 PartitionDescriptor 对象
        PartitionDescriptor partitionDescriptor = PartitionDescriptor.from(partition);
        int maxParallelism = getPartitionMaxParallelism(partition);
        // 构造 ShuffleDescriptor 的 CompletableFuture 对象
        CompletableFuture<? extends ShuffleDescriptor> shuffleDescriptorFuture = vertex
            .getExecutionGraph()
            .getShuffleMaster()
            .registerPartitionWithProducer(partitionDescriptor, producerDescriptor);
        // 构造 ResultPartitionDeploymentDescriptor 的 CompletableFuture 对象
        CompletableFuture<ResultPartitionDeploymentDescriptor> partitionRegistration = shuffleDescriptorFuture
            .thenApply(shuffleDescriptor -> new ResultPartitionDeploymentDescriptor(
                partitionDescriptor,
                shuffleDescriptor,
                maxParallelism,
                sendScheduleOrUpdateConsumersMessage));
        partitionRegistrations.add(partitionRegistration);
    }
    return FutureUtils.combineAll(partitionRegistrations).thenApply(rpdds -> {
        Map<IntermediateResultPartitionID, ResultPartitionDeploymentDescriptor> producedPartitions =
            new LinkedHashMap<>(partitions.size());
        rpdds.forEach(rpdd -> producedPartitions.put(rpdd.getPartitionId(), rpdd));
        return producedPartitions;
    });
}
```

从上面的代码可以看出，整个过程是在构造参数传入 ResultPartitionDeploymentDescriptor 的构造方法中实现的，以此生成 ResultPartitionDeploymentDescriptor 对象。

ResultPartitionDeploymentDescriptor 的字段如下。

- partitionDescriptor：PartitionDescriptor 类型，表示分区的描述符，用于构造 ShuffleDescriptor 对象。
- shuffleDescriptor：ShuffleDescriptor 类型，用于构造输入通道。
- maxParallelism：表示最大并行度。
- sendScheduleOrUpdateConsumersMessage：用于设置任务是否会在分区可被消费时就开始通知作业管理器调度下游任务。

ResultPartitionDeploymentDescriptor 的构造方法如下：

```
public ResultPartitionDeploymentDescriptor(
    PartitionDescriptor partitionDescriptor,
    ShuffleDescriptor shuffleDescriptor,
    int maxParallelism,
    boolean sendScheduleOrUpdateConsumersMessage) {
  this.partitionDescriptor = checkNotNull(partitionDescriptor);
  this.shuffleDescriptor = checkNotNull(shuffleDescriptor);
  this.maxParallelism = maxParallelism;
  this.sendScheduleOrUpdateConsumersMessage = sendScheduleOrUpdateConsumersMessage;
}
```

其中的 PartitionDescriptor 和 ShuffleDescriptor 下文会依次介绍。

## 7.1.3　InputGateDeploymentDescriptor

在 TaskDeploymentDescriptor 的构造过程中，还构造了 InputGateDeploymentDescriptor。InputGateDeploymentDescriptor 用于在任务管理器端构造 InputGate。InputGateDeploymentDescriptor 的构造过程调用了 TaskDeploymentDescriptorFactory 的 createInputGateDeploymentDescriptors()方法：

```
private List<InputGateDeploymentDescriptor> createInputGateDeploymentDescriptors() {
    List<InputGateDeploymentDescriptor> inputGates = new ArrayList<>(inputEdges.length);
    for (ExecutionEdge[] edges : inputEdges) {
        // （1）计算 queueToRequest
        int numConsumerEdges = edges[0].getSource().getConsumers().get(0).size();
        int queueToRequest = subtaskIndex % numConsumerEdges;
        // （2）获取 IntermediateDataSetID 和 ResultPartitionType
        IntermediateResult consumedIntermediateResult = edges[0].getSource().
getIntermediateResult();
        IntermediateDataSetID resultId = consumedIntermediateResult.getId();
        ResultPartitionType partitionType = consumedIntermediateResult.getResultType();
        // （3）构造 InputGateDeploymentDescriptor
        inputGates.add(new InputGateDeploymentDescriptor(
            resultId,
            partitionType,
            queueToRequest,
            getConsumedPartitionShuffleDescriptors(edges)));
    }
    return inputGates;
}
```

从 for 循环遍历的对象可以看出，InputGateDeploymentDescriptor 的构造过程所用到的信息全部来自 inputEdges 字段。该字段的类型为 ExecutionEdge[][]，从前文可了解到 ExecutionEdge 是维护上下游关系的一环。

下面仔细分析 for 循环中的逻辑。

(1) 计算 queueToRequest。

(2) 先获取 IntermediateResult 对象,再从中获取 IntermediateDataSetID 和 ResultPartitionType,并将它们作为构造 InputGateDeploymentDescriptor 的参数。

(3) 构造 InputGateDeploymentDescriptor。

下游每个任务并行实例的所有接收同一个上游输入的 ExecutionEdge 对象,在任务管理器端会属于同一个输入网关(InputGate)。同一个 InputGate 中的输入通道(InputChannel)虽然会从上游不同的任务并行实例读取数据,但是无论读取哪个结果分区(ResultPartition)的数据,读取的都是同一个索引位置的结果子分区(ResultSubpartition)。queueToRequest 表示的就是该索引。

请看下面的计算过程:

```
int numConsumerEdges = edges[0].getSource().getConsumers().get(0).size();
```

这里取 edges[0]没有什么特殊的含义,在数组不越界的情况下,取其他索引值也没有区别,这里主要是想通过 ExecutionEdge 对象获取其 source 字段,即其上游的 IntermediateResultPartition 对象,进而通过该对象获取下游的所有消费者,最终获取消费同一个 IntermediateResultPartition 的消费者的个数。得到这个个数以后,通过该并行实例的索引,就可以计算出在任务管理器端对应的 InputGate 应该读取哪个 ResultSubpartition 的数据。

```
int queueToRequest = subtaskIndex % numConsumerEdges;
```

InputGateDeploymentDescriptor 的字段如下。

- consumedResultId:IntermediateDataSetID 类型,表示 InputGateDeploymentDescriptor 对应的 InputGate 要消费的上游的 IntermediateDataSet(ExecutionGraph 中的 IntermediateResult)。
- consumedPartitionType:ResultPartitionType 类型,表示上游分区的 ResultPartitionType 类型。
- consumedSubpartitionIndex:表示 InputGateDeploymentDescriptor 对应的 InputGate 消费的上游每个 ResultPartition 中的 ResultSubpartition 的索引,这个值就来自上面计算出的 queueToRequest。
- inputChannels:ShuffleDescriptor[]类型,用于在任务管理器端初始化 InputChannel。

InputGateDeploymentDescriptor 的构造方法如下:

```
public InputGateDeploymentDescriptor(
    IntermediateDataSetID consumedResultId,
    ResultPartitionType consumedPartitionType,
    @Nonnegative int consumedSubpartitionIndex,
    ShuffleDescriptor[] inputChannels) {
  this.consumedResultId = checkNotNull(consumedResultId);
  this.consumedPartitionType = checkNotNull(consumedPartitionType);
  this.consumedSubpartitionIndex = consumedSubpartitionIndex;
  this.inputChannels = checkNotNull(inputChannels);
}
```

## 7.1.4 ShuffleDescriptor

在 InputGateDeploymentDescriptor 的构造过程中,还构造了 ShuffleDescriptor。ShuffleDescriptor

用于在任务管理器端构造 InputChannel。ShuffleDescriptor 的构造过程会调用 TaskDeploymentDescriptorFactory 的 getConsumedPartitionShuffleDescriptors() 方法：

```
private ShuffleDescriptor[] getConsumedPartitionShuffleDescriptors(ExecutionEdge[] edges) {
    ShuffleDescriptor[] shuffleDescriptors = new ShuffleDescriptor[edges.length];
    for (int i = 0; i < edges.length; i++) {
        shuffleDescriptors[i] =
            getConsumedPartitionShuffleDescriptor(edges[i], allowUnknownPartitions);
    }
    return shuffleDescriptors;
}
```

可以观察到，ShuffleDescriptor 的构造过程基本上只用到了 ExecutionEdge 的信息。参数 allowUnknownPartitions 的值来自作业的 ScheduleMode。

每一个 ShuffleDescriptor 对象的构造过程都会调用 getConsumedPartitionShuffleDescriptor() 方法：

```
public static ShuffleDescriptor getConsumedPartitionShuffleDescriptor(
    ExecutionEdge edge,
    boolean allowUnknownPartitions) {
    // 获取上游的 IntermediateResultPartition 对象
    IntermediateResultPartition consumedPartition = edge.getSource();
    // 获取上游当前的 Execution 对象
    Execution producer = consumedPartition.getProducer().getCurrentExecutionAttempt();
    // 获取上游任务的执行状态
    ExecutionState producerState = producer.getState();
    // 根据唯一标识获取上游的 ResultPartitionDeploymentDescriptor 的 Optional 对象
    Optional<ResultPartitionDeploymentDescriptor> consumedPartitionDescriptor =
        producer.getResultPartitionDeploymentDescriptor(consumedPartition.getPartitionId());
    // 构造 ResultPartitionID
    ResultPartitionID consumedPartitionId = new ResultPartitionID(
        consumedPartition.getPartitionId(),
        producer.getAttemptId());
    // 根据上面的变量构造 ShuffleDescriptor 对象
    return getConsumedPartitionShuffleDescriptor(
        consumedPartitionId,
        consumedPartition.getResultType(),
        consumedPartition.isConsumable(),
        producerState,
        allowUnknownPartitions,
        consumedPartitionDescriptor.orElse(null));
}
```

之前已详细介绍过 ExecutionGraph 中各个对象的含义及上下游关系。上述代码中 consumedPartitionDescriptor 的值是这样获取的：

```
public Optional<ResultPartitionDeploymentDescriptor> getResultPartitionDeploymentDescriptor(
    IntermediateResultPartitionID id) {
    return Optional.ofNullable(producedPartitions.get(id));
}
```

其值来自之前介绍过的 Execution 中的 producedPartitions 字段。

ResultPartitionID 表示 ResultPartition 的唯一标识，它有如下字段：

```java
public final class ResultPartitionID implements Serializable {
    private final IntermediateResultPartitionID partitionId;
    private final ExecutionAttemptID producerId;
```

**ResultPartitionID** 的构造方法如下：

```java
public ResultPartitionID(IntermediateResultPartitionID partitionId, ExecutionAttemptID producerId) {
    this.partitionId = checkNotNull(partitionId);
    this.producerId = checkNotNull(producerId);
}
```

**getConsumedPartitionShuffleDescriptor()方法如下：**

```java
static ShuffleDescriptor getConsumedPartitionShuffleDescriptor(
    ResultPartitionID consumedPartitionId,
    ResultPartitionType resultPartitionType,
    boolean isConsumable,
    ExecutionState producerState,
    boolean allowUnknownPartitions,
    @Nullable ResultPartitionDeploymentDescriptor consumedPartitionDescriptor) {
    if ((resultPartitionType.isPipelined() || isConsumable) &&
        consumedPartitionDescriptor != null &&
        isProducerAvailable(producerState)) {
        return consumedPartitionDescriptor.getShuffleDescriptor();
    }
    else if (allowUnknownPartitions) {
        return new UnknownShuffleDescriptor(consumedPartitionId);

    }
    else {
        ...
    }
}
```

该方法根据不同的判断条件，或者会从 consumedPartitionDescriptor 中返回一个 NettyShuffleDescriptor 对象，或者会实例化一个 UnknownShuffleDescriptor 对象。这两者都是 ShuffleDescriptor 接口的实现类。

其中，NettyShuffleDescriptor 的构造过程在前文的 registerProducedPartitions() 方法中介绍过：

```java
static CompletableFuture<Map<IntermediateResultPartitionID, ResultPartitionDeployment
Descriptor>> registerProducedPartitions(
    ExecutionVertex vertex,
    TaskManagerLocation location,
    ExecutionAttemptID attemptId,
    boolean sendScheduleOrUpdateConsumersMessage) {
    ...
    for (IntermediateResultPartition partition : partitions) {
        PartitionDescriptor partitionDescriptor = PartitionDescriptor.from(partition);
        int maxParallelism = getPartitionMaxParallelism(partition);
        CompletableFuture<? extends ShuffleDescriptor> shuffleDescriptorFuture = vertex
            .getExecutionGraph()
            .getShuffleMaster()
            .registerPartitionWithProducer(partitionDescriptor, producerDescriptor);
        ...
    }
    ...
}
```

这里从 ExecutionGraph 中获取了 ShuffleMaster 对象，然后调用其 registerPartitionWithProducer() 方法构造 ShuffleDescriptor：

```
public CompletableFuture<NettyShuffleDescriptor> registerPartitionWithProducer(
    PartitionDescriptor partitionDescriptor,
    ProducerDescriptor producerDescriptor) {
  ResultPartitionID resultPartitionID = new ResultPartitionID(
    partitionDescriptor.getPartitionId(),
    producerDescriptor.getProducerExecutionId());
  NettyShuffleDescriptor shuffleDeploymentDescriptor = new NettyShuffleDescriptor(
    producerDescriptor.getProducerLocation(),
    createConnectionInfo(producerDescriptor, partitionDescriptor.getConnectionIndex()),
    resultPartitionID);
  return CompletableFuture.completedFuture(shuffleDeploymentDescriptor);
}
```

下面分别对 NettyShuffleDescriptor 和 UnknownShuffleDescriptor 的字段和方法进行介绍。

（1）NettyShuffleDescriptor 的字段如下。

- producerLocation：ResourceID 类型。
- partitionConnectionInfo：PartitionConnectionInfo 类型，表示与上游分区的连接信息。
- resultPartitionID：ResultPartitionID 类型。

其构造方法如下：

```
public NettyShuffleDescriptor(
    ResourceID producerLocation,
    PartitionConnectionInfo partitionConnectionInfo,
    ResultPartitionID resultPartitionID) {
  this.producerLocation = producerLocation;
  this.partitionConnectionInfo = partitionConnectionInfo;
  this.resultPartitionID = resultPartitionID;
}
```

根据 registerPartitionWithProducer()方法，可观察到 partitionConnectionInfo 的构造会调用 createConnectionInfo()方法：

```
private static PartitionConnectionInfo createConnectionInfo(
    ProducerDescriptor producerDescriptor,
    int connectionIndex) {
  return producerDescriptor.getDataPort() >= 0 ?
    NetworkPartitionConnectionInfo.fromProducerDescriptor(producerDescriptor, connectionIndex) :
    LocalExecutionPartitionConnectionInfo.INSTANCE;
}
```

这里会根据 producerDescriptor 的 dataPort 字段决定返回表示远程连接信息的 NetworkPartitionConnectionInfo 对象或是表示本地连接信息的 LocalExecutionPartitionConnectionInfo 对象。它们都是 NettyShuffleDescriptor 的内部类。

NetworkPartitionConnectionInfo 的构造过程如下：

```
static NetworkPartitionConnectionInfo fromProducerDescriptor(
    ProducerDescriptor producerDescriptor,
    int connectionIndex) {
  InetSocketAddress address =
```

```
            new InetSocketAddress(producerDescriptor.getAddress(), producerDescriptor.
getDataPort());
        return new NetworkPartitionConnectionInfo(new ConnectionID(address, connectionIndex));
    }
```

LocalExecutionPartitionConnectionInfo 中因为不需要进行远程连接，所以没有任何信息。

（2）UnknownShuffleDescriptor 的字段如下。

resultPartitionID：ResultPartitionID 类型。

其构造方法如下：

```
public UnknownShuffleDescriptor(ResultPartitionID resultPartitionID) {
    this.resultPartitionID = resultPartitionID;
}
```

## 7.1.5　ProducerDescriptor 和 PartitionDescriptor

前文介绍了 TaskDeploymentDescriptor、ResultPartitionDeploymentDescriptor、InputGateDeploymentDescriptor 和 ShuffleDescriptor，它们分别对应任务管理器端构造任务、结果分区（ResultPartition）、输入网关（InputGate）和输入通道（InputChannel）的描述符。下面介绍的两个类 ProducerDescriptor 和 PartitionDescriptor 并不会直接对应任务管理器端的某个对象，但它们会参与 ResultPartitionDeploymentDescriptor 和 ShuffleDescriptor 的构造。其中，ProducerDescriptor 大致封装了与上游的连接信息，所以其会参与 ShuffleDescriptor 的构造；PartitionDescriptor 大致封装了上游分区的基本信息，所以其会参与 ResultPartitionDeploymentDescriptor 和 ShuffleDescriptor 的构造。

ProducerDescriptor 的主要字段如下。

- producerLocation：ResourceID 类型。
- producerExecutionId：ExecutionAttemptID 类型。
- address：InetAddress 类型，表示上游连接的地址。
- dataPort：上游连接的端口号。

其构造方法如下：

```
public ProducerDescriptor(
        ResourceID producerLocation,
        ExecutionAttemptID producerExecutionId,
        InetAddress address,
        int dataPort) {
    this.producerLocation = checkNotNull(producerLocation);
    this.producerExecutionId = checkNotNull(producerExecutionId);
    this.address = checkNotNull(address);
    this.dataPort = dataPort;
}
```

ProducerDescriptor 的构造过程在 registerProducedPartitions()方法中实现：

```
    static CompletableFuture<Map<IntermediateResultPartitionID, ResultPartitionDeployment
Descriptor>> registerProducedPartitions(
        ExecutionVertex vertex,
        TaskManagerLocation location,
```

```
      ExecutionAttemptID attemptId,
      boolean sendScheduleOrUpdateConsumersMessage) {
    ProducerDescriptor producerDescriptor = ProducerDescriptor.create(location, attemptId);
  ...
    for (IntermediateResultPartition partition : partitions) {
      ...
      CompletableFuture<? extends ShuffleDescriptor> shuffleDescriptorFuture = vertex
          .getExecutionGraph()
          .getShuffleMaster()
          .registerPartitionWithProducer(partitionDescriptor, producerDescriptor);
      ...
    }
  ...
}
  public static ProducerDescriptor create(TaskManagerLocation producerLocation,
ExecutionAttemptID attemptId) {
    return new ProducerDescriptor(
      producerLocation.getResourceID(),
      attemptId,
      producerLocation.address(),
      producerLocation.dataPort());
}
```

在 registerPartitionWithProducer()方法中传入 producerDescriptor 对象：

```
public CompletableFuture<NettyShuffleDescriptor> registerPartitionWithProducer(
    PartitionDescriptor partitionDescriptor,
    ProducerDescriptor producerDescriptor) {
  ResultPartitionID resultPartitionID = new ResultPartitionID(
    partitionDescriptor.getPartitionId(),
    producerDescriptor.getProducerExecutionId());
  NettyShuffleDescriptor shuffleDeploymentDescriptor = new NettyShuffleDescriptor(
    producerDescriptor.getProducerLocation(),
    createConnectionInfo(producerDescriptor, partitionDescriptor.getConnectionIndex()),
    resultPartitionID);
  return CompletableFuture.completedFuture(shuffleDeploymentDescriptor);
}
```

从上述代码可知，取出了 producerDescriptor 中的各个字段参与 ShuffleDescriptor 的构造。同时还可以看到取出了 partitionDescriptor 中的字段构造了 ShuffleDescriptor。

PartitionDescriptor 的主要字段如下。

- resultId：IntermediateDataSetID 类型。
- totalNumberOfPartitions：表示总的分区数。
- partitionId：IntermediateResultPartitionID 类型。
- partitionType：ResultPartitionType 类型。
- numberOfSubpartitions：表示总的子分区数。

其构造方法如下：

```
public PartitionDescriptor(
    IntermediateDataSetID resultId,
    int totalNumberOfPartitions,
    IntermediateResultPartitionID partitionId,
```

```java
        ResultPartitionType partitionType,
        int numberOfSubpartitions,
        int connectionIndex) {
    this.resultId = checkNotNull(resultId);
    this.totalNumberOfPartitions = totalNumberOfPartitions;
    this.partitionId = checkNotNull(partitionId);
    this.partitionType = checkNotNull(partitionType);
    this.numberOfSubpartitions = numberOfSubpartitions;
    this.connectionIndex = connectionIndex;
}
```

PartitionDescriptor 的构造过程在 registerProducedPartitions()方法中实现:

```java
static CompletableFuture<Map<IntermediateResultPartitionID, ResultPartitionDeploymentDescriptor>> registerProducedPartitions(
        ExecutionVertex vertex,
        TaskManagerLocation location,
        ExecutionAttemptID attemptId,
        boolean sendScheduleOrUpdateConsumersMessage) {
    ...
    for (IntermediateResultPartition partition : partitions) {
        PartitionDescriptor partitionDescriptor = PartitionDescriptor.from(partition);
        int maxParallelism = getPartitionMaxParallelism(partition);
        CompletableFuture<? extends ShuffleDescriptor> shuffleDescriptorFuture = vertex
            .getExecutionGraph()
            .getShuffleMaster()
            .registerPartitionWithProducer(partitionDescriptor, producerDescriptor);
        CompletableFuture<ResultPartitionDeploymentDescriptor> partitionRegistration = shuffleDescriptorFuture
            .thenApply(shuffleDescriptor -> new ResultPartitionDeploymentDescriptor(
                partitionDescriptor,
                shuffleDescriptor,
                maxParallelism,
                sendScheduleOrUpdateConsumersMessage));
        partitionRegistrations.add(partitionRegistration);
    }
    ...
}
public static PartitionDescriptor from(IntermediateResultPartition partition) {
    int numberOfSubpartitions = 1;
    List<List<ExecutionEdge>> consumers = partition.getConsumers();
    if (!consumers.isEmpty() && !consumers.get(0).isEmpty()) {
        // 当前每个分区只支持被一个消费者消费
        if (consumers.size() > 1) {
            throw new IllegalStateException("Currently, only a single consumer group per partition is supported.

");
        }
        numberOfSubpartitions = consumers.get(0).size();
    }
    IntermediateResult result = partition.getIntermediateResult();
    return new PartitionDescriptor(
        result.getId(),
        partition.getIntermediateResult().getNumberOfAssignedPartitions(),
```

```
        partition.getPartitionId(),
        result.getResultType(),
        numberOfSubpartitions,
        result.getConnectionIndex());
}
```

前文已介绍过在 registerPartitionWithProducer()方法中一方面取出 PartitionDescriptor 对象中的字段来构造 ShuffleDescriptor。另一方面，partitionDescriptor 直接作为参数被传入 ResultPartitionDeploymentDescriptor 的构造方法，用于给 ResultPartitionDeploymentDescriptor 的 partitionDescriptor 字段赋值。

### 7.1.6　TaskDeploymentDescriptor 的提交

在 Execution 的 deploy()方法中，构造了 TaskDeploymentDescriptor 对象，通过 Rpc 调用将该对象传输到任务管理器端，最终调用任务管理器的 submitTask()方法：

```
public CompletableFuture<Acknowledge> submitTask(
    TaskDeploymentDescriptor tdd,
    JobMasterId jobMasterId,
    Time timeout) {
  try {
    ...
    // 构造 Task 对象
    Task task = new Task(
        jobInformation,
        taskInformation,
        tdd.getExecutionAttemptId(),
        tdd.getAllocationId(),
        tdd.getSubtaskIndex(),
        tdd.getAttemptNumber(),
        tdd.getProducedPartitions(),
        tdd.getInputGates(),
        tdd.getTargetSlotNumber(),
        memoryManager,
        taskExecutorServices.getIOManager(),
        taskExecutorServices.getShuffleEnvironment(),
        taskExecutorServices.getKvStateService(),
        taskExecutorServices.getBroadcastVariableManager(),
        taskExecutorServices.getTaskEventDispatcher(),
        taskStateManager,
        taskManagerActions,
        inputSplitProvider,
        checkpointResponder,
        aggregateManager,
        blobCacheService,
        libraryCache,
        fileCache,
        taskManagerConfiguration,
        taskMetricGroup,
        resultPartitionConsumableNotifier,
        partitionStateChecker,
        getRpcService().getExecutor());
```

```
        ...
        if (taskAdded) {
            // 启动 Task
            task.startTaskThread();
            ...
        } else {
            ...
        }
    } catch (TaskSubmissionException e) {
        ...
    }
}
```

在该方法中，其中一个主要逻辑就是获取 TaskDeploymentDescriptor 对象中的字段，将其作为参数传入 Task 的构造方法中用于构造 Task 对象。

## 7.2 任务的初始化

Task 表示任务的并行实例。从前面介绍的物理执行图的角度来看，任务的初始化还包括其结果分区（ResultPartition）和输入网关（InputGate）的初始化。

### 7.2.1 Task 的初始化

并行实例的执行需要考虑资源管理、内存管理等复杂问题，但本节将重点放在前面介绍过的 XXXDescriptor 所对应的对象的初始化上。下面是 Task 中的一些字段。

- jobId：JobID 类型，表示作业的唯一标识。
- vertexId：JobVertexID 类型，用于设置该并行实例属于哪个 JobVertex。
- executionId：ExecutionAttemptID 类型，表示该任务对应的 Execution 对象的唯一标识。
- taskInfo：TaskInfo 类型，表示该任务的基本信息，如任务名称等。
- taskNameWithSubtask：表示包含子任务索引的任务名称。
- jobConfiguration：Configuration 类型，表示作业的配置。
- taskConfiguration：Configuration 类型，表示任务的配置。
- nameOfInvokableClass：表示 AbstractInvokable 的实现类的类名。
- consumableNotifyingPartitionWriters：ResultPartitionWriter[]类型，表示并行实例的输出。
- inputGates：InputGate[]类型，表示并行实例的输入。
- invokable：AbstractInvokable 类型，封装该任务的具体执行逻辑。
- executionState：ExecutionState 类型。

可以观察到 Task 依赖 ResultPartitionWriter 和 InputGate，它们分别负责消息的输出和输入。此外，Task 还依赖 AbstractInvokable 类，AbstractInvokable 类的实现类包括 StreamTask 和 BatchTask 等类，它们在某种程度上分别代表流任务和批任务的任务类，其中会封装具体的算子。Task 类与相关类的关系如图 7-2 所示。

# 第 7 章 任务的生命周期

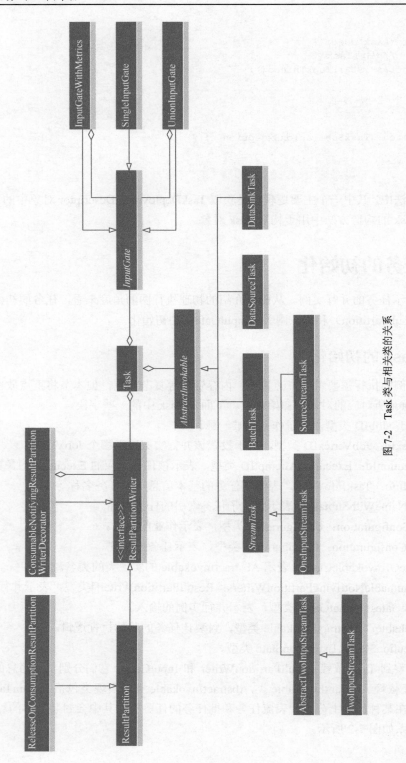

图 7-2 Task 类与相关类的关系

Task 的构造方法主要负责字段的赋值，同时涉及 ResultPartitionWriter 和 InputGate 的构造。

## 7.2.2　ResultPartition 的初始化

ResultPartitionWriter 的构造会用到 ResultPartitionFactory。在当前的架构中，ResultPartitionWriter 的实现类都继承自实现了该接口的抽象类 ResultPartition。

ResultPartitionFactory 的构造在任务管理器的启动过程中实现，可以直接调用其构造方法。ResultPartitionFactory 有以下与构造 ResultPartition 直接相关的字段。

- blockingSubpartitionType：BoundedBlockingSubpartitionType 类型，该类型是枚举类，不同的类型用于构造不同的（ResultSubpartition）。
- forcePartitionReleaseOnConsumption：该字段的值来自配置 taskmanager.network.partition.force-release-on-consumption，用于设置是否强制构造 ReleaseOnConsumptionResultPartition 类型的 ResultPartitionWriter。

ResultPartitionFactory 的构造方法如下：

```
public ResultPartitionFactory(
    ResultPartitionManager partitionManager,
    FileChannelManager channelManager,
    BufferPoolFactory bufferPoolFactory,
    BoundedBlockingSubpartitionType blockingSubpartitionType,
    int networkBuffersPerChannel,
    int floatingNetworkBuffersPerGate,
    int networkBufferSize,
    boolean forcePartitionReleaseOnConsumption,
    boolean blockingShuffleCompressionEnabled,
    String compressionCodec) {
    this.partitionManager = partitionManager;
    this.channelManager = channelManager;
    this.networkBuffersPerChannel = networkBuffersPerChannel;
    this.floatingNetworkBuffersPerGate = floatingNetworkBuffersPerGate;
    this.bufferPoolFactory = bufferPoolFactory;
    this.blockingSubpartitionType = blockingSubpartitionType;
    this.networkBufferSize = networkBufferSize;
    this.forcePartitionReleaseOnConsumption = forcePartitionReleaseOnConsumption;
    this.blockingShuffleCompressionEnabled = blockingShuffleCompressionEnabled;
    this.compressionCodec = compressionCodec;
}
```

上述代码中还涉及一些与内存管理、压缩等有关的字段。

在 ResultPartitionFactory 中定义创建 ResultPartitionWriter 的方法 create()：

```
public ResultPartition create(
        String taskNameWithSubtaskAndId,
        ResultPartitionDeploymentDescriptor desc) {
    // 从 ResultPartitionDeploymentDescriptor 对象中取出一些信息用于构造 ResultPartition 和 ResultSubpartition
    return create(
        taskNameWithSubtaskAndId,
        desc.getShuffleDescriptor().getResultPartitionID(),
        desc.getPartitionType(),
        desc.getNumberOfSubpartitions(),
        desc.getMaxParallelism(),
```

```
        createBufferPoolFactory(desc.getNumberOfSubpartitions(), desc.getPartitionType()));
}
public ResultPartition create(
    String taskNameWithSubtaskAndId,
    ResultPartitionID id,
    ResultPartitionType type,
    int numberOfSubpartitions,
    int maxParallelism,
    FunctionWithException<BufferPoolOwner, BufferPool, IOException> bufferPoolFactory) {
...
    ResultSubpartition[] subpartitions = new ResultSubpartition[numberOfSubpartitions];
    // 创建 ResultPartition
    ResultPartition partition = forcePartitionReleaseOnConsumption || !type.isBlocking()
        ? new ReleaseOnConsumptionResultPartition(
            taskNameWithSubtaskAndId,
            id,
            type,
            subpartitions,
            maxParallelism,
            partitionManager,
            bufferCompressor,
            bufferPoolFactory)
        : new ResultPartition(
            taskNameWithSubtaskAndId,
            id,
            type,
            subpartitions,
            maxParallelism,
            partitionManager,
            bufferCompressor,
            bufferPoolFactory);
    // 创建 ResultSubpartition
    createSubpartitions(partition, type, blockingSubpartitionType, subpartitions);
    return partition;
}
```

在创建 ResultPartition 的过程中，会根据 forcePartitionReleaseOnConsumption 字段和 ResultPartitionType 对象的值决定具体的实现类。

另外，还会调用 createSubpartitions() 方法创建每个 ResultPartition 中的 ResultSubpartition：

```
private void createSubpartitions(
    ResultPartition partition,
    ResultPartitionType type,
    BoundedBlockingSubpartitionType blockingSubpartitionType,
    ResultSubpartition[] subpartitions) {
    if (type.isBlocking()) {
        initializeBoundedBlockingPartitions(
            subpartitions,
            partition,
            blockingSubpartitionType,
            networkBufferSize,
            channelManager);
    } else {
        for (int i = 0; i < subpartitions.length; i++) {
            subpartitions[i] = new PipelinedSubpartition(i, partition);
        }
    }
}
```

```java
    }
    private static void initializeBoundedBlockingPartitions(
        ResultSubpartition[] subpartitions,
        ResultPartition parent,
        BoundedBlockingSubpartitionType blockingSubpartitionType,
        int networkBufferSize,
        FileChannelManager channelManager) {
        int i = 0;
        try {
            for (i = 0; i < subpartitions.length; i++) {
                final File spillFile = channelManager.createChannel().getPathFile();
                subpartitions[i] = blockingSubpartitionType.create(i, parent, spillFile, networkBufferSize);
            }
        }
        catch (IOException e) {
            ...
        }
    }
```

总体而言，要根据 ResultPartitionType 的值和 BoundedBlockingSubpartitionType 的值构造具体的 ResultSubpartition 实现类。

从之前介绍的 Task 的初始化过程可观察到，ResultPartition 的初始化会调用 NettyShuffleEnvironment 的 createResultPartitionWriters() 方法：

```java
public Collection<ResultPartition> createResultPartitionWriters(
        ShuffleIOOwnerContext ownerContext,
        Collection<ResultPartitionDeploymentDescriptor> resultPartitionDeploymentDescriptors) {
    synchronized (lock) {
        ResultPartition[] resultPartitions = new ResultPartition[resultPartitionDeploymentDescriptors.size()];
        int counter = 0;
        for (ResultPartitionDeploymentDescriptor rpdd : resultPartitionDeploymentDescriptors) {
            // 以 ResultPartitionDeploymentDescriptor 对象作为参数，调用 ResultPartitionFactory 的
            //create()方法构造 ResultPartition
            resultPartitions[counter++] = resultPartitionFactory.create(ownerContext.getOwnerName(), rpdd);
        }
        return Arrays.asList(resultPartitions);
    }
}
```

正是在这个方法中，调用了 ResultPartitionFactory 的 create() 方法。

得到 ResultPartition 集合后，还会调用 ConsumableNotifyingResultPartitionWriterDecorator 的 decorate() 方法进行了一层封装：

```java
public static ResultPartitionWriter[] decorate(
        Collection<ResultPartitionDeploymentDescriptor> descs,
        ResultPartitionWriter[] partitionWriters,
        TaskActions taskActions,
        JobID jobId,
        ResultPartitionConsumableNotifier notifier) {
    ResultPartitionWriter[] consumableNotifyingPartitionWriters = new ResultPartitionWriter[partitionWriters.length];
    int counter = 0;
    for (ResultPartitionDeploymentDescriptor desc : descs) {
```

```
            if (desc.sendScheduleOrUpdateConsumersMessage() && desc.getPartitionType().
isPipelined()) {
                consumableNotifyingPartitionWriters[counter] = new ConsumableNotifyingResult
PartitionWriterDecorator(
                    taskActions,
                    jobId,
                    partitionWriters[counter],
                    notifier);
            } else {
                consumableNotifyingPartitionWriters[counter] = partitionWriters[counter];
            }
            counter++;
        }
        return consumableNotifyingPartitionWriters;
    }
```

这里会根据 ResultPartitionDeploymentDescriptor 中的 sendScheduleOrUpdateConsumersMessage 字段值和 ResultPartitionType 的值决定是否再将之前的 ResultPartition 对象封装成 ConsumableNotifyingResultPartitionWriterDecorator 类的对象。

### 7.2.3　InputGate 的初始化

构造 InputGate 会用到 SingleInputGateFactory。SingleInputGateFactory 的构造在任务管理器的启动过程中实现，可以直接调用其构造方法。SingleInputGateFactory 中的许多字段与内存管理、压缩等相关，这里暂不分析。

SingleInputGateFactory 中定义了创建 InputGate 的方法 create()：

```
public SingleInputGate create(
        @Nonnull String owningTaskName,
        @Nonnull InputGateDeploymentDescriptor igdd,
        @Nonnull PartitionProducerStateProvider partitionProducerStateProvider,
        @Nonnull InputChannelMetrics metrics) {

    ...
    // 创建 InputGate
    SingleInputGate inputGate = new SingleInputGate(
        owningTaskName,
        igdd.getConsumedResultId(),
        igdd.getConsumedPartitionType(),
        igdd.getConsumedSubpartitionIndex(),
        igdd.getShuffleDescriptors().length,
        partitionProducerStateProvider,
        bufferPoolFactory,
        bufferDecompressor);
    // 创建 InputChannel
    createInputChannels(owningTaskName, igdd, inputGate, metrics);
    return inputGate;
}
private void createInputChannels(
        String owningTaskName,
        InputGateDeploymentDescriptor inputGateDeploymentDescriptor,
        SingleInputGate inputGate,
        InputChannelMetrics metrics) {
```

```java
        ShuffleDescriptor[] shuffleDescriptors = inputGateDeploymentDescriptor.getShuffle
Descriptors();
        InputChannel[] inputChannels = new InputChannel[shuffleDescriptors.length];
        ChannelStatistics channelStatistics = new ChannelStatistics();
        for (int i = 0; i < inputChannels.length; i++) {
            inputChannels[i] = createInputChannel(
                inputGate,
                i,
                shuffleDescriptors[i],
                channelStatistics,
                metrics);
            ResultPartitionID resultPartitionID = inputChannels[i].getPartitionId();
            inputGate.setInputChannel(resultPartitionID.getPartitionId(), inputChannels[i]);
        }
    }
```

构造 InputChannel 时会调用 createInputChannel()方法，将 ShuffleDescriptor 数组传入其中：

```java
    private InputChannel createInputChannel(
        SingleInputGate inputGate,
        int index,
        ShuffleDescriptor shuffleDescriptor,
        ChannelStatistics channelStatistics,
        InputChannelMetrics metrics) {
        return applyWithShuffleTypeCheck(
            NettyShuffleDescriptor.class,
            shuffleDescriptor,
            unknownShuffleDescriptor -> {
                channelStatistics.numUnknownChannels++;
                return new UnknownInputChannel(
                    inputGate,
                    index,
                    unknownShuffleDescriptor.getResultPartitionID(),
                    partitionManager,
                    taskEventPublisher,
                    connectionManager,
                    partitionRequestInitialBackoff,
                    partitionRequestMaxBackoff,
                    metrics,
                    networkBufferPool);
            },
            nettyShuffleDescriptor ->
                createKnownInputChannel(
                    inputGate,
                    index,
                    nettyShuffleDescriptor,
                    channelStatistics,
                    metrics));
    }
    public static <T, SD extends ShuffleDescriptor> T applyWithShuffleTypeCheck(
        Class<SD> shuffleDescriptorClass,
        ShuffleDescriptor shuffleDescriptor,
        Function<UnknownShuffleDescriptor, T> functionOfUnknownDescriptor,
        Function<SD, T> functionOfKnownDescriptor) {
        if (shuffleDescriptor.isUnknown()) { // 根据 ShuffleDescriptor 的实现类的类型决定构造
                                             // 何种类型的 InputChannel
            return functionOfUnknownDescriptor.apply((UnknownShuffleDescriptor) shuffleDescriptor);
```

```
        } else if (shuffleDescriptorClass.equals(shuffleDescriptor.getClass())) {
            return functionOfKnownDescriptor.apply((SD) shuffleDescriptor);
        } else {
            ...
        }
    }
```

在 applyWithShuffleTypeCheck()方法中，会根据 ShuffleDescriptor 的 isUnknown()方法的返回值决定调用哪个函数去构造 InputChannel。前面介绍过 ShuffleDescriptor 有两个实现类，若为 NettyShuffleDescriptor 则返回值为 false，若为 UnknownShuffleDescriptor 则返回值为 true。如果返回值为 true，则通过构造方法构造出 UnknownShuffleDescriptor，否则调用 createKnownInputChannel() 方法：

```
private InputChannel createKnownInputChannel(
        SingleInputGate inputGate,
        int index,
        NettyShuffleDescriptor inputChannelDescriptor,
        ChannelStatistics channelStatistics,
        InputChannelMetrics metrics) {
    ResultPartitionID partitionId = inputChannelDescriptor.getResultPartitionID();
    if (inputChannelDescriptor.isLocalTo(taskExecutorResourceId)) { // 如果要读取的数据
// 在本地，则创建 LocalInputChannel
        channelStatistics.numLocalChannels++;
        return new LocalInputChannel(
            inputGate,
            index,
            partitionId,
            partitionManager,
            taskEventPublisher,
            partitionRequestInitialBackoff,
            partitionRequestMaxBackoff,
            metrics);
    } else { // 如果要读取的数据不在本地，则创建 RemoteInputChannel
        channelStatistics.numRemoteChannels++;
        return new RemoteInputChannel(
            inputGate,
            index,
            partitionId,
            inputChannelDescriptor.getConnectionId(),
            connectionManager,
            partitionRequestInitialBackoff,
            partitionRequestMaxBackoff,
            metrics,
            networkBufferPool);

    }
}
```

在该方法中，会判断要读取的数据是否在本地，以此决定是创建 LocalInputChannel 还是创建 RemoteInputChannel。其中，进行判断时调用的方法为 isLocalTo()：

```
public boolean isLocalTo(ResourceID consumerLocation) {
    return producerLocation.equals(consumerLocation);
}
```

这里传入的参数为任务管理器的 ResourceID。在并行实例申请槽时，槽会带有对应的任务管理器的 ResourceID，进而被传递到 NettyShuffleDescriptor 对象，赋值给 producerLocation 字段。因此这里可以对比 producerLocation 和 consumerLocation，以判断是否在本地读取数据。

从 Task 的初始化过程可以观察到，InputGate 的创建在 NettyShuffleEnvironment 的 createInputGates() 方法中实现：

```
public Collection<SingleInputGate> createInputGates(
    ShuffleIOOwnerContext ownerContext,
    PartitionProducerStateProvider partitionProducerStateProvider,
    Collection<InputGateDeploymentDescriptor> inputGateDeploymentDescriptors) {
  synchronized (lock) {
    ...
    SingleInputGate[] inputGates = new SingleInputGate[inputGateDeploymentDescriptors.size()];
    int counter = 0;
    for (InputGateDeploymentDescriptor igdd : inputGateDeploymentDescriptors) {
      // 利用 SingleInputGateFactory 的 create()方法创建 InputGate
      SingleInputGate inputGate = singleInputGateFactory.create(
          ownerContext.getOwnerName(),
          igdd,
          partitionProducerStateProvider,
          inputChannelMetrics);
      InputGateID id = new InputGateID(igdd.getConsumedResultId(), ownerContext.getExecutionAttemptID());
      inputGatesById.put(id, inputGate);
      inputGate.getCloseFuture().thenRun(() -> inputGatesById.remove(id));
      inputGates[counter++] = inputGate;
    }
    return Arrays.asList(inputGates);
  }
}
```

## 7.3 任务的执行

在任务的初始化过程中，如果成功创建了 Task 对象并将其添加到了 TaskExecutor 的 taskSlotTable 中，就会调用 Task 的 startTaskThread()方法：

```
public void startTaskThread() {
  executingThread.start();
}
```

Task 类本身实现了 Runnable 接口，所以调用上面的方法相当于调用 Task 的 run()方法。

```
public void run() {
  try {
    doRun();
  } finally {
    terminationFuture.complete(executionState);
  }
}
```

doRun()方法中涵盖一个并行实例的大部分生命周期，其中包括 AbstractInvokable 对象的初始化、任务的执行和任务的结束等。

```java
private void doRun() {
    ...
    try {
        ...
        // 初始化 ResultPartition 和 InputGate
        setupPartitionsAndGates(consumableNotifyingPartitionWriters, inputGates);
        ...
        // 初始化 AbstractInvokable 对象
        invokable = loadAndInstantiateInvokable(userCodeClassLoader, nameOfInvokableClass, env);
        this.invokable = invokable;
        // 将任务状态更新为 RUNNING
        if (!transitionState(ExecutionState.DEPLOYING, ExecutionState.RUNNING)) {
            throw new CancelTaskException();
        }
        taskManagerActions.updateTaskExecutionState(new TaskExecutionState(jobId, executionId, ExecutionState.RUNNING));
        // 任务的执行
        invokable.invoke();
        ...
        // ResultPartition 的结束
        for (ResultPartitionWriter partitionWriter : consumableNotifyingPartitionWriters) {
            if (partitionWriter != null) {
                partitionWriter.finish();
            }
        }
        // 将任务状态更新为 FINISHED
        if (!transitionState(ExecutionState.RUNNING, ExecutionState.FINISHED)) {
            throw new CancelTaskException();
        }
        ...
    }
    catch (Throwable t) {
        ...
    }
    finally {
        try {
            this.invokable = null;
            ...
            // 通知任务结束
            notifyFinalState();
        }
        catch (Throwable t) {
            ...
        }
        ...
    }
}
```

上面的代码保留了关键的 AbstractInvokable 对象的初始化、AbstractInvokable 对象的执行、状态的更新以及与 ResultPartition 和 InputGate 相关的逻辑。接下来对上面的代码添加了注释的部分进行分析。

（1）首先分析 setupPartitionsAndGates()方法：

```
public static void setupPartitionsAndGates(
    ResultPartitionWriter[] producedPartitions, InputGate[] inputGates) throws IOException,
InterruptedException {
    for (ResultPartitionWriter partition : producedPartitions) {
        partition.setup();
    }
    for (InputGate gate : inputGates) {
        gate.setup();
    }
}
```

在之前任务的初始化过程中，ResultPartition 对象和 InputGate 对象都已经进行了实例化。这个方法中会逐一调用 ResultPartition 对象和 InputGate 对象的 setup()方法。

对于 ResultPartition 对象，其 setup()方法如下：

```
public void setup() throws IOException {
    ...
    partitionManager.registerResultPartition(this);
}
```

这里的 setup()方法会将该 ResultPartition 对象注册到该任务管理器的 ResultPartitionManager 中进行管理。

对于 InputGate 对象，其 setup()方法如下：

```
public void setup() throws IOException, InterruptedException {
    ...
    requestPartitions();
}
```

这里的 setup()方法的主要逻辑是发送请求给上游的分区。

另一处 InputGate 与 ResultPartition 相关的地方是其最后会调用 finish()方法：

```
public void finish() throws IOException {
    for (ResultSubpartition subpartition : subpartitions) {
        subpartition.finish();
    }
    isFinished = true;
}
```

ResultSubpartition 有两个实现类——BoundedBlockingSubpartition 和 PipelinedSubpartition。它们的 finish()方法分别如下。

BoundedBlockingSubpartition 的 finish()方法：

```
public void finish() throws IOException {
    isFinished = true;
    flushCurrentBuffer();
    // 添加一条表示结束的数据
    writeAndCloseBufferConsumer(EventSerializer.toBufferConsumer(EndOfPartitionEvent.INSTANCE));
    data.finishWrite();
}
```

PipelinedSubpartition 的 finish()方法：

```java
public void finish() throws IOException {
    // 添加一条表示结束的数据
    add(EventSerializer.toBufferConsumer(EndOfPartitionEvent.INSTANCE), true);
}
```

这里的主要逻辑都是添加一条表示结束的数据。

（2）接下来分析 doRun() 方法中有关状态更新的逻辑。

首先调用 transitionState() 方法将任务的状态由 DEPLOYING 更新为 RUNNING。

之前介绍 Task 类的字段时就讲解过，Task 也有一个 executionState 字段（Execution 类中有同名同类型且含义相同的字段），其是 ExecutionState 类型。这里改动的就是该字段的值。

```java
private boolean transitionState(ExecutionState currentState, ExecutionState newState) {
    return transitionState(currentState, newState, null);
}
private boolean transitionState(ExecutionState currentState, ExecutionState newState,
Throwable cause) {
    if (STATE_UPDATER.compareAndSet(this, currentState, newState)) {
        ...
         return true;
    } else {
        return false;
    }
}
```

更新了任务中该字段的值，那么要通知作业管理器修改相应的值。这时需要调用 taskManagerActions 字段的 updateTaskExecutionState() 方法：

```java
public void updateTaskExecutionState(final TaskExecutionState taskExecutionState) {
    if (taskExecutionState.getExecutionState().isTerminal()) {
        runAsync(() -> unregisterTaskAndNotifyFinalState(jobMasterGateway, taskExecutionState.getID()));
    } else {
        TaskExecutor.this.updateTaskExecutionState(jobMasterGateway, taskExecutionState);
    }
}
```

这里会根据不同的状态进入不同的分支。这里的状态为 RUNNING，所以会进入 else 分支：

```java
private void updateTaskExecutionState(
        final JobMasterGateway jobMasterGateway,
        final TaskExecutionState taskExecutionState) {
    final ExecutionAttemptID executionAttemptID = taskExecutionState.getID();
    CompletableFuture<Acknowledge> futureAcknowledge = jobMasterGateway.updateTaskExecutionState(taskExecutionState);
    ...
}
```

这里通过 Rpc 调用回到了作业管理器端：

```java
public CompletableFuture<Acknowledge> updateTaskExecutionState(
        final TaskExecutionState taskExecutionState) {
    if (schedulerNG.updateTaskExecutionState(taskExecutionState)) {
        return CompletableFuture.completedFuture(Acknowledge.get());
    } else {
        ...
    }
}
```

剩下的工作就交给调度器 schedulerNG 来做了。updateTaskExecutionState()方法在第 6 章已经详细介绍。

同理，在任务结束时再次调用 transitionState()方法将任务的状态更新为 FINISHED。

在 doRun()方法的 finally 代码块中，调用 notifyFinalState()方法用于通知作业管理器端任务已结束：

```
private void notifyFinalState() {
    taskManagerActions.updateTaskExecutionState(new TaskExecutionState(jobId, executionId, executionState, failureCause));
}
```

此时会进入刚才介绍的 updateTaskExecutionState()方法的 if 分支：

```
private void unregisterTaskAndNotifyFinalState(
    final JobMasterGateway jobMasterGateway,
    final ExecutionAttemptID executionAttemptID) {
    // 将 Task 对象从 taskSlotTable 中移除
    Task task = taskSlotTable.removeTask(executionAttemptID);
    if (task != null) {
        ...
        updateTaskExecutionState(
            jobMasterGateway,
            new TaskExecutionState(
                task.getJobID(),
                task.getExecutionId(),
                task.getExecutionState(),
                task.getFailureCause(),
                accumulatorSnapshot,
                task.getMetricGroup().getIOMetricGroup().createSnapshot()));
    } else {
        ...
    }
}
```

这里比之前多了将 Task 对象从 taskSlotTable 中移除的操作。随后，仍然调用 updateTaskExecutionState()方法通知作业管理器端更新状态。

doRun()方法中的重要部分实际上是 AbstractInvokable 对象的初始化和 AbstractInvokable 对象的执行。

AbstractInvokable 类是一个抽象类，它表示在任务管理器端真正执行业务逻辑的任务。它的实现类包括 StreamTask 和 BatchTask，分别代表流处理任务和批处理任务。它的初始化调用的是 Task 类的 loadAndInstantiateInvokable()方法。它的执行调用的是 AbstractInvokable 的 invoke()方法，所有的业务逻辑都在该方法中实现。这里先对 loadAndInstantiateInvokable()方法进行介绍，然后根据不同的实现类分别介绍其他方法。

loadAndInstantiateInvokable()方法如下：

```
private static AbstractInvokable loadAndInstantiateInvokable(
    ClassLoader classLoader,
    String className,
    Environment environment) throws Throwable {
    final Class<? extends AbstractInvokable> invokableClass;
```

```
try {
   invokableClass = Class.forName(className, true, classLoader)
       .asSubclass(AbstractInvokable.class);
} catch (Throwable t) {
   ...
}
Constructor<? extends AbstractInvokable> statelessCtor;
try {
   statelessCtor = invokableClass.getConstructor(Environment.class);
} catch (NoSuchMethodException ee) {
   ...
}
try {
   return statelessCtor.newInstance(environment);
} catch (InvocationTargetException e) {
   ...
}
}
```

其中传入的第二个参数为 Task 的 nameOfInvokableClass 字段。该字段的值最初来自 StreamGraph 生成过程中 addNode()方法设置的一个参数。这部分内容已经介绍过，这里再次回顾相关方法的调用。比如调用 StreamGraph 的 addSource()方法时，有：

```
addNode(vertexID, slotSharingGroup, coLocationGroup, SourceStreamTask.class, operatorFactory,
operatorName);
```

调用一般的 addOperator()方法或者 addSink()方法时，有：

```
addNode(vertexID, slotSharingGroup, coLocationGroup, OneInputStreamTask.class,
operatorFactory, operatorName);
```

调用 addCoOperator()方法时，有：

```
Class<? extends AbstractInvokable> vertexClass = TwoInputStreamTask.class;
addNode(vertexID, slotSharingGroup, coLocationGroup, vertexClass, taskOperatorFactory,
operatorName);
```

这里的 SourceStreamTask、OneInputStreamTask 和 TwoInputStreamTask 类的全名会被传递下去作为 Task 的 nameOfInvokableClass 字段的值。因此，在 loadAndInstantiateInvokable()方法中，变量 invokableClass 就是相应的 Class 对象。接着获取其中的 Constructor 类型的构造器对象 statelessCtor，通过反射的方式实例化一个 AbstractInvokable 对象。其中构造器的参数为 Environment 对象。Environment 是一个接口，具体而言其是 RuntimeEnvironment 类的对象。

### 7.3.1　StreamTask 的初始化

StreamTask 是 AbstractInvokable 类的一个实现类，它也是一个抽象类，是所有流处理任务的基类。它的主要功能就是执行一个或多个链接在一起的 StreamOperator 中的逻辑。StreamOperator 的概念和设计思想在前文已经分析过。在链接在一起的算子中，第一个叫作 head operator，StreamTask 会根据它的类型决定自身的实现类的类型。StreamTask 常用的实现类包括 OneInputStreamTask（表示只有一个输入的任务）、TwoInputStreamTask（表示有两个输入的任务）等。

StreamTask 的定义如下：

```
public abstract class StreamTask<OUT, OP extends StreamOperator<OUT>>
    extends AbstractInvokable
    implements AsyncExceptionHandler
```

这里需注意其中的泛型 OUT 和 OP。

StreamTask 的字段主要包括以下。

- inputProcessor：StreamInputProcessor 类型，表示输入数据的 reader。
- headOperator：泛型 OP 类型，其是 StreamOperator 的子类。
- operatorChain：OperatorChain<OUT, OP>类型，表示链接在一起的算子。
- configuration：StreamConfig 类型，表示该流处理任务的配置。
- stateBackend：StateBackend 类型，表示状态后端。
- timerService：TimerService 类型，用于定时服务。
- recordWriter：RecordWriterDelegate<SerializationDelegate<StreamRecord<OUT>>>类型，表示写出数据的 writer。
- mailboxProcessor：MailboxProcessor 类型，该类封装了逻辑上基于 mailbox 的执行模型。之后会对该类进行分析。

其构造方法如下：

```
protected StreamTask(Environment env) {
    this(env, null);
}
protected StreamTask(Environment env, @Nullable TimerService timerService) {
    this(env, timerService, FatalExitExceptionHandler.INSTANCE);
}
protected StreamTask(
    Environment environment,
    @Nullable TimerService timerService,
    Thread.UncaughtExceptionHandler uncaughtExceptionHandler) {
    this(environment, timerService, uncaughtExceptionHandler, StreamTaskActionExecutor.synchronizedExecutor());
}
protected StreamTask(
    Environment environment,
    @Nullable TimerService timerService,
    Thread.UncaughtExceptionHandler uncaughtExceptionHandler,
    StreamTaskActionExecutor.SynchronizedStreamTaskActionExecutor actionExecutor) {
    this(environment, timerService, uncaughtExceptionHandler, actionExecutor, new TaskMailboxImpl(Thread.currentThread()));
}
protected StreamTask(
    Environment environment,
    @Nullable TimerService timerService,
    Thread.UncaughtExceptionHandler uncaughtExceptionHandler,
    StreamTaskActionExecutor.SynchronizedStreamTaskActionExecutor actionExecutor,
    TaskMailbox mailbox) {
    super(environment);
    this.timerService = timerService;
    this.uncaughtExceptionHandler = Preconditions.checkNotNull(uncaughtExceptionHandler);
    this.configuration = new StreamConfig(getTaskConfiguration());
    this.accumulatorMap = getEnvironment().getAccumulatorRegistry().getUserMap();
    this.recordWriter = createRecordWriterDelegate(configuration, environment);
```

```
        this.actionExecutor = Preconditions.checkNotNull(actionExecutor);
        this.mailboxProcessor = new MailboxProcessor(this::processInput, mailbox, actionExecutor);
        this.asyncExceptionHandler = new StreamTaskAsyncExceptionHandler(environment);
    }
    public AbstractInvokable(Environment environment) {
        this.environment = checkNotNull(environment);
    }
```

从上面代码的调用可以看出，在整个构造方法中，传入的参数 environment 被一直传递到了 AbstractInvokable 的 environment 字段上，其他字段则会通过各种方式被赋予默认值。如 timerService 会根据传入的值来赋值，configuration 会调用 getTaskConfiguration()方法来赋值，recordWriter 会调用 createRecordWriterDelegate()方法来赋值，mailboxProcessor 会调用 MailboxProcessor 类的构造方法来赋值。

下面对 StreamTask 中涉及的重要概念进行介绍。

### 7.3.2　StreamTask 中的重要概念

#### 1. StreamInputProcessor

StreamInputProcessor 是一个接口，它提供了处理数据的方法入口，可以被认为是输入数据的 reader。接口定义如下：

```
public interface StreamInputProcessor extends AvailabilityProvider, Closeable {
    InputStatus processInput() throws Exception;
}
```

返回值为 InputStatus 类型，该类型是一个枚举类：

```
public enum InputStatus {
    // 表示还有数据可供处理，可以立刻调用相应的方法处理数据
    MORE_AVAILABLE,
    // 表示当前没有更多数据可供处理，但未来会有数据被处理
    NOTHING_AVAILABLE,
    // 表示所有数据都被读取
    END_OF_INPUT
}
```

StreamInputProcessor 的实现类有两个——StreamOneInputProcessor 和 StreamTwoInputProcessor，它们分别对应 OneInputStreamTask 和 TwoInputStreamTask。

StreamOneInputProcessor 的定义如下：

```
public final class StreamOneInputProcessor<IN> implements StreamInputProcessor
```

泛型 IN 表示要处理的数据的类型。

StreamOneInputProcessor 的重要字段如下。

- input：StreamTaskInput<In>类型，用于在 StreamOneInputProcessor 中读取数据。
- output：DataOutput<In>类型，用于在 StreamOneInputProcessor 中写出数据。
- operatorChain：OperatorChain 类型，表示链接在一起的算子。

其构造方法如下：

```
public StreamOneInputProcessor(
    StreamTaskInput<IN> input,
```

```
        DataOutput<IN> output,
        Object lock,
        OperatorChain<?, ?> operatorChain) {
    this.input = checkNotNull(input);
    this.output = checkNotNull(output);
    this.lock = checkNotNull(lock);
    this.operatorChain = checkNotNull(operatorChain);
}
```

它对接口方法的实现如下：

```
public InputStatus processInput() throws Exception {
    // 读取接下来的数据并进行处理
    InputStatus status = input.emitNext(output);
    ...
    return status;
}
```

所有的业务逻辑都在 input 的 emitNext()方法中实现，并且可以推测读取数据后会调用 output 中的方法将数据输出。在 emitNext()方法中，涉及从 InputGate 中读取数据，这部分逻辑会在后文详细介绍。

对于一般的数据，被读取出来后会被封装成 StreamRecord 类型，通过 output 中的方法输出后会被算子（StreamOperator）处理。

StreamInputProcessor 的另一个实现类 StreamTwoInputProcessor 的定义如下：

```
public final class StreamTwoInputProcessor<IN1, IN2> implements StreamInputProcessor
```

泛型 IN1 和 IN2 分别表示两个输入源的数据类型。

StreamTwoInputProcessor 的重要字段如下。

- inputSelectionHandler：TwoInputSelectionHandler 类型，用于选择下一个输入的索引。
- input1：StreamTaskInput<IN1>类型，用于在 StreamTwoInputProcessor 中读取其中一个输入源的数据。
- input2：StreamTaskInput<IN2>类型，用于在 StreamTwoInputProcessor 中读取另一个输入源的数据。
- operatorChain：OperatorChain 类型，表示链接在一起的算子。
- output1：DataOutput<IN1>类型，用于在 StreamTwoInputProcessor 中写出其中一个输入源的数据。
- output2：DataOutput<IN2>类型，用于在 StreamTwoInputProcessor 中写出另一个输入源的数据。
- firstInputStatus：InputStatus 类型，表示其中一个输入源的 InputStatus，默认值为 MORE_AVAILABLE。
- secondInputStatus：InputStatus 类型，表示另一个输入源的 InputStatus，默认值为 MORE_AVAILABLE。
- firstStatus：StreamStatus 类型，表示其中一个数据流的状态，默认值为 0。
- secondStatus：StreamStatus 类型，表示另一个数据流的状态，默认值为 0。

StreamStatus 也是一种特殊类型的数据,会被传输到下游。它的字段和构造方法如下:

```java
public final class StreamStatus extends StreamElement {
   public static final int IDLE_STATUS = -1;
   public static final int ACTIVE_STATUS = 0;
   public static final StreamStatus IDLE = new StreamStatus(IDLE_STATUS);
   public static final StreamStatus ACTIVE = new StreamStatus(ACTIVE_STATUS);
   public final int status;
   public StreamStatus(int status) {
      ...
       this.status = status;
   }
```

**StreamTwoInputProcessor** 的构造方法如下:

```java
public StreamTwoInputProcessor(
    CheckpointedInputGate[] checkpointedInputGates,
    TypeSerializer<IN1> inputSerializer1,
    TypeSerializer<IN2> inputSerializer2,
    Object lock,
    IOManager ioManager,
    StreamStatusMaintainer streamStatusMaintainer,
    TwoInputStreamOperator<IN1, IN2, ?> streamOperator,
    TwoInputSelectionHandler inputSelectionHandler,
    WatermarkGauge input1WatermarkGauge,
    WatermarkGauge input2WatermarkGauge,
    OperatorChain<?, ?> operatorChain,
    Counter numRecordsIn) {
   this.lock = checkNotNull(lock);
   this.inputSelectionHandler = checkNotNull(inputSelectionHandler);
   this.output1 = new StreamTaskNetworkOutput<>(
      streamOperator,
      record -> processRecord1(record, streamOperator, numRecordsIn),
      lock,
      streamStatusMaintainer,
      input1WatermarkGauge,
      0);
   this.output2 = new StreamTaskNetworkOutput<>(
      streamOperator,
      record -> processRecord2(record, streamOperator, numRecordsIn),
      lock,
      streamStatusMaintainer,
      input2WatermarkGauge,
      1);
   this.input1 = new StreamTaskNetworkInput<>(
      checkpointedInputGates[0],
      inputSerializer1,
      ioManager,
      new StatusWatermarkValve(checkpointedInputGates[0].getNumberOfInputChannels(), output1),
      0);
   this.input2 = new StreamTaskNetworkInput<>(
      checkpointedInputGates[1],
      inputSerializer2,
      ioManager,
      new StatusWatermarkValve(checkpointedInputGates[1].getNumberOfInputChannels(), output2),
      1);
   this.operatorChain = checkNotNull(operatorChain);
}
```

StreamTwoInputProcessor 构造过程与 StreamOneInputProcessor 构造过程不同的是，Stream OneInputProcessor 在构造方法之外将用于输入和输出的对象（input 和 output）构造好再传入构造方法，而在 StreamTwoInputProcessor 的构造方法中可以看到用于输入和输出的对象的构造方式。这里可以观察到输入会依赖 InputGate 相关对象，输出会依赖 StreamOperator 对象。因此可以推测，整个过程是从 InputGate 中获取输入数据，最终由算子 StreamOperator 处理。

2. OperatorChain

OperatorChain 表示链接在一起的算子。在构建 OperatorChain 的过程中就构建了算子之间的输入输出关系，比如一个 StreamOperator 对象中可能持有一个表示输出的对象，而该表示输出的对象中又持有下一个 StreamOperator 对象，以此将数据传递下去依次进行处理。

OperatorChain 的核心字段如下。

- allOperators：StreamOperator[]类型，表示所有的算子。
- streamOutputs：RecordWriterOutput[]类型，RecordWriterOutput 类实现了 Output 接口，其中有一个 RecordWriter 类型的字段，用于将数据写入 ResultPartition，表示数据的输出。
- chainEntryPoint：WatermarkGaugeExposingOutput 类型，WatermarkGaugeExposingOutput 是一个接口，其继承自 Output 接口，表示链接的入口。
- headOperator：StreamOperator 的子类，表示第一个算子。

其构造方法如下：

```
public OperatorChain(
    StreamTask<OUT, OP> containingTask,
    RecordWriterDelegate<SerializationDelegate<StreamRecord<OUT>>> recordWriterDelegate) {
...
    // 获取 headOperator 的配置信息
    final StreamConfig configuration = containingTask.getConfiguration();
    // 从配置信息中反序列化出 StreamOperatorFactory 对象，即创建 StreamOperator 的工厂类
    StreamOperatorFactory<OUT> operatorFactory = configuration.getStreamOperatorFactory
(userCodeClassloader);
    // 获取链接在一起的所有算子的配置信息（包括 headOperator 的配置信息）
    Map<Integer, StreamConfig> chainedConfigs = configuration.getTransitiveChainedTask
ConfigsWithSelf(userCodeClassloader);
    // 获取链接在一起的算子的最后一个算子的出边，并用该出边信息构造 RecordWriterOutput 对象
    List<StreamEdge> outEdgesInOrder = configuration.getOutEdgesInOrder(userCodeClassloader);
    Map<StreamEdge, RecordWriterOutput<?>> streamOutputMap = new HashMap<>(outEdgesInOrder.
size());
    this.streamOutputs = new RecordWriterOutput<?>[outEdgesInOrder.size()];
    boolean success = false;
    try {
        for (int i = 0; i < outEdgesInOrder.size(); i++) {
            // 构造 RecordWriterOutput 对象
            StreamEdge outEdge = outEdgesInOrder.get(i);
            RecordWriterOutput<?> streamOutput = createStreamOutput(
                recordWriterDelegate.getRecordWriter(i),
                outEdge,
                chainedConfigs.get(outEdge.getSourceId()),
                containingTask.getEnvironment());
            this.streamOutputs[i] = streamOutput;
            streamOutputMap.put(outEdge, streamOutput);
```

```
            }
            ...
            // 构造链接的入口
            this.chainEntryPoint = createOutputCollector(
                containingTask,
                configuration,
                chainedConfigs,
                userCodeClassloader,
                streamOutputMap,
                allOps,
                containingTask.getMailboxExecutorFactory());
            if (operatorFactory != null) {
                WatermarkGaugeExposingOutput<StreamRecord<OUT>> output = getChainEntryPoint();
                // 利用 operatorFactory 构造 headOperator
                headOperator = StreamOperatorFactoryUtil.createOperator(
                    operatorFactory,
                    containingTask,
                    configuration,
                    output);
            } else {
                headOperator = null;
            }
            ...
        }
        finally {
            ...
        }
    }
```

OperatorChain 的构造方法较为复杂，因为并不是单纯的字段赋值操作。在上面的代码中，对比较重要的步骤添加了注释。这里对这些步骤进行详细的分析。

首先从 containingTask 中获取 StreamConfig 对象。在 StreamingJobGraphGenerator 中，生成 JobVertex 对象的方法 createJobVertex() 的返回值的类型就是 StreamConfig。它表示一个算子的配置信息，其中包括 operatorFactory、chainedConfigs 和 outEdgesInOrder 等。operatorFactory 对象用于创建 StreamOperator 对象，这在第 3 章中介绍过。chainedConfigs 包含链接在一起的每个算子的 StreamConfig 信息。outEdgesInOrder 表示物理出边。因为链接在一起的算子内部会有边的连接，而一串链接在一起的算子中的最后一个算子与下一串链接在一起的算子中的第一个算子之间也会有边的连接，后者叫作物理出边。这在第 4 章中已详细介绍。

在构造 OperatorChain 的过程中，需要构造上一个算子到下一个算子的输出。这个输出由接口 Output 表示：

```
public interface Output<T> extends Collector<T> {
    void emitWatermark(Watermark mark);
    <X> void collect(OutputTag<X> outputTag, StreamRecord<X> record);
    void emitLatencyMarker(LatencyMarker latencyMarker);
}
```

该接口定义了发送 Watermark 对象的方法和发送普通消息的方法，其中，发送普通消息的方法中需要传入一个 OutputTag 对象，表示这是 side output。该接口的父接口是 Collector：

```
public interface Collector<T> {
    void collect(T record);
```

```
    void close();
}
```

这里的 collect()方法表示的就是普通的输出,即把消息发送给下一个算子。

在构造 OperatorChain 的过程中,大致会根据算子与算子之间的边是不是物理出边分别构造两种不同的输出对象。如果是物理出边,则会构造 RecordWriterOutput 类的对象,否则会构造其他类型的 Output 实现类的对象,如 CopyingChainingOutput 对象。

回看 OperatorChain 的构造方法,得到物理出边后,会先对其进行遍历,构造 RecordWriterOutput 对象:

```
for (int i = 0; i < outEdgesInOrder.size(); i++) {
   StreamEdge outEdge = outEdgesInOrder.get(i);
   RecordWriterOutput<?> streamOutput = createStreamOutput(
      recordWriterDelegate.getRecordWriter(i),
      outEdge,
      chainedConfigs.get(outEdge.getSourceId()),
      containingTask.getEnvironment());
   this.streamOutputs[i] = streamOutput;
   streamOutputMap.put(outEdge, streamOutput);
}
```

构造出该对象以后,将其放入临时变量 streamOutputMap 和字段 streamOutputs 中进行维护。

接下来,在初始化字段 chainEntryPoint 时,将临时变量 streamOutputMap 作为方法参数传入 createOutputCollector()方法:

```
this.chainEntryPoint = createOutputCollector(
   containingTask,
   configuration,
   chainedConfigs,
   userCodeClassloader,
   streamOutputMap,
   allOps,
   containingTask.getMailboxExecutorFactory());
```

注意,chainEntryPoint 字段的类型是 WatermarkGaugeExposingOutput,其也是一个接口,继承自 Output 接口,紧接着,在初始化 headOperator 时将它作为方法参数传入 StreamOperatorFactoryUtil.createOperator()方法:

```
WatermarkGaugeExposingOutput<StreamRecord<OUT>> output = getChainEntryPoint();
headOperator = StreamOperatorFactoryUtil.createOperator(
   operatorFactory,
   containingTask,
   configuration,
   output);
```

chainEntryPoint 表示的就是 headOperator 的输出。

那么,从 headOperator 开始,一串链接在一起的算子是如何构建上下游输出关系的呢?可以想象,headOperator 的输出 chainEntryPoint 一定以某种形式依赖下一个 StreamOperator,而下一个 StreamOperator 又有其自身的输出对象,该输出对象再依赖它的下一个 StreamOperator。现在来观察 createOutputCollector()方法中的行为:

```java
    private <T> WatermarkGaugeExposingOutput<StreamRecord<T>> createOutputCollector(
        StreamTask<?, ?> containingTask,
        StreamConfig operatorConfig,
        Map<Integer, StreamConfig> chainedConfigs,
        ClassLoader userCodeClassloader,
        Map<StreamEdge, RecordWriterOutput<?>> streamOutputs,
        List<StreamOperator<?>> allOperators,
        MailboxExecutorFactory mailboxExecutorFactory) {
      List<Tuple2<WatermarkGaugeExposingOutput<StreamRecord<T>>, StreamEdge>> allOutputs =
new ArrayList<>(4);
        // 为物理出边构造 RecordWriterOutput 对象
        for (StreamEdge outputEdge : operatorConfig.getNonChainedOutputs(userCodeClassloader)) {
          @SuppressWarnings("unchecked")
          RecordWriterOutput<T> output = (RecordWriterOutput<T>) streamOutputs.get(outputEdge);
          allOutputs.add(new Tuple2<>(output, outputEdge));
        }
        // 为非物理出边构造其他的 Output 对象
        for (StreamEdge outputEdge : operatorConfig.getChainedOutputs(userCodeClassloader)) {
          int outputId = outputEdge.getTargetId();
          StreamConfig chainedOpConfig = chainedConfigs.get(outputId);
          WatermarkGaugeExposingOutput<StreamRecord<T>> output = createChainedOperator(
            containingTask,
            chainedOpConfig,
            chainedConfigs,
            userCodeClassloader,
            streamOutputs,
            allOperators,
            outputEdge.getOutputTag(),
            mailboxExecutorFactory);
          allOutputs.add(new Tuple2<>(output, outputEdge));
        }
        ...
        if (selectors == null || selectors.isEmpty()) {
          if (allOutputs.size() == 1) {
            return allOutputs.get(0).f0;
          }
          else {
            ...
          }
        }
        else {
          ...
        }
    }
```

首先分析传入的参数：

```java
    private <T> WatermarkGaugeExposingOutput<StreamRecord<T>> createOutputCollector(
        StreamTask<?, ?> containingTask, // 表示这个算子所在的 StreamTask 对象
        StreamConfig operatorConfig, // 表示要构造的 Output 对象所对应的算子的 StreamConfig 信息
        Map<Integer, StreamConfig> chainedConfigs, // 链接在一起的每个算子的 StreamConfig 信息
        ClassLoader userCodeClassloader, // 用户代码的类加载器
        Map<StreamEdge, RecordWriterOutput<?>> streamOutputs, // 物理出边的 RecordWriterOutput
                                                              // 对象
        List<StreamOperator<?>> allOperators, // 所有算子
        MailboxExecutorFactory mailboxExecutorFactory // 用于创建 MailboxExecutor 对象的工厂类对象
    ) {
```

此处的 chainedConfigs 就是 OperatorChain 构造方法中获取的临时变量 chainedConfigs，包含链接在一起的所有算子（包括 headOperator）的 StreamConfig 信息。streamOutputs 就是 OperatorChain 构造方法中获取的临时变量 streamOutputMap，它会维护所有物理出边对应的 RecordWriterOutput 对象。operatorConfig 表示 createOutputCollector()方法中即将要构建的 Output 对象所对应的算子的 StreamConfig 信息，在这里就是 headOperator 的 StreamConfig 信息。

了解了参数的含义后，createOutputCollector()方法中的各个步骤就比较好理解了。在构建链接在一起的算子之间的输出对象时，调用 createChainedOperator()方法，该方法的参数有类似的含义。注意，此时传入的 chainedOpConfig 对象已经不是 headOperator 的 StreamConfig 对象，而是中间的算子的 StreamConfig 对象。

```
private <IN, OUT> WatermarkGaugeExposingOutput<StreamRecord<IN>> createChainedOperator(
    StreamTask<OUT, ?> containingTask,
    StreamConfig operatorConfig,
    Map<Integer, StreamConfig> chainedConfigs,
    ClassLoader userCodeClassloader,
    Map<StreamEdge, RecordWriterOutput<?>> streamOutputs,
    List<StreamOperator<?>> allOperators,
    OutputTag<IN> outputTag,
    MailboxExecutorFactory mailboxExecutorFactory) {
    // 调用 createOutputCollector()方法构造下一个 Output 对象
    WatermarkGaugeExposingOutput<StreamRecord<OUT>> chainedOperatorOutput = createOutputCollector(
        containingTask,
        operatorConfig,
        chainedConfigs,
        userCodeClassloader,
        streamOutputs,
        allOperators,
        mailboxExecutorFactory);
    // 将上面构造出的 Output 对象作为参数传入 StreamOperatorFactoryUtil.createOperator()方法来构造
    // StreamOperator 对象
    OneInputStreamOperator<IN, OUT> chainedOperator = StreamOperatorFactoryUtil.createOperator(
        operatorConfig.getStreamOperatorFactory(userCodeClassloader),
        containingTask,
        operatorConfig,
        chainedOperatorOutput);
    allOperators.add(chainedOperator);
    // 由于不是物理出边对应的 Output 对象，因此需要将上面构造出的 StreamOperator 对象封装到当前要返
    // 回的 Output 对象中
    WatermarkGaugeExposingOutput<StreamRecord<IN>> currentOperatorOutput;
    if (containingTask.getExecutionConfig().isObjectReuseEnabled()) {
        currentOperatorOutput = new ChainingOutput<>(chainedOperator, this, outputTag);
    }
    else {
        TypeSerializer<IN> inSerializer = operatorConfig.getTypeSerializerIn1(userCodeClassloader);
        currentOperatorOutput = new CopyingChainingOutput<>(chainedOperator, inSerializer, outputTag, this);
    }
    return currentOperatorOutput;
}
```

这里递归地调用 createOutputCollector() 方法得到了即将要构造的 StreamOperator（可以理解为 headOperator 的下一个 StreamOperator 对象）的输出对象 chainedOperatorOutput，然后将其作为参数传入 StreamOperatorFactoryUtil.createOperator() 方法，从而构造下一个 StreamOperator 对象。这里可以对比先前分析的 chainEntryPoint 的构造和 headOperator 的生成，其过程基本是一致的。最后，将 StreamOperator 对象封装到 Output 对象中并返回，就得到了 chainEntryPoint 对象。由于整个过程是递归调用的，因此整个链接中的上下游算子的输出关系全都被包含在 chainEntryPoint 中，进而被封装到 headOperator 中。

3. StreamConfig

在上面的分析中，频繁出现 StreamConfig 对象。该对象在第 4 章介绍 JobGraph 的生成过程时提及过。该对象中保存了算子的计算逻辑、物理出边、链接在一起的算子的 StreamConfig 等信息。这些信息是在生成 JobGraph 的过程中设置的，有些信息被序列化成 byte 数组，还有一些原始类型的信息会直接以原始类型保存在其中。StreamConfig 中有一个 Configuration 类型的 config 字段，StreamConfig 的所有信息都保存在其中。

下面以 StreamOperator 为例，介绍 StreamConfig 是如何在生成执行图的过程中保存信息并在任务管理器端根据这些信息生成相应对象的。

StreamOperator 在前文已经多次介绍过，其表示的就是算子的计算逻辑。比如在业务代码中定义一个 userFunction，该对象就会被封装到对应的 StreamOperator 实现类中。接着在构造 Transformation 对象时，该对象会被封装到 StreamOperatorFactory 对象中。StreamOperatorFactory 对象会一直被传递到 StreamNode 对象中，在构造 JobGraph 的过程中会生成 StreamConfig 对象，StreamNode 中的 StreamOperatorFactory 对象就会被设置到对应的 StreamConfig 对象中。

```
config.setStreamOperatorFactory(vertex.getOperatorFactory());
public void setStreamOperatorFactory(StreamOperatorFactory<?> factory) {
    if (factory != null) {
        try {
            InstantiationUtil.writeObjectToConfig(factory, this.config, SERIALIZEDUDF);
        } catch (IOException e) {
            ...
        }
    }
}
public static void writeObjectToConfig(Object o, Configuration config, String key)
throws IOException {
    byte[] bytes = serializeObject(o);
    config.setBytes(key, bytes);
}
```

设置的过程中，先将 StreamOperatorFactory 对象序列化，再将其放入 config 字段。

该配置信息最终会随着 TaskDeploymentDescriptor 对象一起被传输到任务管理器端，变成 StreamTask 的 configuration 字段。在构造 OperatorChain 的过程中就利用了该字段，将先前保存的信息取出并构造相应的对象，整个过程在前面分析 OperatorChain 时已经介绍过。

取出 StreamOperatorFactory 对象的过程如下：

```
    StreamOperatorFactory<OUT> operatorFactory = configuration.getStreamOperatorFactory
(userCodeClassloader);
    public <T extends StreamOperatorFactory<?>> T getStreamOperatorFactory(ClassLoader cl) {
        try {
            return InstantiationUtil.readObjectFromConfig(this.config, SERIALIZEDUDF, cl);
        }
        catch (ClassNotFoundException e) {
            ...
        }
    }
    public static <T> T readObjectFromConfig(Configuration config, String key, ClassLoader cl)
throws IOException, ClassNotFoundException {
        byte[] bytes = config.getBytes(key, null);
        if (bytes == null) {
            return null;
        }
        return deserializeObject(bytes, cl);
    }
```

先从 config 字段中取出 byte 数组，再对其进行反序列化。这里需要注意，只有从 userCode ClassLoader 这个类加载器中反序列化才能获取想要的结果。一般在编写业务代码时，很少会用到类加载器，但通常在设计框架时会给用户代码设置一个专门的类加载器。在分析这类框架的源码或是进行二次开发时，一定要注意所涉及的类究竟在哪个类加载器中。

返回 StreamOperatorFactory 对象后，就可以构造 StreamOperator 了。这时会调用 StreamOperatorFactoryUtil 的 createOperator()方法：

```
    headOperator = StreamOperatorFactoryUtil.createOperator(
        operatorFactory,
        containingTask,
        configuration,
        output);
    public static <OUT, OP extends StreamOperator<OUT>> OP createOperator(
        StreamOperatorFactory<OUT> operatorFactory,
        StreamTask<OUT, ?> containingTask,
        StreamConfig configuration,
        Output<StreamRecord<OUT>> output) {
        MailboxExecutorFactory mailboxExecutorFactory = containingTask.getMailboxExecutorFactory();
        if (operatorFactory instanceof YieldingOperatorFactory) {
            MailboxExecutor mailboxExecutor = mailboxExecutorFactory.createExecutor(configuration.
getChainIndex());
            ((YieldingOperatorFactory) operatorFactory).setMailboxExecutor(mailboxExecutor);
        }
        return operatorFactory.createStreamOperator(containingTask, configuration, output);
    }
```

以 SimpleOperatorFactory 为例，其 createStreamOperator()方法返回的就是其中的 operator 字段：

```
    public <T extends StreamOperator<OUT>> T createStreamOperator(StreamTask<?, ?>
containingTask,
        StreamConfig config, Output<StreamRecord<OUT>> output) {
        if (operator instanceof SetupableStreamOperator) {
            ((SetupableStreamOperator) operator).setup(containingTask, config, output);
        }
        return (T) operator;
    }
```

StreamConfig 中的其他信息都是用同样的方式进行保存、传输并在任务管理器端对相应对象进行初始化的。

### 4. RecordWriterDelegate

对于 RecordWriterDelegate，顾名思义，它表示 RecordWriter 的代理。前文介绍每个算子的输出对象时介绍过，物理出边对应的输出对象是 RecordWriterOutput 类型，其中就封装了 RecordWriter 对象。RecordWriter 类表示物理出边的输出，其中封装了 ResultPartition 对象。这部分内容后文会详细介绍。

RecordWriterDelegate 是一个接口，它有 3 个实现类，在 StreamTask 的构造方法中，会调用 createRecordWriterDelegate()方法对其进行初始化，这样刚好可以清晰地观察到 3 个实现类在何种条件下构造：

```
public static <OUT> RecordWriterDelegate<SerializationDelegate<StreamRecord<OUT>>>
createRecordWriterDelegate(
        StreamConfig configuration,
        Environment environment) {
    List<RecordWriter<SerializationDelegate<StreamRecord<OUT>>>> recordWrites =
createRecordWriters(
        configuration,
        environment);
    if (recordWrites.size() == 1) {
      return new SingleRecordWriter<>(recordWrites.get(0));
    } else if (recordWrites.size() == 0) {
      return new NonRecordWriter<>();
    } else {
      return new MultipleRecordWriters<>(recordWrites);
    }
}
```

createRecordWriters()方法会根据物理出边构造 RecordWriter 对象：

```
    private static <OUT> List<RecordWriter<SerializationDelegate<StreamRecord<OUT>>>>
createRecordWriters(
        StreamConfig configuration,
        Environment environment) {
    List<RecordWriter<SerializationDelegate<StreamRecord<OUT>>>> recordWriters = new
ArrayList<>();
    List<StreamEdge> outEdgesInOrder = configuration.getOutEdgesInOrder(environment.
getUserClassLoader());
    Map<Integer, StreamConfig> chainedConfigs = configuration.getTransitiveChained
TaskConfigsWithSelf(environment.getUserClassLoader());
    for (int i = 0; i < outEdgesInOrder.size(); i++) {
      StreamEdge edge = outEdgesInOrder.get(i);
      recordWriters.add(
          createRecordWriter(
              edge,
              i,
              environment,
              environment.getTaskInfo().getTaskName(),
              chainedConfigs.get(edge.getSourceId()).getBufferTimeout()));
    }
    return recordWriters;
}
```

可以认为，上述代码的构造过程主要就是为了构造 RecordWriter 对象。随后，在 OperatorChain 的构造方法中会将 RecordWriterDelegate 对象传入，在构造 Output 对象时，会将其中的 RecordWriter 对象取出：

```
RecordWriterOutput<?> streamOutput = createStreamOutput(
    recordWriterDelegate.getRecordWriter(i),
    outEdge,
    chainedConfigs.get(outEdge.getSourceId()),
    containingTask.getEnvironment());
```

由此，RecordWriter 对象就被封装到 RecordWriterOutput 中。

5. MailboxProcessor

StreamTask 中有一个字段为 mailboxProcessor，它是 MailboxProcessor 类型。当 StreamTask 开始真正执行业务逻辑时，会调用 runMailboxLoop()方法：

```
private void runMailboxLoop() throws Exception {
    mailboxProcessor.runMailboxLoop();
}
```

该方法调用的就是 mailboxProcessor 的 runMailboxLoop()方法。从 MailboxProcessor 类的类名、方法名可以看出，Flink 将真正的业务逻辑的过程抽象成了 mailbox 模型。可以推测，整个处理过程应该会有类似"邮箱"的对象，用于接收各处发送来的"邮件"，进而会根据"邮件"的具体类型触发相应的操作。

MailboxProcessor 中的主要字段如下。

- mailbox：TaskMailbox 类型，表示"邮箱"的抽象。它会接收不同的"邮件"，在不同的方法中会涉及从其中取出"邮件"进行处理。
- mailboxDefaultAction：MailboxDefaultAction 类型，可以将其理解为整个 MailboxProcessor 的默认操作，具体而言就是处理输入数据。
- mainMailboxExecutor：MailboxExecutor 类型，表示"邮箱"的执行器，持有 mailbox 依赖，提供方法用于向"邮箱"中添加"邮件"。
- suspendedDefaultAction：MailboxDefaultAction.Suspension 类型，表示处理过程中的等待状态，提供 resume()方法用于让处理继续进行。

其构造方法如下：

```
public MailboxProcessor(
    MailboxDefaultAction mailboxDefaultAction,
    TaskMailbox mailbox,
    StreamTaskActionExecutor actionExecutor) {
    this(mailboxDefaultAction, actionExecutor, mailbox, new MailboxExecutorImpl(mailbox,
MIN_PRIORITY, actionExecutor));
}
public MailboxProcessor(
    MailboxDefaultAction mailboxDefaultAction,
    StreamTaskActionExecutor actionExecutor,
    TaskMailbox mailbox,
    MailboxExecutor mainMailboxExecutor) {
    this.mailboxDefaultAction = Preconditions.checkNotNull(mailboxDefaultAction);
    this.actionExecutor = Preconditions.checkNotNull(actionExecutor);
```

```
    this.mailbox = Preconditions.checkNotNull(mailbox);
    this.mainMailboxExecutor = Preconditions.checkNotNull(mainMailboxExecutor);
    this.mailboxLoopRunning = true;
    this.suspendedDefaultAction = null;
}
```

该构造方法在 StreamTask 的构造方法中被调用：

```
this.mailboxProcessor = new MailboxProcessor(this::processInput, mailbox, actionExecutor);
```

由于 MailboxDefaultAction 接口只提供了一个方法，因此可以直接用 this::processInput 这种方式来赋值。processInput()方法如下：

```
protected void processInput(MailboxDefaultAction.Controller controller) throws Exception {
    // 对数据进行处理，并返回 InputStatus
    InputStatus status = inputProcessor.processInput();
    ...
}
```

前文分析过 inputProcessor，它会对数据进行读取和处理。

现在回到前面介绍的 runMailboxLoop()方法，观察该方法中的行为，以此来了解 MailboxProcessor 的行为：

```
public void runMailboxLoop() throws Exception {
    final TaskMailbox localMailbox = mailbox;
    final MailboxController defaultActionContext = new MailboxController(this);
    // 循环处理"邮件"，如果返回 true，则执行 MailboxProcessor 默认的操作
    while (processMail(localMailbox)) {
        mailboxDefaultAction.runDefaultAction(defaultActionContext);
    }
}
```

MailboxController 是 MailboxProcessor 的内部类，它持有该 MailboxProcessor 对象的依赖，并提供以下方法：

```
public void allActionsCompleted() {
    mailboxProcessor.allActionsCompleted();
}
public MailboxDefaultAction.Suspension suspendDefaultAction() {
    return mailboxProcessor.suspendDefaultAction();
}
```

在 allActionsCompleted()方法中，会发送"邮件"，以告知不再处理数据：

```
public void allActionsCompleted() {
    mailbox.runExclusively(() -> {
        if (mailbox.getState() == TaskMailbox.State.OPEN) {
            sendControlMail(() -> mailboxLoopRunning = false, "poison mail");
        }
    });
}
private void sendControlMail(RunnableWithException mail, String descriptionFormat,
Object... descriptionArgs) {
    mailbox.putFirst(new Mail(
        mail,
        Integer.MAX_VALUE,
        descriptionFormat,
```

在 suspendDefaultAction()方法中，会给 suspendedDefaultAction 字段赋值，注意，在构造方法中该字段的值为 null：

```
private MailboxDefaultAction.Suspension suspendDefaultAction() {
    if (suspendedDefaultAction == null) {
        suspendedDefaultAction = new DefaultActionSuspension();
        ...
    }
    return suspendedDefaultAction;
}
```

之后在处理"邮件"时，会判断该字段的值是否为 null，以此判定是否能进行默认的操作。因此可以看出，MailboxController 就是 MailboxProcessor 的流程控制器。

processMail()方法如下：

```
private boolean processMail(TaskMailbox mailbox) throws Exception {
    if (!mailbox.createBatch()) {
        return true;
    }
    Optional<Mail> maybeMail;
    while (isMailboxLoopRunning() && (maybeMail = mailbox.tryTakeFromBatch()).isPresent()) {
        maybeMail.get().run();
    }
    while (isDefaultActionUnavailable() && isMailboxLoopRunning()) {
        mailbox.take(MIN_PRIORITY).run();
    }
    return isMailboxLoopRunning();
}
```

在这个过程中，会从"邮箱"mailbox 中取出"邮件"并进行处理。其中，isDefaultActionUnavailable()方法如下：

```
public boolean isDefaultActionUnavailable() {
    return suspendedDefaultAction != null;
}
```

这里，当 suspendedDefaultAction 字段的值不为 null 时，会返回 true，整个流程处于等待状态。回到之前的 processInput()方法：

```
protected void processInput(MailboxDefaultAction.Controller controller) throws Exception {
    InputStatus status = inputProcessor.processInput();
    if (status == InputStatus.MORE_AVAILABLE && recordWriter.isAvailable()) {
        return;
    }
    if (status == InputStatus.END_OF_INPUT) {
        controller.allActionsCompleted();
        return;
    }
    CompletableFuture<?> jointFuture = getInputOutputJointFuture(status);
    MailboxDefaultAction.Suspension suspendedDefaultAction = controller.suspendDefaultAction();
    jointFuture.thenRun(suspendedDefaultAction::resume);
}
```

之前分析这段代码时省略了后面部分。现在观察后面部分的代码可以发现，当状态为 END_OF_INPUT 时会调用之前分析过的 allActionsCompleted() 方法。当状态既不为 MORE_AVAILABLE 又不为 END_OF_INPUT 时，会执行以下的逻辑，最终会调用 suspendedDefaultAction 的 resume() 方法：

```
public void resume() {
   if (mailbox.isMailboxThread()) {
      resumeInternal();
   } else {
      sendControlMail(this::resumeInternal, "resume default action");
   }
}
private void resumeInternal() {
   if (suspendedDefaultAction == this) {
      suspendedDefaultAction = null;
   }
}
```

这里，suspendedDefaultAction 字段的值又被设置为 null，此时 processMail() 方法中的循环会结束，返回 isMailboxLoopRunning() 方法的返回值：

```
private boolean isMailboxLoopRunning() {
   return mailboxLoopRunning;
}
```

该返回值在初始化时就为 true，除非 allActionsCompleted() 方法将它设置为 false。

除了这里的发送"邮件"的例子，还可以再举一个例子。在检查点机制中，当作业管理器结束一次检查点操作后，会触发 TaskExecutor 调用 confirmCheckpoint() 方法，进而会调用 Task 中的 notifyCheckpointComplete() 方法，更进一步地，会调用 StreamTask 的 notifyCheckpointCompleteAsync() 方法：

```
public Future<Void> notifyCheckpointCompleteAsync(long checkpointId) {
   return mailboxProcessor.getMailboxExecutor(TaskMailbox.MAX_PRIORITY).submit(
        () -> notifyCheckpointComplete(checkpointId),
        "checkpoint %d complete", checkpointId);
}
```

此时又会向"邮箱"发送"邮件"，用于执行 notifyCheckpointComplete() 方法：

```
private void notifyCheckpointComplete(long checkpointId) {
   try {
      boolean success = actionExecutor.call(() -> {
         if (isRunning) {
            for (StreamOperator<?> operator : operatorChain.getAllOperators()) {
               if (operator != null) {
                  operator.notifyCheckpointComplete(checkpointId);
               }
            }
            return true;
         } else {
            ...
         }
      });
      ...
   } catch (Exception e) {
      ...
```

            }
        }

该方法又会调用每个 StreamOperator 算子的 notifyCheckpointComplete()方法，显然最终也会执行 userFunction 的相关方法。

在 Kafka 与 Flink 结合应用的架构中，常常会涉及端到端一致性语义的问题，其中有个概念叫作"两阶段提交"，两阶段提交中所用到的类就会继承 TwoPhaseCommitSinkFunction 这个抽象类。该类的 notifyCheckpointComplete()方法会调用相关的 commit()方法进行提交。这也是"邮箱"机制的一个使用场景。

### 7.3.3　StreamTask 的实现类

StreamTask 是一个抽象类，定义了其自身的生命周期，但该类中的一些方法是由具体的实现类实现或重写的。本节对几个常用的实现类进行介绍。

1. SourceStreamTask

SourceStreamTask 表示 source 任务的 StreamTask 实现类。其中包含一个 LegacySourceFunctionThread 类型的 sourceThread 字段，它的 run()方法实现了具体业务逻辑。

在 SourceStreamTask 的构造方法中，会对该字段进行初始化：

```
public SourceStreamTask(Environment env) {
    super(env);
    this.sourceThread = new LegacySourceFunctionThread();
}
```

SourceStreamTask 会重写 processInput()方法：

```
protected void processInput(MailboxDefaultAction.Controller controller) throws Exception {
    ...
    sourceThread.start();
    ...
}
```

其中的关键步骤是启动线程。

线程的 run()方法如下：

```
public void run() {
    try {
        headOperator.run(getCheckpointLock(), getStreamStatusMaintainer(), operatorChain);
        completionFuture.complete(null);
    } catch (Throwable t) {
        ...
    }
}
```

SourceStreamTask 中的 headOperator 是 StreamSource 类型的 StreamOperator，其 run()方法如下：

```
public void run(final Object lockingObject,
        final StreamStatusMaintainer streamStatusMaintainer,
        final OperatorChain<?, ?> operatorChain) throws Exception {
    run(lockingObject, streamStatusMaintainer, output, operatorChain);
}
```

```
public void run(final Object lockingObject,
    final StreamStatusMaintainer streamStatusMaintainer,
    final Output<StreamRecord<OUT>> collector,
    final OperatorChain<?, ?> operatorChain) throws Exception {
  ...
  try {
    userFunction.run(ctx);
    ...
  } finally {
    ...
  }
}
```

其中的关键步骤是调用 userFunction 的 run()方法。另外需注意，run()方法传入了 lockingObject 对象，该对象在检查点的执行过程中会被"上锁"，由此确保检查点的执行过程与数据发送过程不能同时进行，而必须串行化地依次进行。

2. OneInputStreamTask

OneInputStreamTask 表示只有一个输入源的任务。它会重写 init()方法：

```
public void init() throws Exception {
  StreamConfig configuration = getConfiguration();
  int numberOfInputs = configuration.getNumberOfInputs();
  if (numberOfInputs > 0) {
    CheckpointedInputGate inputGate = createCheckpointedInputGate();
    DataOutput<IN> output = createDataOutput();
    StreamTaskInput<IN> input = createTaskInput(inputGate, output);
    inputProcessor = new StreamOneInputProcessor<>(
        input,
        output,
        getCheckpointLock(),
        operatorChain);
  }
}
```

这个过程主要构造一些参数，最终会构造一个 inputProcessor 对象。其中，CheckpointedInputGate 就是对 InputGate 的一层封装。createDataOutput()方法返回的是 PushingAsyncDataInput.Output 对象，它在 OneInputStreamTask 中的实现类是 StreamTaskNetworkOutput，它主要提供这样一个方法：

```
public void emitRecord(StreamRecord<IN> record) throws Exception {
  synchronized (lock) {
    numRecordsIn.inc();
    operator.setKeyContextElement1(record);
    operator.processElement(record);
  }
}
```

即调用 StreamOperator 的 processElement()方法对数据进行处理。

createTaskInput()方法返回的是 StreamTaskInput 对象，可以简单地理解为对 InputGate 的一层封装。这里的实现类为 StreamTaskNetworkInput，其提供了从 InputGate 中读取数据并进行处理的方法。

### 3. TwoInputStreamTask

TwoInputStreamTask 表示有两个输入源的任务。它继承自抽象类 AbstractTwoInputStreamTask。TwoInputStreamTask 类也会重写 init() 方法：

```java
public void init() throws Exception {
    StreamConfig configuration = getConfiguration();
    ClassLoader userClassLoader = getUserCodeClassLoader();
    TypeSerializer<IN1> inputDeserializer1 = configuration.getTypeSerializerIn1(userClassLoader);
    TypeSerializer<IN2> inputDeserializer2 = configuration.getTypeSerializerIn2(userClassLoader);
    int numberOfInputs = configuration.getNumberOfInputs();
    ArrayList<InputGate> inputList1 = new ArrayList<InputGate>();
    ArrayList<InputGate> inputList2 = new ArrayList<InputGate>();
    List<StreamEdge> inEdges = configuration.getInPhysicalEdges(userClassLoader);
    for (int i = 0; i < numberOfInputs; i++) {
        int inputType = inEdges.get(i).getTypeNumber();
        InputGate reader = getEnvironment().getInputGate(i);
        switch (inputType) {
           case 1:
               inputList1.add(reader);
               break;
           case 2:
               inputList2.add(reader);
               break;
           default:
               throw new RuntimeException("Invalid input type number: " + inputType);
        }
    }
    createInputProcessor(inputList1, inputList2, inputDeserializer1, inputDeserializer2);
}
```

在 TwoInputStreamTask 中会实现 createInputProcessor() 方法：

```java
protected void createInputProcessor(
    Collection<InputGate> inputGates1,
    Collection<InputGate> inputGates2,
    TypeSerializer<IN1> inputDeserializer1,
    TypeSerializer<IN2> inputDeserializer2) throws IOException {
    TwoInputSelectionHandler twoInputSelectionHandler = new TwoInputSelectionHandler(
        headOperator instanceof InputSelectable ? (InputSelectable) headOperator : null);
    InputGate unionedInputGate1 = InputGateUtil.createInputGate(inputGates1.toArray(new InputGate[0]));
    InputGate unionedInputGate2 = InputGateUtil.createInputGate(inputGates2.toArray(new InputGate[0]));
    CheckpointedInputGate[] checkpointedInputGates = InputProcessorUtil.createCheckpointedInputGatePair(
        this,
        getConfiguration().getCheckpointMode(),
        getEnvironment().getIOManager(),
        unionedInputGate1,
        unionedInputGate2,
        getEnvironment().getTaskManagerInfo().getConfiguration(),
        getEnvironment().getMetricGroup().getIOMetricGroup(),
        getTaskNameWithSubtaskAndId());
    checkState(checkpointedInputGates.length == 2);
```

```
        inputProcessor = new StreamTwoInputProcessor<>(
            checkpointedInputGates,
            inputDeserializer1,
            inputDeserializer2,
            getCheckpointLock(),
            getEnvironment().getIOManager(),
            getStreamStatusMaintainer(),
            headOperator,
            twoInputSelectionHandler,
            input1WatermarkGauge,
            input2WatermarkGauge,
            operatorChain,
            setupNumRecordsInCounter(headOperator));
    }
```

这里的主要逻辑是构造 inputProcessor 对象。在 StreamTwoProcessor 的构造方法中构造了 StreamTaskInput 和 DataOutput 等对象，整体逻辑与 OneInputStreamTask 的一致。

### 7.3.4　StreamTask 的生命周期

StreamTask 的生命周期在源码的注释中会有所体现：

```
/**
 *  -- invoke()
 *        |
 *        +----> Create basic utils (config, etc) and load the chain of operators
 *        +----> operators.setup()
 *        +----> task specific init()
 *        +----> initialize-operator-states()
 *        +----> open-operators()
 *        +----> run()
 *        +----> close-operators()
 *        +----> dispose-operators()
 *        +----> common cleanup
 *        +----> task specific cleanup()
 *
 *
 */
```

如注释所示，这些操作全部在 invoke() 方法中执行。

invoke()方法如下：

```
public final void invoke() throws Exception {
    try {
        // 执行前的准备工作
        beforeInvoke();
        ...
        // 执行业务逻辑
        runMailboxLoop();
        ...
        // 执行后的收尾工作
        afterInvoke();
    }
    finally {
```

```
        // 清理工作
        cleanUpInvoke();
    }
}
```

上面代码的核心逻辑如注释所示。其中，生命周期中的 Create basic utils (config, etc) and load the chain of operators、operators.setup()、task specific init()、initialize-operator-states()、open-operators() 都在 beforeInvoke()方法中实现，run()在 runMailboxLoop()方法中实现，close-operators()、dispose-operators()在 afterInvoke()方法中实现，common cleanup、task specific cleanup()在 cleanUpInvoke()方法中实现。

下面详细分析前面介绍的生命周期分别对应这些方法中的哪一部分。

### 1. beforeInvoke

beforeInvoke()方法如下：

```
private void beforeInvoke() throws Exception {
    //创建基本的程序（包括配置等）并加载链接在一起的算子
    // 创建 operators.setup()
    ...
    operatorChain = new OperatorChain<>(this, recordWriter);
    headOperator = operatorChain.getHeadOperator();
    // task specific init().
    init();
    actionExecutor.runThrowing(() -> {
        // initialize-operator-states().
        // open-operators().
        initializeStateAndOpen();
    });
}
```

上面的代码只保留了与生命周期直接相关的部分。

生命周期中的第一步和第二步，即 Create basic utils (config, etc) and load the chain of operators 和 operators.setup()，都在构造 OperatorChain 的过程中实现。OperatorChain 的构造方法已经详细分析过，已经介绍过获取配置信息并生成算子的过程。算子的 setup()方法在构造 StreamOperator 对象的过程中实现。

以 SimpleOperatorFactory 为例：

```
public <T extends StreamOperator<OUT>> T createStreamOperator(StreamTask<?, ?> containingTask,
        StreamConfig config, Output<StreamRecord<OUT>> output) {
    if (operator instanceof SetupableStreamOperator) {
        ((SetupableStreamOperator) operator).setup(containingTask, config, output);
    }
    return (T) operator;
}
```

一般自定义的算子类型可实现 SetupableStreamOperator 接口，因此会执行其 setup()方法，其 setup()方法在 AbstractStreamOperator 类中实现：

```
public void setup(StreamTask<?, ?> containingTask, StreamConfig config, Output<StreamRecord<OUT>> output) {
    final Environment environment = containingTask.getEnvironment();
```

```
        this.container = containingTask;
        this.processingTimeService = containingTask.getProcessingTimeService(config.
getChainIndex());
        this.config = config;
        try {
            ...
            this.output = new CountingOutput(output, operatorMetricGroup.getIOMetricGroup().
getNumRecordsOutCounter());
            ...
        } catch (Exception e) {
            ...
        }
        ...
        this.runtimeContext = new StreamingRuntimeContext(this, environment, container.
getAccumulatorMap());
        stateKeySelector1 = config.getStatePartitioner(0, getUserCodeClassloader());
        stateKeySelector2 = config.getStatePartitioner(1, getUserCodeClassloader());
    }
```

在算子执行 setup() 方法的过程中，主要为 container、config、output、runtimeContext、stateKeySelector1 和 stateKeySelector2 这些字段赋值。

AbstractStreamOperator 的子类 AbstractUdfStreamOperator 的 setup() 方法如下：

```
public void setup(StreamTask<?, ?> containingTask, StreamConfig config, Output
<StreamRecord<OUT>> output) {
    super.setup(containingTask, config, output);
    FunctionUtils.setFunctionRuntimeContext(userFunction, getRuntimeContext());
}
public static void setFunctionRuntimeContext(Function function, RuntimeContext context){
    if (function instanceof RichFunction) {
        RichFunction richFunction = (RichFunction) function;
        richFunction.setRuntimeContext(context);
    }
}
```

首先调用父类的 setup() 方法，接着给 userFunction 设置前面初始化过的 runtimeContext。

前面介绍的生命周期中主要是初始化算子，接下来需要初始化 StreamTask。init() 方法是由具体的实现类实现的，前面已分析过 OneInputStreamTask 和 TwoInputStreamTask 的 init() 方法，主要用于初始化 inputProcessor 字段。

在 beforeInvoke() 方法的最后调用了 initializeStateAndOpen() 方法，完成算子执行 initializeState() 和 open() 方法的过程。

```
private void initializeStateAndOpen() throws Exception {
    StreamOperator<?>[] allOperators = operatorChain.getAllOperators();
    for (StreamOperator<?> operator : allOperators) {
        if (null != operator) {
            operator.initializeState();
            operator.open();
        }
    }
}
```

initializeState() 方法涉及状态的恢复，这会在后文详细分析。open() 方法在各个 StreamOperator

实现类中各不相同，其中比较简单的情况是调用 userFunction 的 open()方法。

### 2. runMailboxLoop

runMailboxLoop()方法真正实现了业务逻辑。前面已经分析过 MailboxProcessor 的设计原理和其中的字段的含义。

### 3. afterInvoke

afterInvoke()方法如下：

```
private void afterInvoke() throws Exception {
   actionExecutor.runThrowing(() -> {
      // close-operators().
      closeAllOperators();
      ...
   });
   ...
   // dispose-operators().
   disposeAllOperators(false);
}
```

上面的代码只保留了与生命周期直接相关的部分。

其中 closeAllOperators()方法如下：

```
private void closeAllOperators() throws Exception {
   StreamOperator<?>[] allOperators = operatorChain.getAllOperators();
   for (int i = allOperators.length - 1; i >= 0; i--) {
      StreamOperator<?> operator = allOperators[i];
      if (operator != null) {
         operator.close();
      }
      ...
   }
}
```

该方法为每个 StreamOperator 调用了 close()方法。每个 StreamOperator 实现类的 close()方法都不相同，比如对于 AbstractUdfStreamOperator，其 close()方法会调用 userFunction 的 close()方法。

disposeAllOperators()方法如下：

```
private void disposeAllOperators(boolean logOnlyErrors) throws Exception {
   if (operatorChain != null && !disposedOperators) {
      for (StreamOperator<?> operator : operatorChain.getAllOperators()) {
         if (operator == null) {
            continue;
         }
         if (!logOnlyErrors) {
            operator.dispose();
         }
         else {
            try {
               operator.dispose();
            }
            catch (Exception e) {
               ...
            }
         }
```

```
        }
        disposedOperators = true;
    }
}
```

上面代码的核心逻辑就是调用 StreamOperator 的 dispose()方法。每个 StreamOperator 实现类的 dispose()方法各不相同。以 AbstractStreamOperator 为例，它的 dispose()方法主要用于关闭状态后端：

```
public void dispose() throws Exception {
    ...
    try {
        if (taskCloseableRegistry == null ||
            taskCloseableRegistry.unregisterCloseable(operatorStateBackend)) {
            operatorStateBackend.close();
        }
    } catch (Exception e) {
        ...
    }
    try {
        if (taskCloseableRegistry == null ||
            taskCloseableRegistry.unregisterCloseable(keyedStateBackend)) {
            keyedStateBackend.close();
        }
    } catch (Exception e) {
        ...
    }
    try {
        if (operatorStateBackend != null) {
            operatorStateBackend.dispose();
        }
    } catch (Exception e) {
        ...
    }
    try {
        if (keyedStateBackend != null) {
            keyedStateBackend.dispose();
        }
    } catch (Exception e) {
        ...
    }
}
```

4. cleanUpInvoke

生命周期中最后的清理工作放在 cleanUpInvoke()方法中进行。

该方法如下：

```
private void cleanUpInvoke() throws Exception {
    isRunning = false;
    setShouldInterruptOnCancel(false);
    Thread.interrupted();
    tryShutdownTimerService();
    try {
        cancelables.close();
        shutdownAsyncThreads();
```

```
    } catch (Throwable t) {
      ...
    }
    try {
       cleanup();
    } catch (Throwable t) {
      ...
    }
    disposeAllOperators(true);
    if (operatorChain != null) {
       actionExecutor.run(() -> operatorChain.releaseOutputs());
    } else {
       recordWriter.close();
    }
    mailboxProcessor.close();
}
```

上面代码的核心逻辑就是调用各个组件的 close() 方法。其中 cleanup() 方法如下:

```
protected void cleanup() throws Exception {
   if (inputProcessor != null) {
      inputProcessor.close();
   }
}
```

默认的实现方式是调用 inputProcessor 的 close() 方法。但部分 StreamTask (如 SourceStreamTask) 并没有用到 inputProcessor。在 SourceStreamTask 中重写了该方法,并且在该方法中没有做任何实现。

## 7.3.5　DataSourceTask、BatchTask 和 DataSinkTask

在批任务中,用 DataSourceTask、BatchTask 和 DataSinkTask 等 AbstractInvokable 类的实现类来表示任务。这些类虽然在字段和方法方面与 StreamTask 的有所不同,但是主要的设计思想和生命周期是类似的。下面分别介绍这 3 个实现类中主要字段的含义和 invoke() 方法的实现。

### 1. DataSourceTask

如果是 source 任务,那么 AbstractInvokable 的实现类就是 DataSourceTask。它的定义如下:

```
public class DataSourceTask<OT> extends AbstractInvokable
```

其中的泛型 OT 表示要输出的数据的类型。

DataSourceTask 中的重要字段如下。

- eventualOutputs:List <RecordWriter<?>>类型,表示物理出边的 RecordWriter 列表。RecordWriter 用于将数据写入 ResultPartition,表示数据的输出。RecordWriter 的概念在之前介绍 StreamTask 时已经介绍过。
- output:Collector <OT>类型,表示算子的输出。如果有多个算子链接在一起,则这个链接关系会被封装在 output 中,可类比 OperatorChain 中的 chainEntryPoint 字段。
- format:InputFormat <OT, InputSplit>类型,表示数据的输入方式。
- serializerFactory:TypeSerializerFactory<OT>类型,表示数据的序列化器的工厂类。
- config:TaskConfig 类型,表示算子的配置信息,可类比 StreamTask 类中用于表示算子配置信息的 StreamConfig 类型的对象。
- chainedTasks：ArrayList<ChainedDriver<?, ?>>类型，可类比 OperatorChain 中的 allOperators 字段。

这里有一个新的概念——ChainedDriver。它用于实现 Collector 接口，该接口表示一个算子的输出。在 StreamTask 中常见的 Output 接口就继承自 Collector 接口。另外，ChainedDriver 又表示链接在一起的算子的驱动，其中封装了业务逻辑。

DataSourceTask 的 invoke() 方法如下：

```
public void invoke() throws Exception {
    // 初始化输入与输出
    initInputFormat();
    try {
        initOutputs(getUserCodeClassLoader());
    } catch (Exception ex) {
        ...
    }
    ...
    final TypeSerializer<OT> serializer = this.serializerFactory.getSerializer();
    try {
        // 初始化算子
        BatchTask.openChainedTasks(this.chainedTasks, this);
        // 获取要读取的分片
        final Iterator<InputSplit> splitIterator = getInputSplits();
        // 遍历每个分片
        while (!this.taskCanceled && splitIterator.hasNext())
        {
            final InputSplit split = splitIterator.next();
            final InputFormat<OT, InputSplit> format = this.format;
            format.open(split);
            try {
                final Collector<OT> output = new CountingCollector<>(this.output, numRecordsOut);
                if (objectReuseEnabled) {
                    ...
                } else {
                    // 获取数据并向下游发送
                    while (!this.taskCanceled && !format.reachedEnd()) {
                        OT returned;
                        if ((returned = format.nextRecord(serializer.createInstance())) != null) {
                            output.collect(returned);
                        }
                    }
                }
            } finally {
                // 关闭输入
                format.close();
            }
        }
        // 关闭算子
        BatchTask.closeChainedTasks(this.chainedTasks, this);
        // 关闭输出
        this.output.close();
    }
    catch (Exception ex) {
```

```
            ...
        } finally {
            // 清理工作
            BatchTask.clearWriters(eventualOutputs);
            if (this.format != null && RichInputFormat.class.isAssignableFrom(this.format.getClass())) {
                // 关闭输入
                ((RichInputFormat) this.format).closeInputFormat();
            }
        }
    }
```

该方法中的核心逻辑已体现在注释中。虽然其中的字段、临时变量的类型以及调用的方法与 **StreamTask** 的不同，但是在流程上仍有许多相似之处，比如一开始都会构造算子并进行初始化，然后执行业务逻辑，最后对资源进行关闭、清理。

initInputFormat()方法会初始化 format 字段。它在批处理任务中表示数据源的输入。initOutputs() 方法用于初始化 output 字段、初始化每个算子的输出，以及填充 chainedTasks 字段和 eventualOutputs 字段等，其逻辑与构造 OperatorChain 的类似：

```
    private void initOutputs(ClassLoader cl) throws Exception {
        this.chainedTasks = new ArrayList<ChainedDriver<?, ?>>();
        this.eventualOutputs = new ArrayList<RecordWriter<?>>();
        this.output = BatchTask.initOutputs(this, cl, this.config, this.chainedTasks, this.eventualOutputs,
            getExecutionConfig(), getEnvironment().getAccumulatorRegistry().getUserMap());

    }
    public static <T> Collector<T> initOutputs(AbstractInvokable containingTask,
ClassLoader cl, TaskConfig config,
                            List<ChainedDriver<?, ?>> chainedTasksTarget,
                            List<RecordWriter<?>> eventualOutputs,
                            ExecutionConfig executionConfig,
                            Map<String, Accumulator<?,?>> accumulatorMap)
    throws Exception
    {
        final int numOutputs = config.getNumOutputs();
        final int numChained = config.getNumberOfChainedStubs();
        if (numChained > 0) {
          ...
            @SuppressWarnings("rawtypes")
            Collector previous = null;
            for (int i = numChained - 1; i >= 0; --i)
            {
                final ChainedDriver<?, ?> ct;
                try {
                    // 通过反射机制构造算子驱动
                    Class<? extends ChainedDriver<?, ?>> ctc = config.getChainedTask(i);
                    ct = ctc.newInstance();
                }
                catch (Exception ex) {
                    ...
                }
                // 获取对应算子的 TaskConfig 配置信息
                final TaskConfig chainedStubConf = config.getChainedStubConfig(i);
```

```java
            final String taskName = config.getChainedTaskName(i);
            if (i == numChained - 1) {
                // 获取最后一个算子的输出
                previous = getOutputCollector(containingTask, chainedStubConf, cl,
eventualOutputs, 0, chainedStubConf.getNumOutputs());
            }
            // 算子驱动的 setup()方法
            ct.setup(chainedStubConf, taskName, previous, containingTask, cl, executionConfig,
accumulatorMap);
            chainedTasksTarget.add(0, ct);
            previous = ct;
        }
        return (Collector<T>) previous;
    }
    return getOutputCollector(containingTask , config, cl, eventualOutputs, 0, numOutputs);
}
```

在上面的方法中，核心逻辑可以概括为实例化算子驱动、初始化算子驱动。注意，这是一个循环的过程，每次初始化一个算子驱动时都会将上一个算子驱动作为参数传入，以此维护算子间输入输出的关系。在构造链接中最后一个算子驱动时，调用 getOutputCollector()方法构造了它的输出，该方法的返回值的类型为 OutputCollector，该类不是算子驱动，仅表示数据的输出。getOutputCollector()方法会在后文分析数据传输时详细介绍。

setup()方法如下：

```java
public void setup(TaskConfig config, String taskName, Collector<OT> outputCollector,
        AbstractInvokable parent, ClassLoader userCodeClassLoader, ExecutionConfig
executionConfig,
        Map<String, Accumulator<?,?>> accumulatorMap)
{
    this.config = config;
    this.taskName = taskName;
    this.userCodeClassLoader = userCodeClassLoader;
    this.metrics = parent.getEnvironment().getMetricGroup().getOrAddOperator(taskName);
    this.numRecordsIn = this.metrics.getIOMetricGroup().getNumRecordsInCounter();
    this.numRecordsOut = this.metrics.getIOMetricGroup().getNumRecordsOutCounter();
    this.outputCollector = new CountingCollector<>(outputCollector, numRecordsOut);
    Environment env = parent.getEnvironment();
    if (parent instanceof BatchTask) {
        this.udfContext = ((BatchTask<?, ?>) parent).createRuntimeContext(metrics);
    } else {
        this.udfContext = new DistributedRuntimeUDFContext(env.getTaskInfo(),
userCodeClassLoader,
            parent.getExecutionConfig(), env.getDistributedCacheEntries(), accumulatorMap,
metrics
        );
    }
    this.executionConfig = executionConfig;
    this.objectReuseEnabled = executionConfig.isObjectReuseEnabled();
    setup(parent);
}
```

在该方法中，主要是对 ChainedDriver 中的字段赋值，其中 outputCollector 字段就会维护上下游输入输出的关系。最后的 setup()方法在实现类中实现，以 ChainedMapDriver 为例：

```
public void setup(AbstractInvokable parent) {
    final MapFunction<IT, OT> mapper =
        BatchTask.instantiateUserCode(this.config, userCodeClassLoader, MapFunction.class);
    this.mapper = mapper;
    FunctionUtils.setFunctionRuntimeContext(mapper, getUdfRuntimeContext());
}
```

这里实例化了业务代码中定义的 Function。

DataSourceTask 的 invoke()方法中后面的操作基本上就是实现算子驱动中定义的生命周期，如 BatchTask.openChainedTasks()方法：

```
public static void openChainedTasks(List<ChainedDriver<?, ?>> tasks, AbstractInvokable parent) throws Exception {
    for (int i = 0; i < tasks.size(); i++) {
        final ChainedDriver<?, ?> task = tasks.get(i);
        task.openTask();
    }
}
```

其中调用的就是算子驱动中定义的 openTask()方法。仍以 ChainedMapDriver 为例：

```
public void openTask() throws Exception {
    Configuration stubConfig = this.config.getStubParameters();
    BatchTask.openUserCode(this.mapper, stubConfig);
}
```

这里最终实际上调用了 RichFunction 中的 open()方法。

再如 BatchTask.closeChainedTasks()方法：

```
public static void closeChainedTasks(List<ChainedDriver<?, ?>> tasks, AbstractInvokable parent) throws Exception {
    for (int i = 0; i < tasks.size(); i++) {
        final ChainedDriver<?, ?> task = tasks.get(i);
        task.closeTask();
    }
}
```

其中调用的就是算子驱动中定义的 closeTask()方法。仍以 ChainedMapDriver 为例：

```
public void closeTask() throws Exception {
    BatchTask.closeUserCode(this.mapper);
}
```

这里最终实际上调用了 RichFunction 中的 close()方法。

在向下游算子发送数据时，首先将 output 字段封装成 CountingCollector 对象，然后调用其 collect()方法：

```
public void collect(OUT record) {
    this.numRecordsOut.inc();
    this.collector.collect(record);
}
```

这里的 collector 字段就是传入的 output 字段，因此核心逻辑就是调用 output 的 collect()方法。前文已经介绍过，这里的 output 既是输出，也是算子驱动。以 ChainedMapDriver 为例，这里的 collect()方法进行的是下面的操作：

```
public void collect(IT record) {
   try {
      this.numRecordsIn.inc();
      this.outputCollector.collect(this.mapper.map(record));
   } catch (Exception ex) {
      ...
   }
}
```

这里用到了先前初始化的 mapper，也就是自定义的业务逻辑。经过 mapper 处理后，数据被 outputCollector 继续向下游传递。outputCollector 实际上封装了下一个算子驱动，层层传递，直到最后被写入 ResultPartition。

2. BatchTask

BatchTask，顾名思义，表示批处理任务。BatchTask 的定义如下：

```
public class BatchTask<S extends Function, OT> extends AbstractInvokable implements
TaskContext<S, OT>
```

注意，其中的泛型 S 表示 Function 的子类，OT 表示输出的数据类型。

BatchTask 主要有以下字段。

- driver：Driver<S, OT>类型，表示任务的驱动。不同类型的任务会由不同的 Driver 实现类来执行。
- stub：S 类型，S 表示 Function 的子类，即业务代码中定义的 userFunction。
- output：Collector<OT>类型，表示算子的输出。
- eventualOutputs：List<RecordWriter<?>>类型，与 DataSourceTask 中同名字段的含义相同。
- inputReaders：MutableReader<?>[]类型，表示数据的输入。
- inputIterators：MutableObjectIterator<?>[]类型，表示输入的迭代器。
- localStrategies：CloseableInputProvider<?>[]类型，表示输入数据的本地策略，用于构造数据的输入。
- inputs：MutableObjectIterator<?>[]类型，表示输入的迭代器。它的值可能直接来源于 inputIterators 字段。
- inputSerializers：TypeSerializerFactory<?>[]类型，表示输入数据的序列化器的工厂类。
- inputComparators：TypeComparator<?>[]类型，表示输入数据的比较器。
- config：TaskConfig 类型。
- chainedTasks：ArrayList<ChainedDriver<?, ?>>类型。

其中 MutableReader、MutableObjectIterator、TypeComparator 等概念会在后文介绍数据传输时详细介绍，这里只需知道它们是初始化过程中需要进行初始化的重要对象。Driver 与之前介绍过的 ChainedDriver 的含义类似，只不过它表示第一个算子的驱动。

BatchTask 的 invoke()方法如下：

```
public void invoke() throws Exception {
   ...
   // 初始化第一个算子
   final Class<? extends Driver<S, OT>> driverClass = this.config.getDriver();
```

```java
    this.driver = InstantiationUtil.instantiate(driverClass, Driver.class);
    // 初始化输入
    initInputReaders();
    ...
    // 初始化输出
    initOutputs();
    ...
    try {
        try {
            ...
            // 初始化序列化器和比较器
            initInputsSerializersAndComparators(numInputs, numComparators);
            ...
            // 初始化本地策略
            initLocalStrategies(numInputs);
        }
        catch (Exception e) {
            ...
        }
        // 初始化任务
        initialize();
        ...
        // 执行业务逻辑
        run();
    }
    finally {
        // 清理和关闭资源
        closeLocalStrategiesAndCaches();
        clearReaders(inputReaders);
        clearWriters(eventualOutputs);
    }
}
```

除了初始化一些 BatchTask 独有的字段，其余的操作与前面介绍的 AbstractInvokable 实现类的基本一致。initInputReaders()方法初始化了 inputReaders 字段。initOutputs()方法初始化了输出，与 DataSourceTask 中的过程类似：

```java
protected void initOutputs() throws Exception {
    this.chainedTasks = new ArrayList<ChainedDriver<?, ?>>();
    this.eventualOutputs = new ArrayList<RecordWriter<?>>();
    ClassLoader userCodeClassLoader = getUserCodeClassLoader();
    this.accumulatorMap = getEnvironment().getAccumulatorRegistry().getUserMap();
    this.output = initOutputs(this, userCodeClassLoader, this.config, this.chainedTasks, this.eventualOutputs,
            this.getExecutionConfig(), this.accumulatorMap);
}
```

因此，在这个方法中也构建了算子的输入输出关系。

initInputsSerializersAndComparators()方法初始化了序列化器和比较器，实际上还同时初始化了 inputIterators 字段。initLocalStrategies()方法初始化了本地策略，同时初始化了 inputs 字段。

initialize()方法会调用 driver 的 setup()方法：

```java
protected void initialize() throws Exception {
    try {
```

```java
      this.driver.setup(this);
   }
   catch (Throwable t) {
      ...
   }
   // 初始化自定义函数
   try {
      final Class<? super S> userCodeFunctionType = this.driver.getStubType();
      if (userCodeFunctionType != null) {
         this.stub = initStub(userCodeFunctionType);
      }
   } catch (Exception e) {
      ...
   }
}
```

以 MapDriver 为例，setup()方法如下：

```java
public void setup(TaskContext<MapFunction<IT, OT>, OT> context) {
   this.taskContext = context;
   this.running = true;
   ExecutionConfig executionConfig = taskContext.getExecutionConfig();
   this.objectReuseEnabled = executionConfig.isObjectReuseEnabled();
}
```

getStubType()方法如下：

```java
public Class<MapFunction<IT, OT>> getStubType() {
   final Class<MapFunction<IT, OT>> clazz = (Class<MapFunction<IT, OT>>) (Class<?>) MapFunction.class;
   return clazz;
}
```

最后调用 initStub()方法来实例化自定义函数，原理与 StreamTask 中的类似，这里不再赘述。

在 invoke()方法中调用 run()方法用于执行业务逻辑：

```java
protected void run() throws Exception {
   try {
      // 准备驱动
      try {
         this.driver.prepare();
      }
      catch (Throwable t) {
         ...
      }
      // 初始化算子
      BatchTask.openChainedTasks(this.chainedTasks, this);
      if (this.stub != null) {
         try {
            // 初始化函数
            Configuration stubConfig = this.config.getStubParameters();
            FunctionUtils.openFunction(this.stub, stubConfig);
            stubOpen = true;
         }
         catch (Throwable t) {
            ...
         }
```

```java
        }
        // 执行业务代码
        this.driver.run();
        // 关闭函数
        if (this.running && this.stub != null) {
            FunctionUtils.closeFunction(this.stub);
            stubOpen = false;
        }
        // 关闭任务
        BatchTask.closeChainedTasks(this.chainedTasks, this);
        // 关闭输出
        this.output.close();
    }
    catch (Exception ex) {
        ...
    }
    finally {
        // 清理驱动
        this.driver.cleanup();
    }
}
```

这个方法中包含之前介绍过的许多生命周期。准备驱动的 prepare() 方法中的每个驱动有自己的实现，MapDriver 中没有实现任何逻辑，后文介绍数据传输时，会以其他驱动为例对这个过程进行分析。

driver 的 run() 方法实现了业务逻辑，以 MapDriver 为例，该过程如下：

```java
public void run() throws Exception {
    final MutableObjectIterator<IT> input = this.taskContext.getInput(0);
    final MapFunction<IT, OT> function = this.taskContext.getStub();
    final Collector<OT> output = new CountingCollector<>(this.taskContext.getOutputCollector(), numRecordsOut);
    if (objectReuseEnabled) {
        ...
    }
    else {
        IT record = null;
        // 利用 input 读取数据，利用 output 将数据发送给下游
        while (this.running && ((record = input.next()) != null)) {
            numRecordsIn.inc();
            output.collect(function.map(record));
        }
    }
}
```

这里用到了 MutableObjectIterator 类型的 input 变量来读取数据，然后利用 output 发送数据。output 发送数据的过程与 DataSourceTask 中的类似。input 读取数据的过程会在后文介绍数据传输时详细分析。

最后，invoke() 方法的 finally 代码块会对资源进行清理和关闭。

3. DataSinkTask

如果是 sink 任务，那么 AbstractInvokable 的实现类就是 DataSinkTask。它的定义如下：

```java
public class DataSinkTask<IT> extends AbstractInvokable
```

其中泛型 IT 表示要输出的数据的类型。

DataSinkTask 中的重要字段如下。

- format：OutputFormat<IT>类型，表示数据的输出方式。
- inputReader：MutableReader<?>类型，表示数据的输入。
- inputTypeSerializerFactory：TypeSerializerFactory<IT>类型，表示输入数据的序列化器的工厂类。
- localStrategy：CloseableInputProvider<IT>类型，表示输入数据的本地策略，用于构造数据的输入。
- config：TaskConfig 类型。

这些字段的数据类型和含义都在前文介绍过。

DataSinkTask 的 invoke()方法如下：

```
public void invoke() throws Exception {
    // 初始化输入与输出
    initOutputFormat();
    try {
      initInputReaders();
    } catch (Exception e) {
      ...
    }
    ...
    try {
      // 初始化本地策略、序列化器等
      MutableObjectIterator<IT> input1;
      switch (this.config.getInputLocalStrategy(0)) {
      case NONE:
         localStrategy = null;
         input1 = reader;
         break;
      case SORT:
         try {
            TypeComparatorFactory<IT> compFact = this.config.getInputComparator(0,
               getUserCodeClassLoader());
            if (compFact == null) {
               throw new Exception("Missing comparator factory for local strategy on input " + 0);
            }
            UnilateralSortMerger<IT> sorter = new UnilateralSortMerger<IT>(
               getEnvironment().getMemoryManager(),
               getEnvironment().getIOManager(),
               this.reader, this, this.inputTypeSerializerFactory, compFact.createComparator(),
               this.config.getRelativeMemoryInput(0), this.config.getFilehandlesInput(0),
               this.config.getSpillingThresholdInput(0),
               this.config.getUseLargeRecordHandler(),
               this.getExecutionConfig().isObjectReuseEnabled());
            this.localStrategy = sorter;
            input1 = sorter.getIterator();
         } catch (Exception e) {
            ...
         }
```

```
            break;
        default:
            ...
    }
    final TypeSerializer<IT> serializer = this.inputTypeSerializerFactory.getSerializer();
    final MutableObjectIterator<IT> input = input1;
    final OutputFormat<IT> format = this.format;
    // 开启输出
    format.open(this.getEnvironment().getTaskInfo().getIndexOfThisSubtask(), this.getEnvironment().getTaskInfo().getNumberOfParallelSubtasks());
        if (objectReuseEnabled) {
            ...
        } else {
            IT record;
            // 执行业务逻辑
            while (!this.taskCanceled && ((record = input.next()) != null)) {
                numRecordsIn.inc();
                format.writeRecord(record);
            }
        }
        // 关闭输出
        if (!this.taskCanceled) {
            this.format.close();
            this.format = null;
        }
    }
    catch (Exception ex) {
        ...
    }
    finally {
        // 清理和关闭资源
        ...
    }
}
```

分析完 DataSourceTask 和 BatchTask 后，再分析 DataSinkTask 的 invoke()方法就会发现其流程基本是一致的。一开始也是初始化输入和输出。因为后面没有链接在一起的算子了，所以初始化输出就是初始化 format，初始化输入就是初始化 inputReader。接着初始化本地策略、序列化器等。后面开启输出、执行业务逻辑、关闭输出以及 finally 代码块中清理和关闭资源的逻辑也与 DataSourceTask 和 BatchTask 中的基本一致。

## 7.4 总结

本章以作业管理器端构造出来的描述符为起点，分析了这些描述符在任务管理器端对应对象的初始化逻辑。初始化完成后，任务就会开始执行，其中重要的环节是 AbstractInvokable 对象的初始化和运行。本章重点介绍了整个过程中流处理相关实现类的重要概念和生命周期。

# 第 8 章

# 数据传输

第 7 章分析了 Task 和 AbstractInvokable 实现类的生命周期，本章分析数据如何在任务之间流转。

数据的流转可以分为两个过程，一个过程是数据在链接在一起的算子中流转，即在同一个任务中流转，这部分内容在第 7 章已经介绍过，本章在必要时会再次提及；另一个过程是数据在不同的任务间传输，这些任务可能处于同一个任务管理器，也可能在不同的物理节点上，需要分别考虑本地传输和远程传输两种不同的实现。

对于数据在不同的任务间传输，还涉及分布式计算框架中的一个重要概念——混洗。这是所有基于 MapReduce 计算模型设计的分布式计算框架都必须要考虑的。本章会详细分析 Flink 对于这个过程的设计思想和实现原理。

第 7 章介绍了两个重要的概念——ResultPartition 和 InputGate。其中 ResultPartition 负责任务的输出，InputGate 负责任务的输入，Flink 中的数据传输的设计几乎都是基于这两者实现的。本章将介绍这两者在数据传输中发挥的作用。

希望在学习完本章后，读者能够了解：
- ResultPartition 和 InputGate 的数据结构以及其中的主要字段和方法；
- 数据在本地传输与远程传输的区别；
- 流处理任务和批处理任务在数据传输方式上的差别；
- 混洗的本质；
- 反压机制的原理。

## 8.1 基本概念与设计思想

要理解 Flink 的数据传输方式，不仅需要了解 ResultPartition 和 InputGate 的字段和方法，还需要"跳出" Flink 框架本身，从更"高"的视角来思考分布式计算框架在进行数据传输时需要考虑什么。本节首先会回顾从 JobGraph 到物理执行图的转换过程，但会将重点放在 ResultPartition 和 InputGate 上。读者可以重新熟悉 ResultPartition 和 InputGate 的由来。回顾该转换过程，可从中观察到批处理任务和流处理任务"走"的是相同的流程，但有一些变量使不同的任务有了不同的调度方式、数据传输方式，并且这些不同之处都是可插拔的。随后，本节会以 Hadoop 和 Spark 为例，分析混洗的本质。读者可以从中看到不同的分布式计算框架在混洗过程中的异曲同工之处，进而在学习 8.1.3 节中介绍的 Flink 的混洗机制时，就可以迅速理解其内在的设计思想，并且可以推测出 Flink 未来的优化方向。最后，本节会介绍 Flink 中常被提及的一个概念——反压。

## 8.1.1 从逻辑执行图到物理执行图

第 4 章详细分析了批处理任务和流处理任务执行图的转换流程。无论是批处理任务的执行图还是流处理任务的执行图，最终都会经由 JobGraph 转换成 ExecutionGraph，进而在调度层面被转换成调度层面的对象并被部署到任务管理器端执行。虽然在任务管理器端并没有一个类来表示整体的执行图，但是从第 7 章可了解到，任务管理器端与任务执行、输入输出直接相关的对象都是由作业管理器端执行图中的对象转换而来的，并且它们有一一对应的关系，因此可以将任务管理器端的这些对象所构成的逻辑结构称为"物理执行图"。这个过程中批处理任务和流处理任务的执行计划用同样的数据结构表示，区别仅是某个字段的值不同。

从 JobGraph 到 ExecutionGraph 的过程如图 8-1 所示。

在 JobGraph 中，用 IntermediateDataSet 表示上游 JobVertex 的输出，用 JobEdge 表示下游 JobVertex 的输入。在 JobGraph 转换成 ExecutionGraph 后，一个 IntermediateDataSet 对应一个 IntermediateResult，而此时每个 JobVertex 会按照并行度生成对应个数的 ExecutionVertex，每个 ExecutionVertex 都对应一个 IntermediateResultPartition。下游的每个 ExecutionVertex 实例针对每个上游的输入，会根据其 IntermediateResultPartition 的个数（上游并行度）生成对应个数的 ExecutionEdge，并分别与上游的每个 IntermediateResultPartition 连接。

从图 8-1 可以观察到，每个上游的 IntermediateResultPartition 可能会连接多个下游的 ExecutionEdge。在从 ExecutionGraph 到物理执行图的转换中，连接关系会进一步细化，如图 8-2 所示。

在物理执行图中，每个 ExecutionEdge 对应一个 InputChannel，同一个并行实例中的针对同一个上游输入的所有 InputChannel 属于同一个 InputGate。由于在物理执行图层面，每个并行实例都可能运行在不同的进程中，因此不方便使用一个对象表示整个 IntermediateResult，而每一个 IntermediateResultPartition 则会对应一个 ResultPartition。在 ExecutionGraph 中，每个 IntermediateResultPartition 可能会连接多个 ExecutionEdge，在物理执行图中则会生成对应个数的 ResultSubpartition 与下游的 InputChannel 连接。数据的重分区，实际上就发生在每个并行实例将输出数据写入每个 ResultSubpartition 的过程中。

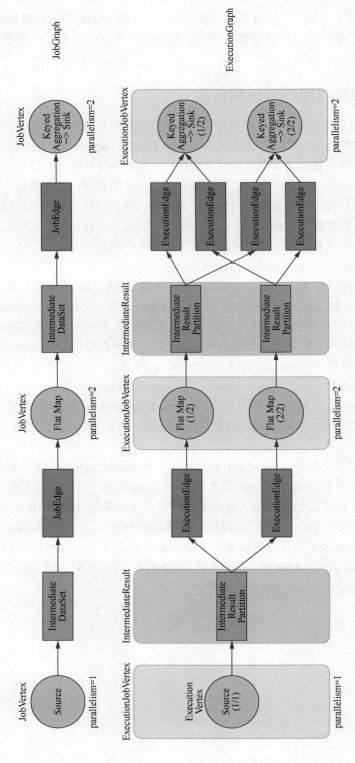

图 8-1 从 JobGraph 到 ExecutionGraph 的转换过程

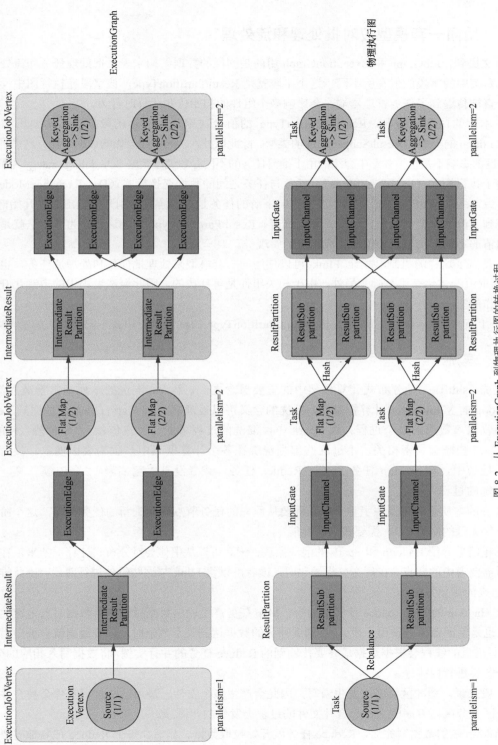

图 8-2 从 ExecutionGraph 到物理执行图的转换过程

## 8.1.2 用同一套模型应对批处理和流处理

前文提到，JobGraph 和 ExecutionGraph 用的是同样的数据结构来表示批处理任务和流处理任务，只是其中的字段的值有所不同。这个字段就是 ResultPartitionType，该字段在执行图中反复出现，一直被传递到了任务管理器端，会影响多个组件的初始化和运行时行为。

在本章的讨论范围内，ResultPartitionType 的值会决定任务的输出数据是否需要压缩以及 ResultPartition 的类型和 ResultSubpartition 的类型。除此以外，它的值还会影响数据传输。这里讲的对任务调度的影响不是指第 6 章中介绍的由于批流任务的不同导致调度策略 SchedulingStrategy 不同，而是指由于执行模式（ExecutionMode）、任务与任务之间的数据交换模式（DataExchangeMode）的不同导致 ResultPartitionType 不同，进而影响下游的任务是在上游数据刚开始输出时就开始部署，还是等到上游数据全部输出完毕后再开始部署。ResultPartitionType 影响这些行为的方法就是根据自身的值的不同，决定相关组件生成相应的实现类。

正是有了这样的机制，才让 Flink 可以使用同一套模型来处理批任务和处理流任务，也使得 Flink 的批流一体化有了基础。另外，由于相关组件是可插拔的，因此混洗过程中的定制化开发有了可操作的空间。

8.2.1 节介绍源码时，会详细分析 ResultPartitionType 的赋值过程和其不同值带来的影响。

## 8.1.3 混洗

混洗（shuffle）是分布式计算框架中的重要概念之一。然而，混洗的本质却常常被人混淆。在 Hadoop 的 MapReduce 计算框架中，混洗的含义可以被理解为从 Map 任务输出数据到 Reduce 任务接收输入数据的整个过程；在 Spark 中，混洗的过程可能与"宽依赖""窄依赖""阶段"（Stage）这些概念紧密相关。不过这些都是混洗在各个计算框架中产生的表面现象。换言之，在其他框架中，即便 Map 任务后不是 Reduce 任务，或者没有"宽依赖""窄依赖"等，也会出现混洗的过程。

混洗的本质是数据发生了重分区，需要从原先的任务节点移动到新的任务节点。这个阶段往往会发生大量的网络 I/O，需要极高的性能。

这里简单介绍 Hadoop 和 Spark 的混洗原理。读者可以从中体会对分布式计算框架来说在混洗阶段所需要考虑的要点。在后文分析源码时，读者可以将 Flink 与这些组件进行对比，由此推测出 Flink 未来的优化方向。

在 Hadoop 的 MapReduce 计算框架中，Map 任务产生输出数据时会先将数据写入内存的缓冲区（这也是保证效率的常用手段）。等到缓冲区的数据达到某个阈值时，这些数据就会被溢写到磁盘。溢写磁盘的线程会根据数据最终要传输到的 Reduce 任务的并行实例，将数据写入相应的分区。每个分区会进行内排序。

一般来说，缓冲区会多次发生溢写，因此会产生多个文件。最后这些溢写文件会被合并成一个已分区且分区内有序的文件。合并文件的过程会发生归并排序。

当 Map 端的数据写完后，Reduce 任务就开始拉取数据。显然，每个 Reduce 任务都需要到各个 Map 任务的输出文件中拉取自己分区的数据。拉取数据的过程也会进行归并排序。当所有数据在 Reduce 端被排好序后就可以开始聚合处理。Hadoop 混洗如图 8-3 所示。

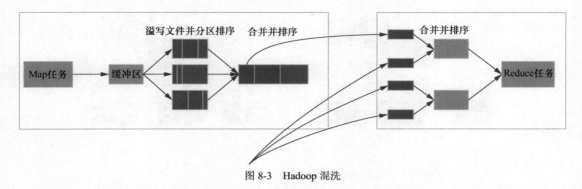

图 8-3　Hadoop 混洗

图 8-3 展示了 Hadoop 的 MapReduce 计算框架中混洗阶段的核心过程，从这个过程中可以看到所谓的 3 次排序。第一次是 Map 端的分区内排序，第二次是 Map 端合并文件时发生的归并排序，第三次是 Reduce 任务拉取到各个 Map 任务的输出数据后进行合并时发生的归并排序。这一设计可以说是后来所有分布式计算框架混洗机制实现的基石。

再来看 Spark 的混洗机制。在 Map 端没有对数据进行排序，而仅分区输出数据。这种机制在 Spark 中被称为哈希混洗（如图 8-4 所示）。它的优点是上游输出时不需要进行任何排序和合并，因此速度非常快，而且从实现上来说也更为简单。

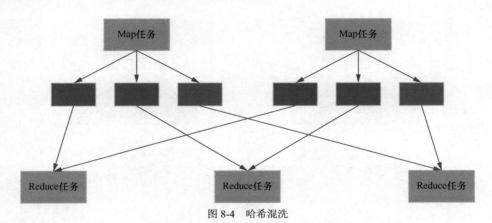

图 8-4　哈希混洗

在哈希混洗过程中，上游任务的每个并行实例会根据数据将要输出到的下游任务的并行实例，将数据写入相应的中间文件。下游任务的并行实例会找到上游任务每个并行实例中的与其自身相对应的文件并进行数据拉取。如果上游任务并行度为 $M$，下游任务并行度为 $R$，那么混洗过程产生的文件数就是 $M \times R$。在生产环境中，这会是一个相当大的数。这么多的小文件不仅会占用相当大的文件句柄数，而且在磁盘 I/O 上会消耗巨大性能。

针对这个问题，Spark 提出了优化方案，即 FileConsolidation（如图 8-5 所示），旨在将由同一个 CPU 核心处理的 Map 任务的输出数据按分区输出到同一个大文件中。这样，中间文件的数量就会从 $M×R$ 降低到 $C$（CPU 核数）$×R$，其中 $C$ 为 CPU 核数。

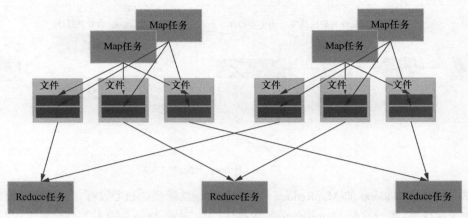

图 8-5　FileConsolidation

不过，这只能缓解而并不能彻底解决中间文件数过多的问题。Spark 1.2 以后的版本推出了与上面介绍的 MapReduce 同样的混洗机制，这在 Spark 中被称为基于排序的混洗（sort-based shuffle），如图 8-6 所示。这样，中间过程产生的文件数就减少到了 $2×$（一个数据文件+一个索引文件）$×C$。在后续版本的迭代中，Spark 还多次对混洗过程进行过优化，但总的来说还是延续了上述思想。

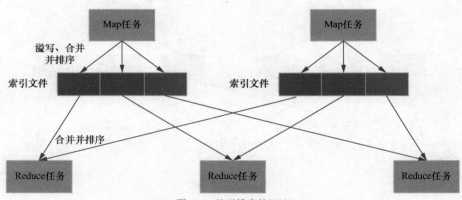

图 8-6　基于排序的混洗

那么，Flink 的混洗机制会采用哪一种方式呢？

Flink 的计算模型与 Hadoop 的 MapReduce 和 Spark 有着本质的区别。Hadoop 的 MapReduce 只能用于批处理，Spark 虽然可以处理无界数据流，但它底层的数据抽象仍然是数据集。Flink 是纯流处理引擎，处理数据的最小单位是单条数据（或者说事件）。对于流处理任务，上游刚刚生产数据时就会通知下游来消费，因此也不可能进行排序、合并，可以理解为哈希混洗。对于批处理任务，从中间过程产生的文件数来看，也应属于哈希混洗。不过，针对某些算子的语义，Flink 还是会在上游数据输出前，在内存中对数据进行排序和聚合，但最终数据仍然是按分区输出到不同文件。这种排序和聚合并不针对对应的并行实例的所有输出数据，也不会针对最终的输出文件，因此与上文介绍的基于排序的混洗有较大的差别。

Flink 之所以采用这种混洗方式，主要是因为其自身的流处理引擎定位。对于处理数据流，显然是用哈希混洗更为自然。

不过在生产环境中，有大量的业务场景需要用 Flink 进行批处理。在这种情况下使用哈希混洗或许就不是最好的选择。目前社区已经提出将基于排序的混洗引入 Flink 的计划，阿里巴巴公司开发的 Blink 分支早已做了这样的尝试。也许在未来 Flink 的批处理中，基于排序的混洗会逐步取代哈希混洗。

## 8.1.4 流量控制

上游向下游传输数据时"天然"存在一个问题，那就是下游消费数据的速率与上游生产数据的速率不一定完全一致。当下游消费数据的速率跟不上上游生产数据的速率时，上游进程中的数据就会在内存中越积越多。为了防止这种情况发生，需要引入流量控制机制，使得当下游任务难以处理时，会以某种方式通知上游任务，最终通知 source 任务，以减慢数据生产的速率。这个机制在 Flink 中叫作反压或背压（back pressure）。

反压机制如图 8-7 所示。

要实现有效的反压机制，就需要做到图 8-7 中的两点。

（1）下游任务的反压需要传递到上游任务中。

（2）反压需要在单个任务中从输出（ResultPartition）传递到输入（InputGate），进而再向上游传递。

Flink 中的反压机制简单而有效。Flink 在网络层有单独的固定大小的缓冲池，当下游任务的缓冲池用满时，就会通知上游任务使其不再发送消息。由于上游任务无法向外发送消息，其网络内存会很快被占满，导致它无法从它的上游读取数据，因此反压传递到了 source 端，会影响从数据源读取数据的速率。

这种下游通知上游的方式在不同版本中有不同的实现。在 Flink 1.5 之前的版本中，这种机制被称为"TCP-based Flow Control"，1.5 版本及之后版本的机制被称为"Credit-based Flow Control"。前者利用的是 TCP 本身的流量控制机制，后者借鉴了前者的思路，在上层做了类似的实现。

TCP 用滑动窗口（sliding window）的方式来实现流量控制，图 8-8～图 8-13 用一个简单的例子阐释了其流量控制机制。

发送端有一个窗口，表示接下来可以发送的数据。这个窗口的大小是变动的，其会由接下来接收端发送回的响应中的字段值决定，字段初始值为 3。接收端有一个固定大小的窗口作为缓冲区，大小为 5。假设发送端发送数据的速率为 3，接收端处理数据的速率为 1，如图 8-8 所示。

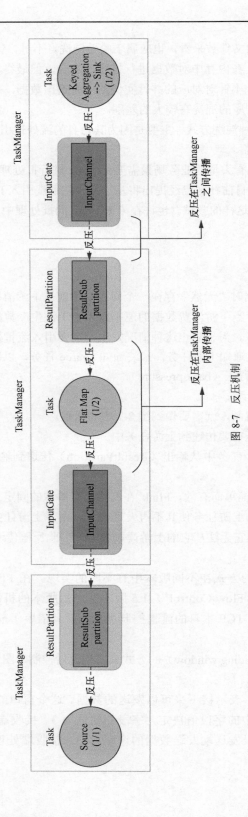

图 8-7 反压机制

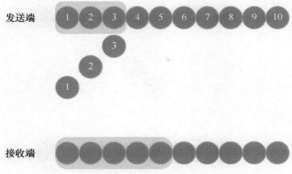

图 8-8　TCP-based Flow Control1

当发送端发送 3 个数据包后，这 3 个数据包被接收端接收，放入缓冲区。此时接收端并没有发送任何响应给发送端，如图 8-9 所示。

图 8-9　TCP-based Flow Control2

接着，接收端消费了一个数据包，其窗口向前移动一个位置。此时接收端发送响应给发送端，响应主要包含两条消息——接下来从哪里开始消费（之前消费到了 3，接下来需要从 4 开始消息）以及发送端的窗口大小（缓冲区剩余大小），如图 8-10 所示。

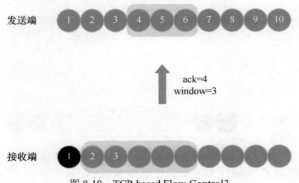

图 8-10　TCP-based Flow Control3

于是如图 8-11 所示，发送端继续将 3 个数据包发送给接收端。

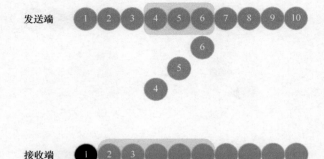

图 8-11　TCP-based Flow Control4

接收端再次消费一个数据包，窗口向前移动。这时响应中需要告知发送端从 7 开始消费，窗口大小为 1，如图 8-12 所示。

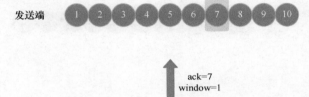

图 8-12　TCP-based Flow Control5

这时就可以看到，因为接收端的消费速率跟不上发送端的发送速率，所以经过几次响应后，发送端的发送速率降低了，如图 8-13 所示。

图 8-13　TCP-based Flow Control6

图 8-14～图 8-16 展示了 Flink 1.5 之前的流量控制机制。

如图 8-14 所示，每个任务管理器（TaskManager）会从堆外申请一定大小的网络缓冲池（network buffer pool），每个任务的结果子分区（ResultSubpartition）和输入通道（InputChannel）都会从中申请内存形成本地缓冲池（local buffer pool）。底层通过套接字进行数据传输。

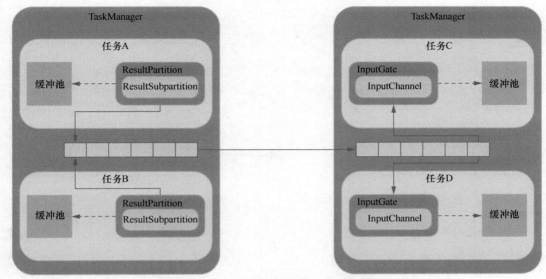

图 8-14　Flink 1.5 版本之前的流控机制（TaskManager 之间）1

如图 8-15 所示，假设数据生产速率大于消费速率，则在任务 C 中 InputChannel 的数据不能及时被消费，因而会申请更多的内存存放数据，直至缓冲池被耗尽。当缓冲池被耗尽时，上层会停止从套接字中读取数据。

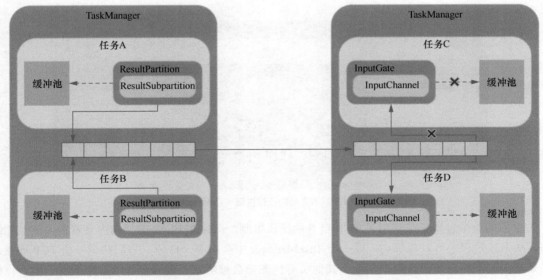

图 8-15　Flink 1.5 版本之前的流控机制（TaskManager 之间）2

如图 8-16 所示，根据前文介绍过的 TCP 流量控制机制可知，上游逐渐无法给下游发送数据，进而造成上游套接字缓冲区被填满。当套接字缓冲区被填满时，ResultSubpartition 无法继续将数据写入其中,消息便停止发送。

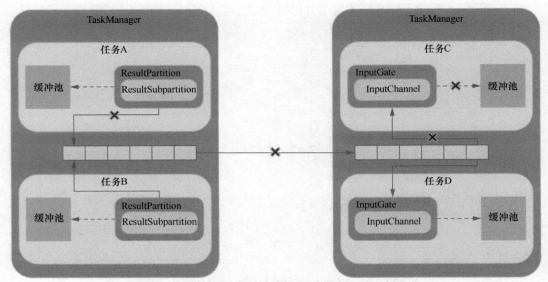

图 8-16　Flink 1.5 版本之前的流控机制(TaskManager 之间)3

从 TaskManager 内部来看,最终 ResultSubpartition 被填满,上游的缓冲池被耗尽,进一步导致 InputChannel 无法继续申请内存存放数据,由此,反压继续传递到了更上游。整个过程如图 8-17～图 8-19 所示。

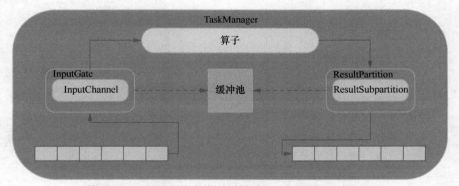

图 8-17　Flink 1.5 版本之前的流控机制(TaskManager 内部)1

上面的整个过程正是利用了 TCP 自身的反压机制,但它在 Flink 的架构内有着很大的性能缺陷。从上述过程中可以观察到,同一个 TaskManager 中会有多个任务,它们共用一个 TCP 连接。如果其中一个任务使链路阻塞,那么其他所有任务都会被阻塞,进而会导致资源闲置、检查点机制无法顺利进行等。另外,由于该机制利用的是底层 TCP 的流量控制机制,因此反压的链路较长,可发现反压的耗时也会更长。

为了解决这个问题,从 Flink 1.5 开始引入了 Credit-based Flow Control。这种方式将流量控制放在了应用层来实现。图 8-20 和图 8-21 展示了该过程。

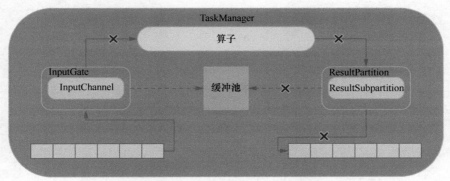

图 8-18　Flink 1.5 版本之前的流控机制（TaskManager 内部）2

图 8-19　Flink 1.5 版本之前的流控机制（TaskManager 内部）3

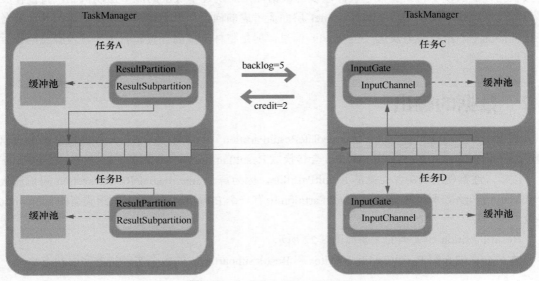

图 8-20　Credit-based Flow Control1

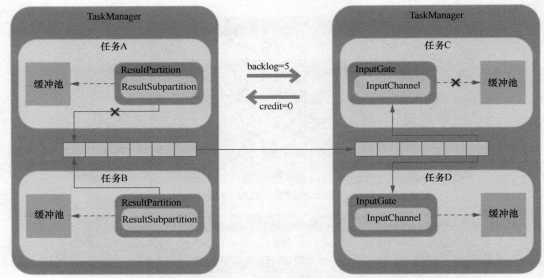

图 8-21　Credit-based Flow Control2

如图 8-20 所示，此时不再通过底层来进行流量控制，而是上游每次会告知下游将要传输的数据量 backlog size，下游根据自身缓冲区的大小进行判断，如果可以接收这些数据，则其会在请求中带上 credit 值，表示它能接收的数据量。

如图 8-21 所示，如果 InputChannel 不能接收数据，则下游发送给上游的 credit 的值为 0。上游接收到该值后便不再给套接字缓冲区发送数据，由此跳过了底层的流量控制机制，直接在应用层形成反压。这时反压变成 InputChannel 层面需考虑的问题，单个通道阻塞不会影响其他任务。在这种情况下，虽然会多发送一些 credit 信息，但是整体上提高了资源利用率，检查点的执行过程也更加顺利。

## 8.2　数据的输出

在 ExecutionGraph 中，每个 IntermediateResultPartition 可以连接多个 ExecutionEdge。在物理执行的层面，每个 IntermediateResultPartition 会转换成 ResultPartitionWriter 对象。ResultPartitionWriter 是一个接口，通常使用的实现类就是 ResultPartition。因为每个 IntermediateResultPartition 可以连接多个 ExecutionEdge，所以在实现类 ResultPartition 中有一个 ResultSubpartition 数组类型的 subpartitions 字段。

ResultPartition 相关类的关系如图 8-22 所示。

本节会详细介绍 ResultPartitionWriter 和 ResultSubpartition 的各个实现类的字段和方法。

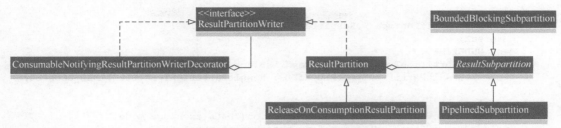

图 8-22　ResultPartition 相关类的关系

## 8.2.1　ResultPartitionType

ResultPartitionWriter 的初始化过程在第 7 章中已经介绍过。每个 ResultPartitionWriter 具体如何初始化都由作业管理器端传来的 ResultPartitionDeploymentDescriptor 决定，ResultPartitionWriter 类封装了 ResultPartitionType 对象。

ResultPartitionType 是枚举类，它包括如下字段。
- isPipelined：用于设置输出的数据是否可以边生产边消费。
- hasBackPressure：用于设置是否产生反压。
- isBounded：用于设置分区是否使用有限个 network buffer。
- isPersistent：当其值为 true 时，分区被消费后不会立即被释放。

对于这些字段，一般主要关注 isPipelined 的值。ResultPartitionType 包括如下几个枚举值：

```
BLOCKING(false, false, false, false),
BLOCKING_PERSISTENT(false, false, false, true),
PIPELINED(true, true, false, false),
PIPELINED_BOUNDED(true, true, true, false);
```

ResultPartitionType 的构造方法如下：

```
ResultPartitionType(boolean isPipelined, boolean hasBackPressure, boolean isBounded,
boolean isPersistent) {

    this.isPipelined = isPipelined;
    this.hasBackPressure = hasBackPressure;
    this.isBounded = isBounded;
    this.isPersistent = isPersistent;
}
```

注意，BLOCKING 和 BLOCKING_PERSISTENT 的 isPipelined 的值为 false，PIPELINED 和 PIPELINED_BOUNDED 的 isPipelined 的值为 true。

ResultPartitionType 的值来自 JobGraph 拓扑中 IntermediateDataSet 类中的同类型字段。由于流处理作业和批处理作业生成 JobGraph 的过程不一样，因此赋值的逻辑需要分开讨论。

流处理作业中，赋值逻辑如下：

```
ResultPartitionType resultPartitionType;
switch (edge.getShuffleMode()) {
    case PIPELINED:
        resultPartitionType = ResultPartitionType.PIPELINED_BOUNDED;
        break;
```

```
    case BATCH:
        resultPartitionType = ResultPartitionType.BLOCKING;
        break;
    case UNDEFINED:
        resultPartitionType = streamGraph.isBlockingConnectionsBetweenChains() ?
            ResultPartitionType.BLOCKING : ResultPartitionType.PIPELINED_BOUNDED;
        break;
    default:
        throw new UnsupportedOperationException("Data exchange mode " +
            edge.getShuffleMode() + " is not supported yet.");
}
```

从上面的代码可以看到，resultPartitionType 变量的值取决于 shuffleMode 的值。在流处理中，该值一般默认为 UNDEFINED。streamGraph 中的 blockingConnectionsBetweenChains 的值默认为 false，一般不会被修改。因此，resultPartitionType 的值在流处理中一般为 PIPELINED_BOUNDED。

批处理作业中，赋值逻辑如下：

```
final ResultPartitionType resultType;
switch (channel.getDataExchangeMode()) {
    case PIPELINED:
        resultType = ResultPartitionType.PIPELINED;
        break;
    case BATCH:
        resultType = channel.getSource().isOnDynamicPath()
            ? ResultPartitionType.PIPELINED
            : ResultPartitionType.BLOCKING;
        break;
    case PIPELINE_WITH_BATCH_FALLBACK:
        throw new UnsupportedOperationException("Data exchange mode " +
            channel.getDataExchangeMode() + " currently not supported.");
    default:
        throw new UnsupportedOperationException("Unknown data exchange mode.");
}
```

此时，可以认为 resultType 的值取决于 channel 对象的 dataExchangeMode 的值。dataExchangeMode 的类型为 DataExchangeMode，它也是枚举类，表示节点间数据交换的模式。它的枚举值有：

```
PIPELINED,
BATCH,
PIPELINE_WITH_BATCH_FALLBACK;
```

第一种模式的含义是数据边生产边消费，接收端向发送端提供反压机制。第二种模式的含义是生产完再消费。第三种模式暂时还没有相应的实现。

该枚举类提供了 3 个静态字段和 1 个静态代码块：

```
private static final DataExchangeMode[] FORWARD = new DataExchangeMode[ExecutionMode.values().length];

private static final DataExchangeMode[] SHUFFLE = new DataExchangeMode[ExecutionMode.values().length];

private static final DataExchangeMode[] BREAKING = new DataExchangeMode[ExecutionMode.values().length];

static {
```

```
    FORWARD[ExecutionMode.PIPELINED_FORCED.ordinal()] = PIPELINED;
    SHUFFLE[ExecutionMode.PIPELINED_FORCED.ordinal()] = PIPELINED;
    BREAKING[ExecutionMode.PIPELINED_FORCED.ordinal()] = PIPELINED;

    FORWARD[ExecutionMode.PIPELINED.ordinal()] = PIPELINED;
    SHUFFLE[ExecutionMode.PIPELINED.ordinal()] = PIPELINED;
    BREAKING[ExecutionMode.PIPELINED.ordinal()] = BATCH;

    FORWARD[ExecutionMode.BATCH.ordinal()] = PIPELINED;
    SHUFFLE[ExecutionMode.BATCH.ordinal()] = BATCH;
    BREAKING[ExecutionMode.BATCH.ordinal()] = BATCH;

    FORWARD[ExecutionMode.BATCH_FORCED.ordinal()] = BATCH;
    SHUFFLE[ExecutionMode.BATCH_FORCED.ordinal()] = BATCH;
    BREAKING[ExecutionMode.BATCH_FORCED.ordinal()] = BATCH;
}
```

注意，这里出现了另一个枚举类 ExecutionMode，表示执行模式。ExecutionMode 枚举类有如下 4 个枚举值：

```
PIPELINED,
PIPELINED_FORCED,
BATCH,
BATCH_FORCED
```

对于前面代码中的 3 个静态字段 FORWARD、SHUFFLE 和 BREAKING，可以理解为上下游任务之间的数据交换方式。FORWARD 表示没有重分区的数据传输，SHUFFLE 表示发生了重分区，BREAKING 表示因为某些原因导致的必须先等上游写完下游才能消费的数据交换方式。

由此可以看出，前面的静态代码块的含义就是指当数据交换方式分别为 FORWARD、SHUFFLE 和 BREAKING 时，在不同执行模式下所对应的 DataExchangeMode 的值。读者应该可以比较容易地理解各种情况的含义。

DataExchangeMode 的值是在调用下面的方法时确定的：

```java
public static DataExchangeMode select(ExecutionMode executionMode, ShipStrategyType shipStrategy,
                                      boolean breakPipeline) {
    if (shipStrategy == null || shipStrategy == ShipStrategyType.NONE) {
        throw new IllegalArgumentException("shipStrategy may not be null or NONE");
    }
    if (executionMode == null) {
        throw new IllegalArgumentException("executionMode may not me null");
    }
    if (breakPipeline) {
        return getPipelineBreakingExchange(executionMode);
    }
    else if (shipStrategy == ShipStrategyType.FORWARD) {
        return getForForwardExchange(executionMode);
    }
    else {
        return getForShuffleOrBroadcast(executionMode);
    }
}
public static DataExchangeMode getForForwardExchange(ExecutionMode mode) {
    return FORWARD[mode.ordinal()];
```

```java
}
public static DataExchangeMode getForShuffleOrBroadcast(ExecutionMode mode) {
    return SHUFFLE[mode.ordinal()];
}
public static DataExchangeMode getPipelineBreakingExchange(ExecutionMode mode) {
    return BREAKING[mode.ordinal()];
}
```

从上面的赋值过程可以看出，批处理任务中的 ResultPartition 的 ResultPartitionType 的值未必都是 BLOCKING。它的值与指定的 ExecutionMode 有关，也与任务之间的数据交换方式有关，而且它的值会影响 ResultPartitionWriter 和 ResultSubpartition 的具体行为。

## 8.2.2　ResultPartitionWriter

前文介绍过，无论是批处理任务还是流处理任务，一个任务中都可能有一个或多个链接在一起的算子。其中最后一个算子的输出会依赖 RecordWriter 类型的字段。RecordWriter 类中依赖了 ResultPartitionWriter 类型的 targetPartition 字段，由此可以将数据写入 ResultPartitionWriter。

这里会对 RecordWriter 进行简单的介绍，可帮助读者了解数据是如何从 RecordWriter 转换到 ResultPartitionWriter 中的。

RecordWriter 是一个抽象类，它有两个实现类——BroadcastRecordWriter 和 ChannelSelectorRecordWriter。这两个实现类分别表示广播数据的输出器和将数据写入特定子分区的输出器。下面是一些 RecordWriter 中的主要字段。

- targetPartition：ResultPartitionWriter 类型，用于将数据真正写入分区。
- numberOfChannels：子分区的个数。
- serializer：RecordSerializer 类型，用于将数据序列化。

RecordWriter 中定义的重要方法是 emit() 方法，表示将数据发送到子分区：

```java
protected void emit(T record, int targetChannel) throws IOException, InterruptedException {
    // 将数据序列化到序列化器中的缓冲区
    serializer.serializeRecord(record);
    // 将数据从序列化器中的缓冲区复制到目标缓冲区
    if (copyFromSerializerToTargetChannel(targetChannel)) {
        // 清理序列化器的缓冲区
        serializer.prune();
    }
}
```

整个过程主要就是利用序列化器提供的方法先将数据序列化到序列化器中的缓冲区，再将数据从其中复制到目标缓冲区。copyFromSerializerToTargetChannel() 方法如下：

```java
protected boolean copyFromSerializerToTargetChannel(int targetChannel) throws IOException,
InterruptedException {
    ...
    // 获取目标缓冲区
    BufferBuilder bufferBuilder = getBufferBuilder(targetChannel);
    // 将数据复制到目标缓冲区
    SerializationResult result = serializer.copyToBufferBuilder(bufferBuilder);
    ...
}
```

将数据写入目标缓冲区的过程还涉及判断缓冲区是否写满等步骤。

getBufferBuilder()是实现类需要实现的方法,表示根据目标子分区索引获取一个 BufferBuilder 对象。BufferBuilder 可以理解为一个用于写入数据的目标缓冲区。

以 ChannelSelectorRecordWriter 为例,getBufferBuilder()方法如下:

```
public BufferBuilder getBufferBuilder(int targetChannel) throws IOException,
InterruptedException {
    if (bufferBuilders[targetChannel] != null) {
      return bufferBuilders[targetChannel];
    } else {
      return requestNewBufferBuilder(targetChannel);
    }
}
```

如果 BufferBuilder 对象之前赋过值,则返回之前的值,否则调用 requestNewBufferBuilder() 方法申请新的 BufferBuilder 对象。

```
public BufferBuilder requestNewBufferBuilder(int targetChannel) throws IOException,
InterruptedException {
    // 获取 BufferBuilder 对象
    BufferBuilder bufferBuilder = targetPartition.getBufferBuilder();
    // 构造 BufferConsumer 对象,将其添加到子分区中
    targetPartition.addBufferConsumer(bufferBuilder.createBufferConsumer(), targetChannel);
    bufferBuilders[targetChannel] = bufferBuilder;
    return bufferBuilder;
}
```

上面的方法中,第一步会申请内存,封装成一个 BufferBuilder 对象并返回,后文会在介绍 ResultPartition 时讲解该过程。随后,BufferBuilder 对象被转换成 BufferConsumer 对象。BufferConsumer 与 BufferBuilder 持有的是相同的数据,但是 BufferBuilder 用于写入数据,BufferConsumer 可以理解为数据的消费者,它可以直接被转换成 Buffer,从而被下游消费。在 addBufferConsumer()方法中,数据被添加进了相应的子分区。

如果一个数据需要被广播,那么在 BroadcastRecordWriter 中,emit()方法的实现如下:

```
public void emit(T record) throws IOException, InterruptedException {
    broadcastEmit(record);
}
public void broadcastEmit(T record) throws IOException, InterruptedException {
    emit(record, 0);
}
```

由此可以看到,在广播中并不会产生多份数据,而是在调用 emit()方法时只向一个子分区发送数据。其中的索引可以为任意一个索引,这里取 0 是为了保证不会造成数组越界等异常。

BroadcastRecordWriter 中的 requestNewBufferBuilder()方法如下:

```
public BufferBuilder requestNewBufferBuilder(int targetChannel) throws IOException,
InterruptedException {
    BufferBuilder builder = targetPartition.getBufferBuilder();
    if (randomTriggered) {
        targetPartition.addBufferConsumer(randomTriggeredConsumer = builder.createBuffer-
Consumer(), targetChannel);
    } else {
```

```
            try (BufferConsumer bufferConsumer = builder.createBufferConsumer()) {
                for (int channel = 0; channel < numberOfChannels; channel++) {
                    targetPartition.addBufferConsumer(bufferConsumer.copy(), channel);
                }
            }
        }
        bufferBuilder = builder;
        return builder;
    }
```

从 else 分支可观察到，发送到每个子分区的 BufferConsumer 都是由同一个 BufferBuilder 对象构造的。

从上面的过程可以了解到，数据从最后一个任务的 RecordWriter 流转到了 ResultPartitionWriter 中。ResultPartitionWriter 是一个接口。上面的过程主要调用了两个方法，一个是 getBufferBuilder() 方法，用于申请一个用于写入数据的 BufferBuilder 对象；另一个是 addBufferConsumer() 方法，用于将构造出的 BufferConsumer 对象添加到子分区。另外，ResultPartitionWriter 还定义了初始化的 setup() 方法、刷写数据用的 flush() 方法和 flushAll() 方法等。下面对 RecordPartitionWriter 的实现类和其中实现的方法加以分析。

1. ResultPartition

ResultPartition 是 ResultPartitionWriter 的基本实现类，它持有 ResultSubpartition 对象。

下面是它的一些重要字段。

- owningTaskName：任务名称。
- partitionId：ResultPartitionID 类型，分区的唯一标识。
- partitionType：ResultPartitionType 类型。
- subpartitions：ResultSubpartition[] 类型。
- partitionManager：ResultPartitionManager 类型，分区管理器。
- bufferPool：BufferPool 类型，内存池。
- bufferPoolFactory：FunctionWithException 类型，表示初始化内存池的工厂类。
- bufferCompressor：BufferCompressor 类型，用于减少 I/O 的压缩器。

其构造方法用于对这些字段进行初始化赋值。

ResultPartition 对象是在 ResultPartitionFactory 的 create() 方法中被创建出来的，前文已经介绍过。这里简单回顾：

```
ResultPartition partition = forcePartitionReleaseOnConsumption || !type.isBlocking()
    ? new ReleaseOnConsumptionResultPartition(
        taskNameWithSubtaskAndId,
        id,
        type,
        subpartitions,
        maxParallelism,
        partitionManager,
        bufferCompressor,
        bufferPoolFactory)
    : new ResultPartition(
        taskNameWithSubtaskAndId,
```

```
            id,
            type,
            subpartitions,
            maxParallelism,
            partitionManager,
            bufferCompressor,
            bufferPoolFactory);
```

创建对象时进行了判断，如果没有指定 forcePartitionReleaseOnConsumption 的值为 true，且 ResultPartitionType 为 BLOCKING 或 BLOCKING_PERSISTENT，那么会创建 ResultPartition 对象，否则会创建 ReleaseOnConsumptionResultPartition 对象。ForcePartitionReleaseOnConsumption 的值是配置中指定的值，如果它的值为 true，则表示子分区被消费完时立刻释放分区。

ResultPartition 的 setup()方法如下：

```
public void setup() throws IOException {
    // 初始化内存池
    BufferPool bufferPool = checkNotNull(bufferPoolFactory.apply(this));
    this.bufferPool = bufferPool;
    // 在分区管理器中注册该分区
    partitionManager.registerResultPartition(this);
}
```

ResultPartition 的 getBufferBuilder()方法如下：

```
public BufferBuilder getBufferBuilder() throws IOException, InterruptedException {
    return bufferPool.requestBufferBuilderBlocking();
}
```

这里直接利用内存池获取内存并构造 BufferBuilder 对象。

ResultPartition 的 addBufferConsumer()方法如下：

```
public boolean addBufferConsumer(BufferConsumer bufferConsumer, int subpartitionIndex)
throws IOException {
    ResultSubpartition subpartition;
    try {
        subpartition = subpartitions[subpartitionIndex];
    }
    catch (Exception ex) {
      ...
    }
    return subpartition.add(bufferConsumer);
}
```

该方法获取了对应的子分区，并将 BufferConsumer 对象添加到了该子分区。

ResultPartitionr 的 flushAll()、flush()、finish()、release()等方法都是调用子分区的方法：

```
public void flushAll() {
    for (ResultSubpartition subpartition : subpartitions) {
        subpartition.flush();
    }
}
public void flush(int subpartitionIndex) {
    subpartitions[subpartitionIndex].flush();
}
```

```
public void finish() throws IOException {
   for (ResultSubpartition subpartition : subpartitions) {
      subpartition.finish();
   }
   isFinished = true;
}
public void release(Throwable cause) {
   if (isReleased.compareAndSet(false, true)) {
      ...
      for (ResultSubpartition subpartition : subpartitions) {
         try {
            subpartition.release();
         }
         catch (Throwable t) {
            ...
         }
      }
   }
}
```

ResultPartition 还提供了 onConsumedSubpartition() 方法，该方法会在子分区被消费完释放分区时被调用：

```
void onConsumedSubpartition(int subpartitionIndex) {
   if (isReleased.get()) {
      return;
   }
}
```

该方法会在 ResultSubpartition 的生命周期中被调用，在 ResultPartition 中并没有做什么实现。

2. ReleaseOnConsumptionResultPartition

ReleaseOnConsumptionResultPartition 有以下字段。

- consumedSubpartitions：boolean 数组类型，用于标记每个子分区是否被消费。
- numUnconsumedSubpartitions：表示还没有被消费的子分区的个数。

ReleaseOnConsumptionResultPartition 的构造过程在前面已经介绍过。它继承自 ResultPartition 类：

```
public class ReleaseOnConsumptionResultPartition extends ResultPartition
```

因此，它的大部分方法是调用的 ResultPartition 类中的实现方法。只不过，它会在子分区被消费完时释放分区，因此它重写了 onConsumedSubpartition() 方法：

```
void onConsumedSubpartition(int subpartitionIndex) {
   ...
   if (remainingUnconsumed == 0) {
      partitionManager.onConsumedPartition(this);
   } else if (remainingUnconsumed < 0) {
      ...
   }
}
```

上述代码的核心逻辑就是当所有子分区被消费完时，调用分区管理器的 onConsumedPartition() 方法。该方法会将分区从分区管理器中移除并调用 ResultPartition 的 release() 方法来释放分区。

3. ConsumableNotifyingResultPartitionWriterDecorator

在 ResultPartitionWriter 的初始化过程中，还需要调用 ConsumableNotifyingResultPartitionWriter-

Decorator 的 decorate()方法对其进行装饰。该过程在前文已经介绍过：

```
if (desc.sendScheduleOrUpdateConsumersMessage() && desc.getPartitionType().isPipelined()) {
    consumableNotifyingPartitionWriters[counter] = new ConsumableNotifyingResultPartition-
WriterDecorator(
        taskActions,
        jobId,
        partitionWriters[counter],
        notifier);
} else {
    consumableNotifyingPartitionWriters[counter] = partitionWriters[counter];
}
```

如果 ResultPartitionType 为 PIPELINED 或 PIPELINED_BOUNDED，且设置了当上游一生产数据就开始调度和部署或更新下游消费者，会将 ResultPartition 对象封装成 ConsumableNotifyingResultPartitionWriterDecorator 对象。

下面是 ConsumableNotifyingResultPartitionWriterDecorator 的一些重要字段。

- partitionWriter：ResultPartitionWriter 类型。
- partitionConsumableNotifier：ResultPartitionConsumableNotifier 类型，用于通知分区可消费。

其中，partitionConsumableNotifier 的实现类为 RpcResultPartitionConsumableNotifier。该对象的作用是通知作业管理器端分区可消费，最终会调用 SchedulerBase 的 scheduleOrUpdateConsumers() 方法。SchedulerBase 的行为在前文已经详细介绍。

ConsumableNotifyingResultPartitionWriterDecorator 的大部分方法调用的是其持有的 partitionWriter 字段的对应方法，只有 addBufferConsumer()方法和 finish()方法有所不同：

```
public boolean addBufferConsumer(BufferConsumer bufferConsumer, int subpartitionIndex)
throws IOException {
    boolean success = partitionWriter.addBufferConsumer(bufferConsumer, subpartitionIndex);
    if (success) {
        notifyPipelinedConsumers();
    }
    return success;
}
public void finish() throws IOException {
    partitionWriter.finish();
    notifyPipelinedConsumers();
}
```

在这两个方法中，都调用了 notifyPipelinedConsumers()方法，以通知消费者分区可消费。

```
private void notifyPipelinedConsumers() {
    if (!hasNotifiedPipelinedConsumers) {
        partitionConsumableNotifier.notifyPartitionConsumable(jobId, partitionWriter.
getPartitionId(), taskActions);
        hasNotifiedPipelinedConsumers = true;
    }
}
```

由此可以串联起调度流程。如果分区为 ConsumableNotifyingResultPartitionWriterDecorator 类型，那么会在分区可消费时就通知下游任务可以部署，最终在整个任务执行完后，还会再次通知。

### 8.2.3 ResultSubpartition

可在前面介绍 ResultPartition 类中了解到，它的许多方法最终调用的都是结果子分区（ResultSubpartition）的对应方法。ResultSubpartition 是一个抽象类，它有如下字段。

- index：子分区索引。
- parent：ResultPartition 类型，表示该子分区属于哪个 ResultPartition。

在介绍 ResultPartition 类时已经介绍过它定义的一些主要方法，如添加 BufferConsumer 的 add() 方法、刷写时调用的 flush() 方法、子分区输出完成时调用的 finish() 方法、释放子分区时调用的 release() 方法等。这些方法仅在实现类中实现。

```
public abstract boolean add(BufferConsumer bufferConsumer) throws IOException;
public abstract void flush();
public abstract void finish() throws IOException;
public abstract void release() throws IOException;
public abstract ResultSubpartitionView createReadView(BufferAvailabilityListener availabilityListener) throws IOException;
```

createReadView() 方法用于创建子分区的视图，会在读取数据时用到。

另外，ResultSubpartition 类中实现了 onConsumedSubpartition() 方法，会在子分区被消费后调用。正是在这个方法中调用了 ResultPartition 的 onConsumedSubpartition() 方法：

```
protected void onConsumedSubpartition() {
    parent.onConsumedSubpartition(index);
}
```

ResultSubpartition 中有一个内部类 BufferAndBacklog。这个类封装了包含数据的 buffer 与 backlog。backlog 表示子分区中剩余"non-event buffer"的个数。这个类会在读取数据时用到。

ResultSubpartition 有两个实现类——PipelinedSubpartition 和 BoundedBlockingSubpartition。它们的初始化在 ResultPartitionFactory 的 createSubpartitions() 方法中实现，前文已经介绍过：

```
if (type.isBlocking()) {
    initializeBoundedBlockingPartitions(
        subpartitions,
        partition,
        blockingSubpartitionType,
        networkBufferSize,
        channelManager);
} else {
    for (int i = 0; i < subpartitions.length; i++) {
        subpartitions[i] = new PipelinedSubpartition(i, partition);
    }
}
```

当 ResultPartitionType 为 BLOCKING 或 BLOCKING_PERSISTENT 时，会创建 BoundedBlockingSubpartition，否则会创建 PipelinedSubpartition。

下面介绍这两个实现类，其中着重介绍与数据输出有关的方法。

#### 1. PipelinedSubpartition

PipelinedSubpartition 表示以 pipeline 的方式进行数据的生产和消费。在 PipelinedSubpartition

中，数据只存在于内存中，并且只会被消费一次。

下面是 PipelinedSubpartition 的一些重要字段。
- buffers：ArrayDeque\类型，表示该子分区的所有 BufferConsumer 队列。
- buffersInBacklog：表示目前该子分区中 "non-event buffer" 的个数。
- readView：PipelinedSubpartitionView 类型，表示该子分区的视图，用于读取数据。

当添加 BufferConsumer 时会调用 add()方法：

```
private boolean add(BufferConsumer bufferConsumer, boolean finish) {
    final boolean notifyDataAvailable;
    synchronized (buffers) {
        ...
        // 将 BufferConsumer 添加到 buffers 队列中
        buffers.add(bufferConsumer);
        // 增加 BufferInBacklog 的值
        increaseBuffersInBacklog(bufferConsumer);
        notifyDataAvailable = shouldNotifyDataAvailable() || finish;
    }
    // 通知数据可被消费
    if (notifyDataAvailable) {
        notifyDataAvailable();
    }
    return true;
}
```

其中，increaseBuffersInBacklog()方法如下：

```
private void increaseBuffersInBacklog(BufferConsumer buffer) {
    if (buffer != null && buffer.isBuffer()) {
        buffersInBacklog++;
    }
}
```

这里的 isBuffer()方法即用于判断是否为 "non-event buffer"。

如果可以通知数据可被消费，则会调用 notifyDataAvailable()方法：

```
private void notifyDataAvailable() {
    if (readView != null) {
        readView.notifyDataAvailable();
    }
}
```

这里调用的是 readView 的对应方法。ResultSubpartition 的 flush()方法也会调用 notifyDataAvailable()方法通知数据可被消费。可以感受到，在 PipelinedSubpartition 的设计中，只要生产了数据，就可能会触发通知机制通知消费者消费数据。

PipelinedSubpartition 的 finish()方法如下：

```
public void finish() throws IOException {
    add(EventSerializer.toBufferConsumer(EndOfPartitionEvent.INSTANCE), true);
}
```

这里添加了一个事件，表示数据流的结束。这个事件被封装成了 BufferConsumer，并被添加到了 buffers 队列中等待被消费。

## 2. BoundedBlockingSubpartition

BoundedBlockingSubpartition 表示结果为有界的数据，数据先全部被生产出来，然后被消费。结果数据可被多次消费。

下面是 BoundedBlockingSubpartition 的一些重要字段。

- currentBuffer：BufferConsumer 类型，表示当前正在被填充的 BufferConsumer 对象。
- data：BoundedData 类型，表示真正写入数据的入口。其多个实现类代表多种不同的数据存储方式。
- readers：<BoundedBlockingSubpartitionReader>类型，表示所有创建的并且尚未被释放的子分区数据读取器。
- numDataBuffersWritten：表示写入的"non-event buffer"的个数。

PipelinedSubpartition 中的 readView 字段仅表示一个读取子分区的视图，这里的 readers 字段是一个集合，表示可以有多个视图来读取该子分区。numDataBuffersWritten 字段与 PipelinedSubpartition 中的 buffersInBacklog 字段含义相同。

BoundedBlockingSubpartition 的 add()方法如下：

```
public boolean add(BufferConsumer bufferConsumer) throws IOException {
    ...
    flushCurrentBuffer();
    currentBuffer = bufferConsumer;
    return true;
}
```

每次添加一个新的 BufferConsumer 时，都会调用 flushCurrentBuffer()方法刷写当前的 BufferConsumer，然后把新添加的 BufferConsumer 设置为 currentBuffer。

```
private void flushCurrentBuffer() throws IOException {
    if (currentBuffer != null) {
        writeAndCloseBufferConsumer(currentBuffer);
        currentBuffer = null;
    }
}
private void writeAndCloseBufferConsumer(BufferConsumer bufferConsumer) throws IOException {
    try {
        final Buffer buffer = bufferConsumer.build();
        try {
            if (canBeCompressed(buffer)) {
                ...
            } else {
                // 写入数据
                data.writeBuffer(buffer);
            }
            numBuffersAndEventsWritten++;
            if (buffer.isBuffer()) {
                numDataBuffersWritten++;
            }
        }
        finally {
            ...
        }
```

```
        }
        finally {
            …
        }
    }
```

在写入数据的过程中，完成了 numDataBuffersWritten 的计数。这里与 PipelinedSubpartition 的有所不同，BufferConsumer 被转换成了只读的 Buffer，真正写入数据时是调用 data 的 writeBuffer() 方法将只读的 Buffer 写入具体的存储媒介的。

这里可与 PipelinedSubpartition 进行对比。在 PipelinedSubpartition 中是在读取数据时才将 BufferConsumer 转换成只读的 Buffer 的，这是因为在 PipelinedSubpartition 的流程中，数据边写边读且一直都在内存中，所以在数据被读取时再转换会更加合理。

data 字段的类型为 BoundedData。BoundedData 是一个接口，会定义数据最终被写入哪里。既然会定义数据的写入方法，自然也会定义数据的读取方法，读取的部分会在后文介绍，这里将重点介绍数据的写入。

BoundedData 中与数据写入有关的方法有：

```
void writeBuffer(Buffer buffer) throws IOException;
void finishWrite() throws IOException;
```

BoundedData 的实现类有 3 个——FileChannelBoundedData、FileChannelMemoryMappedBoundedData 和 MemoryMappedBoundedData。具体构造哪个实现类由配置 taskmanager.network.blocking-shuffle.type 决定。默认为构造 FileChannelBoundedData。

FileChannelBoundedData 会将数据写入 FileChannel，读取数据时读取 FileChannel。FileChannelMemoryMappedBoundedData 在写入数据时会将数据写入 FileChannel，写完后将文件映射到内存，读取时直接读取内存。MemoryMappedBoundedData 的写入和读取都在内存中进行。

以 FileChannelBoundedData 为例，其 writeBuffer()方法如下：

```
public void writeBuffer(Buffer buffer) throws IOException {
    size += BufferReaderWriterUtil.writeToByteChannel(fileChannel, buffer,
headerAndBufferArray);
}
```

该方法将数据序列化到 FileChannel，并且会返回写入的数据的大小，然后将其累加到 size 字段。finishWrite()方法就是用于关闭 FileChannel 的：

```
public void finishWrite() throws IOException {
    fileChannel.close();
}
```

BoundedBlockingSubpartition 的 flush()、finish()方法会调用 flushCurrentBuffer()和 writeAndClose BufferConsumer()方法。

## 8.3 数据的读取

在 ExecutionGraph 中，输入由 ExecutionEdge 表示。每个 ExecutionEdge 在物理执行层面对应

一个 InputChannel，每个 InputChannel 都会从一个 ResultSubpartition 中读取数据。所有从同一个上游任务读取数据的 InputChannel 属于同一个 InputGate。

InputGate 相关类的关系如图 8-23 所示。

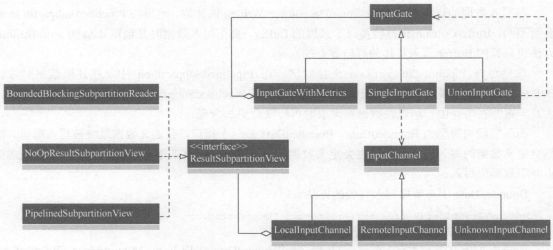

图 8-23　InputGate 相关类的关系

本节会详细介绍 InputGate 和 InputChannel 的各个实现类的字段和方法。

## 8.3.1　ResultSubpartitionView

下游要读取上游的数据时，会给上游发送请求。此时上游会先创建一个视图对象去读取数据，这个对象对应的类就是 ResultSubpartitionView。利用这个视图对象读取出数据后，再传递给下游。

在 ResultSubpartitionView 中定义的方法中，有两个方法十分关键：

```
// 获取下一个 BufferAndBacklog 对象
BufferAndBacklog getNextBuffer() throws IOException, InterruptedException;
// 通知数据可被消费
void notifyDataAvailable();
```

前文已经介绍过，BufferAndBacklog 类是 ResultSubpartition 的内部类，其封装了 buffer 和 buffersInBacklog。

针对 PipelinedSubpartition 和 BoundedBlockingSubpartition，ResultSubpartitionView 有两个对应的实现类——PipelinedSubpartitionView 和 BoundedBlockingSubpartitionReader。

### 1. PipelinedSubpartitionView

PipelinedSubpartition 的 createReadView() 方法会返回一个 PipelinedSubpartitionView 对象：

```
public PipelinedSubpartitionView createReadView(BufferAvailabilityListener availabilityListener)
throws IOException {
    final boolean notifyDataAvailable;
    synchronized (buffers) {
        readView = new PipelinedSubpartitionView(this, availabilityListener);
```

```
        notifyDataAvailable = !buffers.isEmpty();
    }
    if (notifyDataAvailable) {
        notifyDataAvailable();
    }
    return readView;
}
```

下面是 PipelinedSubpartitionView 中的一些重要字段。
- parent：PipelinedSubpartition 类型，表示该视图要消费的子分区。
- availabilityListener：BufferAvailabilityListener 类型，表示监听器，专门用于监听数据是否可被消费。

当要获取下一个 BufferAndBacklog 对象时，可调用 getNextBuffer()方法：

```
public BufferAndBacklog getNextBuffer() {
    return parent.pollBuffer();
}
```

实际上就是调用 PipelinedSubpartition 中的 pollBuffer()方法。该方法在前文介绍数据输出时并未提及，因为该方法专门用于供视图读取数据：

```
BufferAndBacklog pollBuffer() {
    synchronized (buffers) {
        ...
        while (!buffers.isEmpty()) {
            BufferConsumer bufferConsumer = buffers.peek();
            buffer = bufferConsumer.build();
            ...
        }
        ...
        return new BufferAndBacklog(
            buffer,
            isAvailableUnsafe(),
            getBuffersInBacklog(),
            nextBufferIsEventUnsafe());
    }
}
```

消费数据时，就是从 PipelinedSubpartition 中的 buffers 队列中取出第一个 BufferConsumer 并将其转换成只读的 Buffer，再将其与 buffersInBacklog 信息一起封装成 BufferAndBacklog 对象并返回。

前文介绍 PipelinedSubpartition 时，在多个方法中体现了 notifyDataAvailable()方法的调用，这些方法会调用 readView 字段的 notifyDataAvailable()方法，也就是 PipelinedSubpartitionView 的该方法：

```
public void notifyDataAvailable() {
    availabilityListener.notifyDataAvailable();
}
```

这里的关键在于这个监听器的实现类是什么。这个监听器为 BufferAvailabilityListener 类型，它是一个接口，有两个实现类——CreditBasedSequenceNumberingViewReader 和 LocalInputChannel，后文会介绍它们在被通知数据可消费时做了什么。

### 2. BoundedBlockingSubpartitionReader

BoundedBlockingSubpartition 的 createReadView()方法会构造 BoundedBlockingSubpartitionReader 对象：

```
public ResultSubpartitionView createReadView(BufferAvailabilityListener availability) throws IOException {
    synchronized (lock) {
        availability.notifyDataAvailable();
        final BoundedBlockingSubpartitionReader reader = new BoundedBlockingSubpartitionReader(
            this, data, numDataBuffersWritten, availability);
        readers.add(reader);
        return reader;
    }
}
```

下面是 BoundedBlockingSubpartitionReader 的一些重要字段。

- parent：BoundedBlockingSubpartition 类型，表示该视图要消费的子分区。
- availabilityListener：BufferAvailabilityListener 类型，与 PipelinedSubpartitionView 中的同名字段含义相同。
- nextBuffer：Buffer 类型，表示下一个要被消费的 Buffer。
- dataReader：BoundedData.Reader 类型，表示数据的读取器。
- dataBufferBacklog：剩余的"non-event buffer"的个数。

BoundedBlockingSubpartitionReader 的 getNextBuffer()方法如下：

```
public BufferAndBacklog getNextBuffer() throws IOException {
    final Buffer current = nextBuffer;
    if (current.isBuffer()) {
        dataBufferBacklog--;
    }
    nextBuffer = dataReader.nextBuffer();
    return BufferAndBacklog.fromBufferAndLookahead(current, nextBuffer, dataBufferBacklog);
}
```

这里的核心逻辑是用 dataReader 获取下一个 Buffer，然后将其与 backlog 信息一并封装成 BufferAndBacklog 对象。

dataReader 是 BoundedData.Reader 类型。它在 BoundedBlockingSubpartitionReader 的构造方法中由传入的 BoundedData 对象的 createReader()方法所创建。前文介绍过，由于写入数据时有多种不同的实现，因此在读取数据时也会有不同的读取方式。以 FileChannelBoundedData 为例，其 createReader()方法如下：

```
public Reader createReader(ResultSubpartitionView subpartitionView) throws IOException {
    final FileChannel fc = FileChannel.open(filePath, StandardOpenOption.READ);
    return new FileBufferReader(fc, memorySegmentSize, subpartitionView);
}
```

此时构造的读取器实现类是 FileBufferReader。它的 nextBuffer()方法如下：

```
public Buffer nextBuffer() throws IOException {
    final MemorySegment memory = buffers.pollFirst();
    if (memory == null) {
```

```
        return null;
    }
    final Buffer next = BufferReaderWriterUtil.readFromByteChannel(fileChannel, headerBuffer,
memory, this);
    if (next == null) {
        isFinished = true;
        recycle(memory);
    }
    return next;
}
```

该方法从 FileChannel 中读取数据并将其封装到 Buffer 对象中返回。

BoundedBlockingSubpartitionReader 也实现了 notifyDataAvailable() 方法。在 nextBuffer() 方法中可以观察到，在从 FileChannel 中读取数据之前，会先从 buffers 中获取一部分内存，然后将 FileChannel 中的数据写入其中。如果 buffers 中没有内存可用，显然就不能再读取数据。FileChannelBoundedData 中定义了 recycle() 方法，当别处用完了内存想要归还时，可以调用该方法归还内存到 buffers 中：

```
public void recycle(MemorySegment memorySegment) {
    buffers.addLast(memorySegment);
    if (!isFinished) {
        subpartitionView.notifyDataAvailable();
    }
}
```

此时如果数据没有被消费完，则会调用 BoundedBlockingSubpartitionReader 的 notifyDataAvailable() 方法：

```
public void notifyDataAvailable() {
    if (nextBuffer == null) {
        try {
            nextBuffer = dataReader.nextBuffer();
        } catch (IOException ex) {
            ...
        }
        if (nextBuffer != null) {
            availabilityListener.notifyDataAvailable();
        }
    }
}
```

该方法继续用读取器读取数据，并且通知监听器数据可被消费。

### 8.3.2　InputGate

InputGate 是下游任务消费上游分区数据的入口。它是一个抽象类，主要定义了两个方法：

```
public abstract Optional<BufferOrEvent> getNext() throws IOException, InterruptedException;
public abstract Optional<BufferOrEvent> pollNext() throws IOException, InterruptedException;
```

其中，getNext() 方法的调用为阻塞调用，pollNext() 方法的调用为非阻塞调用。

下面分析各个实现类的构造以及对这些方法的实现。

**1. SingleInputGate**

SingleInputGate 是基本的 InputGate 的实现类。

下面是 SingleInputGate 的一些重要字段。

- owningTaskName：表示任务名称。
- consumedResultId：IntermediateDataSetID 类型，唯一标识该 InputGate 消费的 IntermediateResult。
- consumedPartitionType：ResultPartitionType 类型，表示该 InputGate 消费的 ResultPartition 的 ResultPartitionType。
- consumedSubpartitionIndex：表示该 InputGate 消费的子分区的索引。一个 InputGate 虽然会消费多个 ResultPartition，但是其消费的 ResultSubpartition 的索引是相同的。
- numberOfInputChannels：表示该 InputGate 中 InputChannel 的个数。因为一个 InputChannel 对应一个 ResultSubpartition，所以 InputChannel 的个数与消费的 ResultSubpartition 的个数相同。
- inputChannels：Map<IntermediateResultPartitionID, InputChannel>类型，以每个 InputChannel 消费的 IntermediateResultPartition 的唯一标识 IntermediateResultPartitionID 为键，以 InputChannel 对象为值，构建 Map。
- inputChannelsWithData：ArrayDeque<InputChannel>类型，表示有数据的 InputChannel 队列。
- bufferPool：BufferPool 类型，表示内存池。
- bufferPoolFactory：SupplierWithException 类型，表示初始化内存池的工厂类。
- bufferDecompressor：BufferDecompressor 类型，由于 ResultPartition 输出数据时可能对数据进行了压缩，因此在读取数据时可能需要根据情况对数据进行解压。

SingleInputGate 对象的创建在第 7 章已经介绍过，其是在 SingleInputGateFactory 的 create()方法中利用其构造方法直接实例化的。

SingleInputGate 的 setup()方法如下：

```
public void setup() throws IOException, InterruptedException {
    // 给 InputChannel 分配内存
    assignExclusiveSegments();
    // 初始化内存池
    BufferPool bufferPool = bufferPoolFactory.get();
    setBufferPool(bufferPool);
    // 请求分区
    requestPartitions();
}
```

在初始化方法中，对内存做了多次初始化操作。最后该方法调用 requestPartitions()方法请求分区。请求分区的过程实际上就是遍历 InputChannel 去请求对应的子分区，在这个过程中会在上游任务创建视图，即 ResultSubpartitionView 对象。这个过程根据不同的 InputChannel 实现类会有不同的实现。

```
void requestPartitions() throws IOException, InterruptedException {
    synchronized (requestLock) {
        if (!requestedPartitionsFlag) {
            ...
            for (InputChannel inputChannel : inputChannels.values()) {
                inputChannel.requestSubpartition(consumedSubpartitionIndex);
            }
```

```
        }
        ...
    }
}
```

**SingleInputGate** 的 getNext()方法和 pollNext()方法如下：

```
public Optional<BufferOrEvent> getNext() throws IOException, InterruptedException {
    return getNextBufferOrEvent(true);
}
public Optional<BufferOrEvent> pollNext() throws IOException, InterruptedException {
    return getNextBufferOrEvent(false);
}
```

真正的实现都在 getNextBufferOrEvent()方法中，只是传入的参数不一样，true 表示阻塞，false 表示非阻塞。

```
private Optional<BufferOrEvent> getNextBufferOrEvent(boolean blocking) throws IOException, InterruptedException {
    ...
    Optional<InputWithData<InputChannel, BufferAndAvailability>> next = waitAndGetNextData(blocking);
    ...
    InputWithData<InputChannel, BufferAndAvailability> inputWithData = next.get();
    return Optional.of(transformToBufferOrEvent(
        inputWithData.data.buffer(),
        inputWithData.moreAvailable,
        inputWithData.input));
}
```

这里的核心逻辑在 waitAndGetNextData()方法中。相关代码如下：

```
private Optional<InputWithData<InputChannel, BufferAndAvailability>> waitAndGetNextData(boolean blocking)
        throws IOException, InterruptedException {
    while (true) {
        // 从 inputChannelsWithData 队列中获取有数据的 InputChannel
        Optional<InputChannel> inputChannel = getChannel(blocking);
        ...
        // 调用 InputChannel 的 getNextBuffer()方法获取数据
        Optional<BufferAndAvailability> result = inputChannel.get().getNextBuffer();
        synchronized (inputChannelsWithData) {
            if (result.isPresent() && result.get().moreAvailable()) {
                // 如果该 InputChannel 还有数据没被消费完，则将其继续添加到 inputChannelsWithData 队列中
                inputChannelsWithData.add(inputChannel.get());
                enqueuedInputChannelsWithData.set(inputChannel.get().getChannelIndex());
            }
            ...
            if (result.isPresent()) {
                return Optional.of(new InputWithData<>(
                    inputChannel.get(),
                    result.get(),
                    !inputChannelsWithData.isEmpty()));
            }
        }
    }
}
```

在上面的方法中，首先从 inputChannelsWithData 队列中获取有数据的 InputChannel。相关代码如下：

```
private Optional<InputChannel> getChannel(boolean blocking) throws InterruptedException {
    synchronized (inputChannelsWithData) {
        // 当队列中有元素时跳出循环
        while (inputChannelsWithData.size() == 0) {
            ...
            if (blocking) {
                inputChannelsWithData.wait();
            }
            else {
                ...
            }
        }
        // 从队列中取出刚刚被其他线程添加进去的有数据的 InputChannel
        InputChannel inputChannel = inputChannelsWithData.remove();
        enqueuedInputChannelsWithData.clear(inputChannel.getChannelIndex());
        return Optional.of(inputChannel);
    }
}
```

inputChannelsWithData 队列也会被其他线程操作，将有数据的 InputChannel 添加进其中。8.3.3 节分析 InputChannel 时会介绍其何时被添加进队列。

2. InputGateWithMetrics

在任务的初始化阶段，SingleInputGate 对象被创建后，又被封装成了 InputGateWithMetrics 对象。InputGateWithMetrics 也继承自 InputGate，它仅是对 SingleInputGate 的简单封装，只会提供一些统计信息。它持有一个 InputGate 类型的 inputGate 字段，其所有继承自 InputGate 的方法都调用了 inputGate 字段的对应方法。

```
public void setup() throws IOException, InterruptedException {
    inputGate.setup();
}
public Optional<BufferOrEvent> getNext() throws IOException, InterruptedException {
    return inputGate.getNext().map(this::updateMetrics);
}
public Optional<BufferOrEvent> pollNext() throws IOException, InterruptedException {
    return inputGate.pollNext().map(this::updateMetrics);
}
```

3. UnionInputGate

UnionInputGate 中封装了多个 InputGate，也继承自 InputGate。它的一些重要字段如下。

- inputGates：InputGate[]类型。
- inputGatesWithData：LinkedHash<InputGate>类型，表示有数据的 InputGate 集合。
- totalNumberOfInputChannels：所有 InputGate 中的 InputChannel 的总数。

UnionInputGate 的方法的实现思路与 SingleInputGate 的类似，只不过 SingleInputGate 底层会调用其 InputChannel 的方法，而 UnionInputGate 会调用其持有的 InputGate 的相关方法。仍然以 getNext()方法和 pollNext()方法为例：

```
public Optional<BufferOrEvent> getNext() throws IOException, InterruptedException {
    return getNextBufferOrEvent(true);
```

```
}
public Optional<BufferOrEvent> pollNext() throws IOException, InterruptedException {
    return getNextBufferOrEvent(false);
}
```

UnionInputGate 也定义并实现了 getNextBufferOrEvent()方法：

```
private Optional<BufferOrEvent> getNextBufferOrEvent(boolean blocking) throws IOException,
InterruptedException {
    ...
    Optional<InputWithData<InputGate, BufferOrEvent>> next = waitAndGetNextData(blocking);
    ...
    InputWithData<InputGate, BufferOrEvent> inputWithData = next.get();
    handleEndOfPartitionEvent(inputWithData.data, inputWithData.input);
    return Optional.of(adjustForUnionInputGate(
        inputWithData.data,
        inputWithData.input,
        inputWithData.moreAvailable));
}
```

从实现上来看，其与 SingleInputGate 在结构上是极为类似的，只是数据类型有所区别。

```
private Optional<InputWithData<InputGate, BufferOrEvent>> waitAndGetNextData(boolean blocking)
        throws IOException, InterruptedException {
    while (true) {
        // 从 inputGatesWithData 集合中获取有数据的 InputGate
        Optional<InputGate> inputGate = getInputGate(blocking);
        ...
        // 调用 InputGate 的 pollNext()方法获取数据
        Optional<BufferOrEvent> bufferOrEvent = inputGate.get().pollNext();
        synchronized (inputGatesWithData) {
            // 如果该 InputGate 还有数据没被消费完，则将其继续添加到 inputGatesWithData 集合中
            if (bufferOrEvent.isPresent() && bufferOrEvent.get().moreAvailable()) {
                inputGatesWithData.add(inputGate.get());
            } else if (!inputGate.get().isFinished()) {
                inputGate.get().getAvailableFuture().thenRun(() ->
queueInputGate(inputGate.get()));
            }
            ...
            if (bufferOrEvent.isPresent()) {
                return Optional.of(new InputWithData<>(
                    inputGate.get(),
                    bufferOrEvent.get(),
                    !inputGatesWithData.isEmpty()));
            }
        }
    }
}
```

在 SingleInputGate 中，是获取有数据的 InputChannel，而这里是获取有数据的 InputGate。获取 InputGate 对象后，调用 pollNext()方法获取数据，这就回到了 SingleInputGate 的逻辑。

### 8.3.3　InputChannel

每个 InputChannel 都会消费一个 ResultSubpartition 中的数据。InputChannel 是一个抽象类，

下面是它的一些重要字段。

- **channelIndex**：表示 InputChannel 的索引。
- **partitionId**：ResultPartitionID 类型，表示 ResultPartition 的唯一标识。表示该 InputChannel 要消费哪个 ResultPartition。
- **inputGate**：SingleInputGate 类型，表示该 InputChannel 属于哪个 InputGate。

InputChannel 中主要定义了下面两个方法：

```
abstract void requestSubpartition(int subpartitionIndex) throws IOException,
InterruptedException;
    abstract Optional<BufferAndAvailability> getNextBuffer() throws IOException,
InterruptedException;
```

requestSubpartition()方法主要用于请求对应的子分区，子分区会创建视图对象用以准备读取数据。getNextBuffer()方法会返回接下来可供消费的数据。

InputChannel 的实现类有两个——LocalInputChannel 和 RemoteInputChannel。前者用于读取本地数据，后者用于读取远程数据。

**1. LocalInputChannel**

首先可注意到，LocalInputChannel 继承自 InputChannel，并且实现了 BufferAvailabilityListener 接口：

```
public class LocalInputChannel extends InputChannel implements BufferAvailabilityListener
```

LocalInputChannel 本身就是一个监听器，用于监听数据是否可供消费，当数据可供消费时就会触发操作。

下面是 LocalInputChannel 的一些重要字段。

- **partitionManager**：ResultPartitionManager 类型表示分区管理器。
- **subpartitionView**：ResultSubpartitionView 类型。

这里介绍一下结果分区管理器（ResultPartitionManager）。它是任务管理器级的管理器，管理着所有任务中的分区。它是在任务管理器初始化的过程中直接通过构造方法构造出来的。

ResultPartitionManager 持有一个 Map<ResultPartitionID, ResultPartition>类型的 registeredPartitions 字段。每个分区在初始化时都会将自身注册在分区管理器中。

```
public void registerResultPartition(ResultPartition partition) {
    synchronized (registeredPartitions) {
        ResultPartition previous = registeredPartitions.put(partition.getPartitionId(),
partition);
        ...
    }
}
```

既然维护了所有的分区，那么下游任务在请求上游子分区时，就会利用 ResultPartitionManager 找到对应的子分区并创建视图。其中创建视图的过程如下：

```
public ResultSubpartitionView createSubpartitionView(
    ResultPartitionID partitionId,
    int subpartitionIndex,
    BufferAvailabilityListener availabilityListener) throws IOException {
```

```
synchronized (registeredPartitions) {
  final ResultPartition partition = registeredPartitions.get(partitionId);
  ...
  return partition.createSubpartitionView(subpartitionIndex, availabilityListener);
}
```

调用 ResultPartition 的方法创建视图：

```
public ResultSubpartitionView createSubpartitionView(int index, BufferAvailabilityListener
availabilityListener) throws IOException {
  ResultSubpartitionView readView = subpartitions[index].createReadView(availabilityListener);
  return readView;
}
```

此外，在介绍 ReleaseOnConsumptionResultPartition 时介绍过在子分区被消费时会调用其 onConsumedSubpartition()方法，该方法会调用 ResultPartitionManager 的 onConsumedPartition()方法。下面是 ResultPartitionManager 的 onConsumedPartition()方法：

```
void onConsumedPartition(ResultPartition partition) {
  synchronized (registeredPartitions) {
    final ResultPartition previous = registeredPartitions.remove(partition.getPartitionId());
    if (partition == previous) {
      partition.release();
    }
  }
}
```

上述代码调用的是分区的 release()方法，用于释放分区。

回到 LocalInputChannel。它的 requestSubpartition()方法的实现如下：

```
void requestSubpartition(int subpartitionIndex) throws IOException, InterruptedException {
  synchronized (requestLock) {
    if (subpartitionView == null) {
      try {
        // 利用分区管理器创建视图
        ResultSubpartitionView subpartitionView = partitionManager.createSubpartitionView(
            partitionId, subpartitionIndex, this);
        this.subpartitionView = subpartitionView;
        ...
      } catch (PartitionNotFoundException notFound) {
        ...
      }
    }
  }
}
```

在 InputGate 的 setup()方法中调用了 requestPartitions()方法，在该方法中调用了每个 InputChannel 的 requestSubpartition()方法。

前面已经介绍过分区管理器的 createSubpartitionView()方法。这里其实就是利用分区管理器，找到同一个任务管理器中相应的子分区，然后返回视图。注意，创建视图调用 createSubpartitionView()方法时，将 LocalInputChannel 自身作为参数传入，最后就赋值给了视图对象中的 availabilityListener 字段（两种视图中都有该字段）。因此，在有数据可供消费时，最后就会调用 LocalInputChannel

的 notifyDataAvailable()方法：

```
public void notifyDataAvailable() {
    notifyChannelNonEmpty();
}
protected void notifyChannelNonEmpty() {
    inputGate.notifyChannelNonEmpty(this);
}
void notifyChannelNonEmpty(InputChannel channel) {
    queueChannel(checkNotNull(channel));
}
```

最终调用了 InputGate 的 queueChannel()方法，将该 LocalInputChannel 对象放入了 InputGate 的 inputChannelsWithData 队列中。

在 getNextBuffer()方法中，其实就是利用视图对象来获取数据：

```
Optional<BufferAndAvailability> getNextBuffer() throws IOException, InterruptedException {
    ResultSubpartitionView subpartitionView = this.subpartitionView;
    ...
    // 利用视图对象的 getNextBuffer()方法获取数据
    BufferAndBacklog next = subpartitionView.getNextBuffer();
    ...
    return Optional.of(new BufferAndAvailability(next.buffer(), next.isMoreAvailable(), next.buffersInBacklog()));
}
```

获取数据后，经过简单的封装后返回。

2．RemoteInputChannel

RemoteInputChannel 的逻辑要复杂一些，因为要考虑远程通信。

下面是 RemoteInputChannel 的一些重要字段。

- connectionId：ConnectionID 类型，表示网络连接的唯一标识。
- connectionManager：ConnectionManager 类型，用于创建网络连接的客户端。
- receivedBuffers：ArrayDeque\类型，将远程获取的 Buffer 放入该队列，并被消费者消费。
- partitionRequestClient：PartitionRequestClient 类型，表示网络连接的客户端。

RemoteInputChannel 的 requestSubpartition()方法如下：

```
public void requestSubpartition(int subpartitionIndex) throws IOException, InterruptedException {
    if (partitionRequestClient == null) {
        try {
            // 利用 ConnectionManager 创建网络连接的客户端
            partitionRequestClient = connectionManager.createPartitionRequestClient(connectionId);
        } catch (IOException e) {
            ...
        }
        // 请求子分区
        partitionRequestClient.requestSubpartition(partitionId, subpartitionIndex, this, 0);
    }
}
```

partitionRequestClient 的 requestSubpartition()方法会发送一个 PartitionRequest 请求给上游，该

请求中包含 ResultPartitionID 等信息，上游接收到请求后就知道该去哪个子分区获取数据。

上游接收到请求后，会在 Netty 的调用链中执行这样的逻辑：

```
NetworkSequenceViewReader reader;
reader = new CreditBasedSequenceNumberingViewReader(
    request.receiverId,
    request.credit,
    outboundQueue);
reader.requestSubpartitionView(
    partitionProvider,
    request.partitionId,
    request.queueIndex);
```

这里创建了一个 CreditBasedSequenceNumberingViewReader 类的对象。该类提供创建视图、获取数据等的方法。创建视图时调用的就是 requestSubpartitionView() 方法：

```
public void requestSubpartitionView(
    ResultPartitionProvider partitionProvider,
    ResultPartitionID resultPartitionId,
    int subPartitionIndex) throws IOException {
  synchronized (requestLock) {
      if (subpartitionView == null) {
          this.subpartitionView = partitionProvider.createSubpartitionView(
              resultPartitionId,
              subPartitionIndex,
              this);
      } else {
          ...
      }
  }
}
```

创建视图的方式与本地模式的一致。同样，获取数据的方式——getNextBuffer() 方法，也与之一致：

```
public BufferAndAvailability getNextBuffer() throws IOException, InterruptedException {
    BufferAndBacklog next = subpartitionView.getNextBuffer();
    if (next != null) {
        ...
        return new BufferAndAvailability(
            next.buffer(), isAvailable(next), next.buffersInBacklog());
    } else {
        return null;
    }
}
```

因此，在远程模式下，下游给上游发送了要读取数据的请求后，上游就会创建 CreditBased-SequenceNumberingViewReader 对象进而创建视图来获取上游的数据。

另外，CreditBasedSequenceNumberingViewReader 还实现了 BufferAvailabilityListener 接口，因此有数据可供消费时，就会通知 CreditBasedSequenceNumberingViewReader 调用其 notifyDataAvailable() 方法，将数据发往下游。

下游接收到数据后，会将其维护在 RemoteInputChannel 的 receivedBuffers 队列中。在 RemoteChannel 的 getNextBuffer() 方法中，会从该队列中获取数据进行消费：

```java
Optional<BufferAndAvailability> getNextBuffer() throws IOException {
    final Buffer next;
    final boolean moreAvailable;
    synchronized (receivedBuffers) {
        next = receivedBuffers.poll();
        moreAvailable = !receivedBuffers.isEmpty();
    }
    ...
    return Optional.of(new BufferAndAvailability(next, moreAvailable, getSenderBacklog()));
}
```

## 8.4　反压机制的原理

在 8.2 和 8.3 节中分析了数据输出和数据读取的主要方法，但是在介绍远程数据传输时并未将所有环节串联起来。本节会完整地分析远程数据传输的整个链路，并且会在这个过程中讲解反压的原理。

在下游 RemoteInputChannel 初始化时，会发送 PartitionRequest 请求给上游。上游接收到该请求后，会初始化 CreditBasedSequenceNumberingViewReader 对象并创建视图。

RemoteInputChannel 中有一个 initialCredit 字段，表示初始情况下下游可以接收的 Buffer 个数。这个值会被封装到 PartitionRequest 中一起发送到上游，赋值给 CreditBasedSequenceNumberingViewReader 对象的 numCreditsAvailable 字段。

```java
PartitionRequest(ResultPartitionID partitionId, int queueIndex, InputChannelID receiverId, int credit) {
    this.partitionId = checkNotNull(partitionId);
    this.queueIndex = queueIndex;
    this.receiverId = checkNotNull(receiverId);
    this.credit = credit;
}
CreditBasedSequenceNumberingViewReader(
        InputChannelID receiverId,
        int initialCredit,
        PartitionRequestQueue requestQueue) {
    this.receiverId = receiverId;
    this.numCreditsAvailable = initialCredit;
    this.requestQueue = requestQueue;
}
```

但此时不会响应数据给下游，因为上游的数据可能还没有生产，不能被消费。当有数据可供消费时，调用 notifyDataAvailable() 方法：

```java
public void notifyDataAvailable() {
    requestQueue.notifyReaderNonEmpty(this);
}
void notifyReaderNonEmpty(final NetworkSequenceViewReader reader) {
    ctx.executor().execute(() -> ctx.pipeline().fireUserEventTriggered(reader));
}
```

随后，在 userEventTriggered() 方法中，会针对相应的数据类型进行处理：

```java
public void userEventTriggered(ChannelHandlerContext ctx, Object msg) throws Exception {
    if (msg instanceof NetworkSequenceViewReader) {
```

```
        enqueueAvailableReader((NetworkSequenceViewReader) msg);
    } else if (msg.getClass() == InputChannelID.class) {
        ...
    } else {
        ...
    }
}
private void enqueueAvailableReader(final NetworkSequenceViewReader reader) throws Exception
  {
    ...
    if (triggerWrite) {
        writeAndFlushNextMessageIfPossible(ctx.channel());
    }
}
```

在 writeAndFlushNextMessageIfPossible()方法中，会获取数据并将其发送给下游：

```
private void writeAndFlushNextMessageIfPossible(final Channel channel) throws IOException {
    ...
    BufferAndAvailability next = null;
    try {
        while (true) {
            // 从 availableReaders 队列中获取 CreditBasedSequenceNumberingViewReader 对象
            NetworkSequenceViewReader reader = pollAvailableReader();
            ...
            // 获取数据
            next = reader.getNextBuffer();
            if (next == null) {
                ...
            } else {
                // 如果能继续从该 CreditBasedSequenceNumberingViewReader 对象中读取数据，则将其
                // 放入 availableReaders 队列
                if (next.moreAvailable()) {
                    registerAvailableReader(reader);
                }
                // 封装数据并发送给下游
                BufferResponse msg = new BufferResponse(
                    next.buffer(),
                    reader.getSequenceNumber(),
                    reader.getReceiverId(),
                    next.buffersInBacklog());
                channel.writeAndFlush(msg).addListener(writeListener);
                return;
            }
        }
    } catch (Throwable t) {
        …
    }
}
```

注意，封装数据时将 backlog 信息也封装了进去。将数据发送给下游时还添加了监听器，该监听器会在响应成功后再次触发writeAndFlushNextMessageIfPossible()方法继续读取新的数据并发送给下游。每次读取数据后，numCreditsAvailable 字段的值都会减 1：

```
public BufferAndAvailability getNextBuffer() throws IOException, InterruptedException {
    BufferAndBacklog next = subpartitionView.getNextBuffer();
    if (next != null) {
```

```
        sequenceNumber++;
        if (next.buffer().isBuffer() && --numCreditsAvailable < 0) {
            ...
        }
        return new BufferAndAvailability(
            next.buffer(), isAvailable(next), next.buffersInBacklog());
    } else {
        return null;
    }
}
```

而返回的 BufferAndAvailability 对象的 moreAvailable()方法就会返回上面构造方法中传入的第二个参数：

```
public BufferAndAvailability(Buffer buffer, boolean moreAvailable, int buffersInBacklog) {
    this.buffer = checkNotNull(buffer);
    this.moreAvailable = moreAvailable;
    this.buffersInBacklog = buffersInBacklog;
}
public boolean moreAvailable() {
    return moreAvailable;
}
```

isAvailable()方法如下：

```
private boolean isAvailable(BufferAndBacklog bufferAndBacklog) {
    if (numCreditsAvailable > 0) {
        return bufferAndBacklog.isMoreAvailable();
    }
    else {
        return bufferAndBacklog.nextBufferIsEvent();
    }
}
```

也就是说，当 numCreditsAvailable 的值减少到 0 时，就会用 nextBufferIsEvent()方法来判定是否可以继续读取数据。当接下来的是事件时才能继续读取数据。由此可以控制从上游读取的数据。

响应发送到下游后，会在 Netty 调用链中被处理：

```
NettyMessage.BufferResponse bufferOrEvent = (NettyMessage.BufferResponse) msg;
RemoteInputChannel inputChannel = inputChannels.get(bufferOrEvent.receiverId);
...
decodeBufferOrEvent(inputChannel, bufferOrEvent);
```

根据 id 获取接收数据的 InputChannel。

```
private void decodeBufferOrEvent(RemoteInputChannel inputChannel, NettyMessage.BufferResponse bufferOrEvent) throws Throwable {
    try {
        ByteBuf nettyBuffer = bufferOrEvent.getNettyBuffer();
        final int receivedSize = nettyBuffer.readableBytes();
        if (bufferOrEvent.isBuffer()) {
            ...
            Buffer buffer = inputChannel.requestBuffer();
            if (buffer != null) {
                ...
                inputChannel.onBuffer(buffer, bufferOrEvent.sequenceNumber, bufferOrEvent.backlog);
```

```
            } else if (inputChannel.isReleased()) {
                ...
            } else {
                ...
            }
        } else {
            ...
        }
    } finally {
        ...
    }
}
```

最终会调用 RemoteInputChannel 的 onBuffer()方法，将数据放在 receivedBuffers 队列中：

```
public void onBuffer(Buffer buffer, int sequenceNumber, int backlog) throws IOException {
    boolean recycleBuffer = true;
    try {
        final boolean wasEmpty;
        synchronized (receivedBuffers) {
            ...
            receivedBuffers.add(buffer);
        }
        ...
        if (backlog >= 0) {
            onSenderBacklog(backlog);
        }
    } finally {
        ...
    }
}
```

因为先前传回了 backlog 信息，所以这里会在该值大于或等于 0 时调用 onSenderBacklog()方法。该值大于 0 表示上游还有 Buffer 没被消费，那么可以推测，onSenderBacklog()方法可能会再次向上游发送请求以消费数据。

```
void onSenderBacklog(int backlog) throws IOException {
    synchronized (bufferQueue) {
        ...
        // 计算需要接收的 Buffer 个数
        numRequiredBuffers = backlog + initialCredit;
        while (bufferQueue.getAvailableBufferSize() < numRequiredBuffers && !isWaitingForFloatingBuffers) {
            Buffer buffer = inputGate.getBufferPool().requestBuffer();
            if (buffer != null) {
                bufferQueue.addFloatingBuffer(buffer);
                numRequestedBuffers++;
            } else if (inputGate.getBufferProvider().addBufferListener(this)) {
                isWaitingForFloatingBuffers = true;
                break;
            }
        }
    }
    // 通知有空间可以用于消费数据
    if (numRequestedBuffers > 0 && unannouncedCredit.getAndAdd(numRequestedBuffers) == 0) {
        notifyCreditAvailable();
    }
}
```

在这个过程中，会计算要接收的 Buffer 个数。首先会用初始的 credit 值加上 backlog 的值，然后根据内存的分配对 numRequiredBuffers 进行累加。最后调用 notifyCreditAvailable()方法通知有空间可以用于消费数据。

```
private void notifyCreditAvailable() {
    partitionRequestClient.notifyCreditAvailable(this);
}
public void notifyCreditAvailable(RemoteInputChannel inputChannel) {
    clientHandler.notifyCreditAvailable(inputChannel);
}
public void notifyCreditAvailable(final RemoteInputChannel inputChannel) {
    ctx.executor().execute(() -> ctx.pipeline().fireUserEventTriggered(inputChannel));
}
```

随后会在 userEventTriggered()方法中对其进行处理：

```
public void userEventTriggered(ChannelHandlerContext ctx, Object msg) throws Exception {
    if (msg instanceof RemoteInputChannel) {
        boolean triggerWrite = inputChannelsWithCredit.isEmpty();
        inputChannelsWithCredit.add((RemoteInputChannel) msg);
        if (triggerWrite) {
            writeAndFlushNextMessageIfPossible(ctx.channel());
        }
    } else {
        ctx.fireUserEventTriggered(msg);
    }
}
private void writeAndFlushNextMessageIfPossible(Channel channel) {
    while (true) {
        RemoteInputChannel inputChannel = inputChannelsWithCredit.poll();
        ...
        if (!inputChannel.isReleased()) {
            AddCredit msg = new AddCredit(
                inputChannel.getPartitionId(),
                inputChannel.getAndResetUnannouncedCredit(),
                inputChannel.getInputChannelId());
            channel.writeAndFlush(msg).addListener(writeListener);
            return;
        }
    }
}
```

这里构造了 AddCredit 请求，将 credit 值放入了其中。发送请求后添加了监听器。这个监听器会在请求发送成功后再次调用 writeAndFlushNextMessageIfPossible()方法继续发送 AddCredit 请求。

上游接收到 AddCredit 请求后，会将 credit 值添加到对应的 CreditBasedSequenceNumberingViewReader 对象中。

```
AddCredit request = (AddCredit) msg;
outboundQueue.addCredit(request.receiverId, request.credit);
```

addCredit()方法如下：

```
void addCredit(InputChannelID receiverId, int credit) throws Exception {
    NetworkSequenceViewReader reader = allReaders.get(receiverId);
    if (reader != null) {
```

```
            reader.addCredit(credit);
            enqueueAvailableReader(reader);
        } else {
            ...
        }
    }
    public void addCredit(int creditDeltas) {
        numCreditsAvailable += creditDeltas;
    }
```
由此串联起了整个远程传输和流量控制的过程。

## 8.5 总结

本章介绍了 Flink 数据传输、流量控制的设计思想，对混洗机制进行了深入探讨。数据传输的过程主要涉及数据的输出和输入。输出逻辑主要由 ResultPartition（或者 ResultPartitionWriter）相关实现类负责，输入逻辑主要由 InputGate 相关实现类负责。数据传输过程中的流量控制则是基于 Credit 机制实现的。

# 第 9 章

# 时间与窗口

通常，接触过 Spark 或 Flink 等框架的读者一定知道，在这类可以进行流处理的计算引擎中，一般可以对数据流进行加窗操作，然后可以在各个窗口内对数据进行聚合或其他计算。在进行加窗操作时，一般会指定间隔多长时间创建新的窗口，以及窗口本身涵盖的时间范围等。

Flink 提供了接口用于对数据流进行 Window/Window All 加窗操作，加窗后可以继续进行 Window Reduce/Window Fold 聚合处理。对数据流进行加窗处理后，返回的数据流为 WindowedStream/AllWindowedStream 对象，在该数据流上可以进行与窗口有关的操作。

本章会剖析 Flink 中的窗口相关的操作的设计思想，指出它在批流一体化中的重要意义，并且会对 Flink 中的相关实现进行详细的介绍。

希望在学习完本章后，读者能够了解：
- 流处理中窗口操作的产生背景、意义和发展方向；
- 数据流模型（dataflow model）的设计思想；
- Flink 中与窗口操作相关的核心概念，如时间类型、水位线等；
- Flink 中与窗口相关的组件的具体实现。

## 9.1 基本概念和设计思想

在 Flink 中，如果要对数据进行窗口操作，必须指定时间类型（事件时间、处理时间等）、窗口分配器（WindowAssigner）、窗口函数等，有可能还需要指定窗口触发器（Trigger）和窗口剔除器（Evictor）等。

当得到数据流对象后，可以对其进行窗口相关操作。下面是一段典型的窗口操作业务代码：

```
stream
        .window(...)   // 指定窗口分配器
        .trigger(...)  // 指定窗口触发器（可选）
        .evictor(...)  // 指定窗口剔除器（可选）
        .sum(1);       // 指定窗口函数
```

本节会介绍 Flink 是基于什么思想设计了这样一套 API 以及它解决了什么问题。

### 9.1.1 从批处理到流处理

从业务角度来看，实际生产场景中对数据处理的"低延迟"这一特性的要求越来越高，且对于无界数据的处理场景越来越多,因此将数据看成无界数据流(而非有界数据集)来处理的方式变得愈发常见。从技术角度来看,用流处理的方式可以让资源更平均地被利用,其不像批处理需要集中计算。不过目前大多数流处理系统并不成熟。这不难理解,从直观的感受来说,在不断运行的数据处理系统中,任何异常都可能发生,数据本身也可能存在乱序的情况,在这种情况下同时兼顾低延迟和结果的正确性绝非一件易事。

因此,长久以来流处理一直被业界认为是一种提供低延迟但结果的正确率低的处理方式。它常被用于 Nathan Marz 提出的 Lambda 架构中。Lambda 架构的核心思想是同时开发两套逻辑相同的系统:一套是流处理系统,用于提供低延迟但正确率相对较低的结果;另一套是批处理系统,用于保证数据的最终一致性。Lambda 架构如图 9-1 所示。

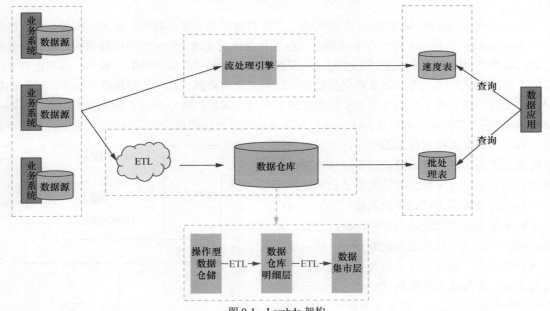

图 9-1 Lambda 架构

显而易见,使用 Lambda 架构有诸多不便,其中主要的问题就是开发和维护两套系统的成本过高。如何解决这个问题呢?一种方案是提供一种更高层次的抽象,使得开发人员可以编写一套代码,由框架将其"翻译"成批处理和流处理两套不同的实现。这种方案的弊端在于:批处理和流处理能够抽象出来的交叉部分很少(读者可以回忆 Flink 的 DataStream API 和 DataSet API 的差别并思考其中的原因),这种高层抽象实现起来十分复杂;另外,即便开发出了这种框架,运维和调试的成本仍然不可避免。

在当今 Lambda 架构被广泛使用的背景下,不妨思考一个问题:为什么要使用 Lambda 架构?根本原因在于业务上既需要保证低延迟,又需要保证正确性。人们认为流处理系统无法保证 100% 的正确率,这才需要另一套批处理系统来保证数据的最终一致性。那么,是否可以设计一个延迟较低的批处理系统,或是一个能够提供更高正确率的流处理系统,以此来降低整个架构的开发和维护的成本呢？

在不考虑架构改进（如从 Lambda 架构演进到 Kappa 架构）的情况下，开发一个兼顾实时性和正确性的数据处理系统是可行的。2013 年谷歌公司发表了一篇论文"MillWheel: Fault-Tolerant Stream Processing at Internet Scale"，介绍了 Google MillWheel 系统。该系统在当时是先进的流处理系统之一。

一方面，Google MillWheel 系统提供了状态管理机制、肯定应答（acknowledgement，ACK）机制等保证了数据的一致性。它的上游计算单元给下游计算单元发送数据时，会利用 ACK 机制保证数据一定能够发送给下游进行处理，同时会利用数据的唯一 ID 保证数据只被处理一次。每个计算单元对数据进行处理时，会产生一些中间结果，这些中间结果会被持久化到后台数据库中进行维护。

另一方面，Google MillWheel 系统提供了多种时间类型（事件时间、处理时间等）和低水位（low watermark）机制用于对乱序和晚到的数据进行合理的处理。处理时间指数据被处理时的自然时间，而事件时间指事件发生时的时间，一般被记录在数据自身的字段中。事件时间和处理时间显然不可能完全相同，那么如果希望按照事件时间来处理数据，就无法避免数据乱序和晚到的情况。过去许多流处理系统正是因为不能很好地处理这一问题，所以导致正确率低的结果。Google MillWheel 系统提出的低水位实际上可以理解为通过某种算法提取出来的时间戳，从上游发往下游时，每个计算单元根据传来的时间戳对自身的低水位进行更新。低水位的值表示系统认为事件时间在该值表示的时间戳之前的数据都已到达，可统一对这部分数据进行有序的处理。比如，可以基于低水位触发窗口的聚合操作。在此之后，如果有事件时间低于低水位的数据到达，就会被视为真正"晚到"的数据，会对其另作处理，比如将其丢弃。在谷歌公司的测试中，丢弃的数据只占整体数据的极少部分，这被认为是可以接受的。

图 9-2 是 Google MillWheel 给出的低水位示意，从上至下表示自然时间的流动，从左至右移动的竖线表示低水位。每幅图中位于横轴上方浅色的方块表示未处理的消息，横轴下方黑色的方块表示已处理的消息。消息处理的顺序是不确定的。对于图 9-2，从上至下可以观察到已处理的消息的数量依次增多。在图 9-2 中，低水位的含义就是"该值表示的时间戳之前的消息已经全部处理完成"。

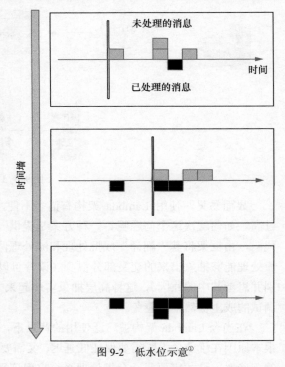

图 9-2　低水位示意[①]

Google MillWheel 系统并不能解决所有的问题，但是它在解决流处理系统的数据一致性问题

---

① 引自论文 *MillWheel: Fault-Tolerant Stream Processing at Internet Scale*.

上给人们提供了宝贵的思路和实践。

## 9.1.2 数据流模型的设计思想

在 9.1.1 节中，我们了解了业务对于流处理系统的迫切需求。从 Google MillWheel 系统的实践还能看出流处理系统在保证低延迟特性的同时也能够较好地处理数据一致性的问题。

在论文"MillWheel: Fault-Tolerant Stream Processing at Internet Scale"发表后两年，谷歌公司发表了一篇论文"The Dataflow Model: A Practical Approach to Balancing Correctness, Latency, and Cost in Massive-Scale, Unbounded, Out-of-Order Data Processing"。这篇论文的"抽象程度"更高，提出了批流一体化的计算模型。在该模型中，对于传统批处理和流处理的选择变成了同一个模型中对于结果的正确性、延迟性和计算成本这三者的权衡。该模型的核心，就是窗口的相关设计。

这篇论文首先进行了概念上的澄清。它没有采用"流处理"和"批处理"的概念来区分不同的数据处理方式。因为人们往往会将"流处理""批处理"这样的概念与具体的计算引擎关联在一起，反而忽视了这些概念的本质。实际上，通过反复运行批处理计算引擎可以处理无界的数据，达到流处理的效果，而流处理系统自然也可以处理有界的数据集。从数据流模型的视角来看，重要的是数据"有界"或"无界"，而数据的处理方式是流处理还是批处理并没有本质的差别。可以认为批处理是流处理的一种特例。这正是批流一体化的出发点。而批流一体化的"桥梁"，便是窗口。有了窗口，才可以将无界、乱序的数据作为一个个批次来进行有序的处理。

有了窗口的概念，首先需要思考的问题就是 9.1.1 节提及的对于多种时间类型（主要是处理时间和事件时间）的支持，因为什么时候创建、触发、销毁窗口，以及窗口本身代表的时间范围是需要通过某种时间机制来确定和表示的。如果只考虑处理时间，问题就会变得非常容易解决。然而，业务上常常需要以事件时间为标准进行计算。由于网络阻塞、计算耗时等原因，事件时间和处理时间不可能完全相同，因此从事件时间的维度来看，数据就会出现乱序、晚到等情况。在这样的背景下，对窗口来说，一个很困难的问题就是如何知道这个窗口的数据已经到齐。进一步思考就会发现这是永远无法知道的，因此更先进的模型需要解决的问题不是"明确数据何时到齐"，而是接受数据会乱序、晚到的事实，并提供完备的窗口触发机制和对处理结果的更新、撤回机制。

基于这样的思考，数据流模型被拆解为以下 3 个模型。

- 窗口模型（window model），主要用于描述窗口如何创建、何时创建、如何合并等。
- 触发模型（triggering model），主要用于描述窗口在何种条件下触发计算。
- 增量处理模型（incremental processing model），主要用于描述如何对先前的结果进行更新、撤回等。

数据流模型要解决的问题，在论文中被归纳成 4 个维度，分别用 what、where、when 和 how 来描述。用这几个词来进行归纳总结是英语中的习惯，结合论文上下文将它们翻译成中文大致可以总结为以下几个问题：

- 对于窗口的计算结果是什么；
- 如何利用事件时间构建窗口；
- 何时触发窗口中的计算；
- 对于计算结果如何进行更新、撤回等操作。

下面来看数据流模型如何解决这几个问题。

在关于有界数据的处理中，即在传统的批处理中，可以认为处理的对象是(key,value)二元组。处理的过程大致可以分为两部分，一部分是数据的转换操作，另一部分是数据的分区聚合。由于对无界数据来说不知道数据何时结束，因此无法聚合。由此引入了窗口的概念。分区聚合操作就变成了在窗口内的分区聚合操作。

数据流模型中定义了以下 3 种窗口。
- Fixed Window（Tumbling Window）：固定大小的窗口。
- Sliding Window：固定大小的窗口和滑动周期。滑动周期小于窗口大小时，窗口就会重叠。Fixed Window 就是窗口大小与滑动周期相等的 Sliding Window。
- Session Window：在设定的超时时间内的数据都属于同一个窗口。若超过一定的时间，窗口没有收到新的数据，则该窗口结束。

数据流模型中的 3 种窗口如图 9-3 所示。

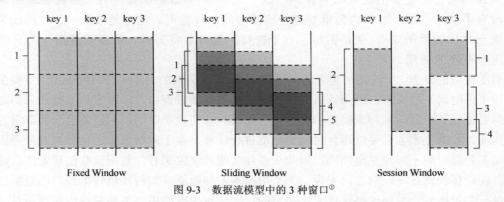

图 9-3　数据流模型中的 3 种窗口[①]

既然引入了窗口的概念，那么需要知道如何给数据加窗。加窗操作包括分配窗口（将数据分配到零个或多个窗口中）和合并窗口（将多个窗口合并成一个窗口）两部分。为了支持这种操作，数据流模型处理的对象从(key,value)二元组变成了(key,value,event_time,window)四元组。其中 event_time 表示数据的事件时间，window 表示数据被分配到了哪个窗口。如果所有元素都被分配到了 global window 中，模型就会变成传统的所谓的批处理模型。

图 9-4 展示了分配窗口的过程。

$(k, v_1, 12:00, [0, \infty))$, $(k, v_2, 12:00, [0, \infty))$

↓ AssignWindows(Sliding(2$m$, 1$m$))

$(k, v_1, 12:00, [11:59, 12:01))$,
$(k, v_1, 12:00, [12:00, 12:02))$,
$(k, v_2, 12:01, [12:00, 12:02))$,
$(k, v_2, 12:01, [12:01, 12:03))$

图 9-4　分配窗口

---

① 图 9-3～图 9-5 引自论文 *The Dataflow Model: A Practical Approach to Balancing Correctness, Latency, and Cost in Massive-Scale, Unbounded, Out-of-Order Data Processing*.

图 9-5 展示了合并窗口的过程。

这就基本确定了窗口模型，接下来需要考虑的主要问题是何时触发窗口中定义的计算。由于窗口模型以事件时间为基础，数据存在乱序或晚到的情况，因此不能直接根据事件时间来判断数据是否已经到齐，还需要采用另外的机制来通知窗口可以对数据进行处理。比较容易想到的是水位线（watermark）机制。前文介绍过，水位线表示的含义就是"系统认为该时间戳之前的数据已经到齐"。然而，"系统认为"并不代表真实的情况，如果水位线定义得"太高"，则会有大量数据在水位线之后才到达，这时窗口被触发得"太快"。另外，如果完全依赖水位线来触发窗口中的计算，有可能会因为某一条数据来得过慢导致计算被触发得"太慢"。

回顾 Lambda 架构的基本思想。Lambda 架构希望通过流处理保证低延迟，用批处理保证数据的最终一致性。那么在数据流模型中如何同时做到这两点呢？其思路是单个窗口可以快速地被触发，以此提供低延迟；同时单个窗口又可以在这之后多次被触发，以此保证数据的最终一致性。提供灵活的触发机制就是触发模型的基本内涵。

当窗口触发计算后，还需要提供多种对之前的计算结果进行更正的模式。数据流模型中有以下 3 种模式。

- Discarding：触发后，窗口中的内容被清空。后续添加进窗口的数据与之前的数据没有关系。每次计算的结果相互独立。
- Accumulating：触发后，窗口中的内容被保留。该窗口下次的触发结果在上一次结果的基础上进行更新。
- Accumulating & Retracting：触发后，与 Accumulating 模式一样会保留窗口中的内容。但该窗口下一次触发计算时，会撤回上一次的计算结果并重新计算。

相信在未来的业务场景中，对无界数据的处理会逐渐

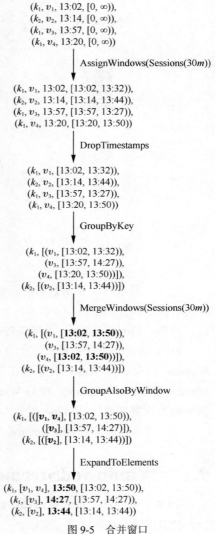

图 9-5 合并窗口

成为主流。数据流模型告诉人们，对有界数据的处理其实用无界数据处理引擎也可处理得很好，所谓批处理不过是流处理的一种特例。它通过窗口模型、触发模型和增量处理模型实现了批流一体化。开发人员可以根据实际情况选择流处理、批处理或者微批处理，在延迟性、正确性和计算成本间进行取舍与权衡。

## 9.1.3 Flink 中与窗口操作相关的核心概念

数据流模型的设计思想深深影响了 Flink 中与窗口操作相关的概念的实现。

与数据流模型类似，Flink 中定义了时间类型。但 Flink 中除了处理时间、事件时间，还定义了摄取时间。时间语义如图 9-6 所示。

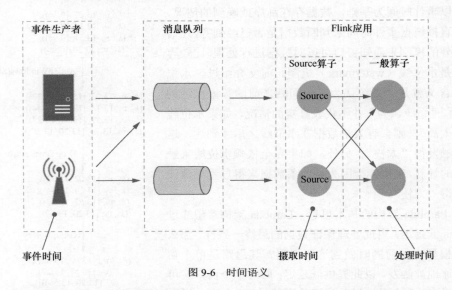

图 9-6　时间语义

其中，摄取时间表示数据进入系统的时间，实际场景中使用得较少，系统对它的处理与事件时间的类似。

同时，Flink 中也设计了与事件时间联系紧密的水位线。

Flink 中还定义了多种类型的窗口。大体上也是 Fixed Window（Tumbling Window）、Sliding Window 和 Session Window。不过在 Fixed Window 和 Sliding Window 中，Flink 还分别设计了 Count Window 和 Time Window。Time Window 的含义与数据流模型中的一致，Count Window 则是以元素个数为依据来划分 Window 的，与时间无关。

本节开头代码片段中的 window、trigger 和 evictor 则可以被认为是对窗口相关模型的具体实现。

## 9.2　WindowedStream

如 9.1 节所述，引入窗口操作后，对数据的处理大致可以分成数据的转换操作和分区后的窗口内的分区聚合操作。在 Flink 中，对数据流执行 keyBy() 操作后，再调用 window() 方法，就会返回 WindowedStream，表示分区后又加窗的数据流。如果数据流没有经过分区，直接调用 window() 方法则会返回 AllWindowedStream。AllWindowedStream 与 WindowedStream 的字段在方法上没有太大区别，可以认为 AllWindowedStream 是把所有数据放在了一个分区中，是 WindowedStream 的一个特例。下面只介绍 WindowedStream。

WindowedStream 的定义如下：

public class WindowedStream<T, K, W extends Window>

其中 T、K、W 为泛型，分别表示输入数据的类型、键的类型和 Window 的实现类。Window

是 Flink 中对窗口这一概念的抽象，是一个抽象类。

下面是 WindowedStream 中的一些重要字段。

- input：KeyedStream 类型，表示被加窗的输入流。
- windowAssigner：WindowAssigner<? super T, W>类型，表示窗口分配器。
- trigger：Trigger<? super T, ? super W>类型，表示窗口触发器。
- evictor：Evictor<? super T, ? super W>类型，表示窗口剔除器。

WindowedStream 中提供了方法对 trigger 和 evictor 字段赋值。除此以外，WindowedStream 中的方法主要是关于定义窗口内的聚合计算的。以 reduce()方法为例：

```
public SingleOutputStreamOperator<T> reduce(ReduceFunction<T> function) {
    ...
    return reduce(function, new PassThroughWindowFunction<K, W, T>());
}
public <R> SingleOutputStreamOperator<R> reduce(
    ReduceFunction<T> reduceFunction,
    WindowFunction<T, R, K, W> function) {
    TypeInformation<T> inType = input.getType();
    TypeInformation<R> resultType = getWindowFunctionReturnType(function, inType);
    return reduce(reduceFunction, function, resultType);
}
```

一般来说，开发人员会定义 ReduceFunction 来表示窗口中的聚合操作。WindowedStream 内部会调用其重载方法，传入默认的 WindowFunction——PassThroughWindowFunction。得到返回的 TypeInformation 后，继续调用重载的 reduce()方法：

```
public <R> SingleOutputStreamOperator<R> reduce(
    ReduceFunction<T> reduceFunction,
    WindowFunction<T, R, K, W> function,
    TypeInformation<R> resultType) {
    ...
    if (evictor != null) { // 当定义了窗口剔除器时
        @SuppressWarnings({"unchecked", "rawtypes"})
        TypeSerializer<StreamRecord<T>> streamRecordSerializer =
            (TypeSerializer<StreamRecord<T>>) new StreamElementSerializer(input.getType().createSerializer(getExecutionEnvironment().getConfig()));
        ListStateDescriptor<StreamRecord<T>> stateDesc =
            new ListStateDescriptor<>("window-contents", streamRecordSerializer);
        operator =
            new EvictingWindowOperator<>(windowAssigner,
                windowAssigner.getWindowSerializer(getExecutionEnvironment().getConfig()),
                keySel,
                input.getKeyType().createSerializer(getExecutionEnvironment().getConfig()),
                stateDesc,
                new InternalIterableWindowFunction<>(new ReduceApplyWindowFunction<>(reduceFunction, function)),
                trigger,
                evictor,
                allowedLateness,
                lateDataOutputTag);
    } else { // 当没有定义窗口剔除器时
        ReducingStateDescriptor<T> stateDesc = new ReducingStateDescriptor<>("window-contents",
            reduceFunction,
```

```
            input.getType().createSerializer(getExecutionEnvironment().getConfig())));
        operator =
            new WindowOperator<>(windowAssigner,
                windowAssigner.getWindowSerializer(getExecutionEnvironment().getConfig()),
                keySel,
                input.getKeyType().createSerializer(getExecutionEnvironment().getConfig()),
                stateDesc,
                new InternalSingleValueWindowFunction<>(function),
                trigger,
                allowedLateness,
                lateDataOutputTag);
    }
    return input.transform(opName, resultType, operator);
}
```

该方法主要根据 evictor 是否被定义而进入两个相应的分支。两个分支分别先构造了状态描述符，再将其作为参数构造了 StreamOperator 对象。最后，调用了 DataStream 中的 transform()方法，将返回类型、StreamOperator 对象传入其中，构造 Transformation 对象。

由此可以了解到，在 WindowedStream 中定义计算逻辑后的流程与其他数据流的流程一致，都会构造 StreamOperator 对象，进而构造出 Transformation 对象。在构造 StreamOperator 对象时，可观察到 trigger 和 evictor 等字段作为参数传入其中。于是，现在的问题为这些不同的 StreamOperator 实现类的内部逻辑是怎样的，以及 trigger、evictor 和传入的状态描述符在内部发挥了怎样的作用。这些内容会在接下来的几节详细分析。

## 9.3 窗口相关模型的实现

对数据流加窗后，会形成 WindowedStream。WindowedStream 中包含窗口分配器、窗口触发器、窗口剔除器等对应的字段，这些对应数据流模型中窗口、触发器等模型的具体实现，最终它们会被封装到具体的 StreamOperator 实现类中来发挥作用。本节主要介绍这些类中的字段和方法。

### 9.3.1 Window 类

Window 类是 Flink 中对窗口的抽象。它是一个抽象类，包含抽象方法 maxTimestamp()，用于获取属于该窗口的最大时间戳。

其中重要的实现类就是 TimeWindow，它有两个 long 类型的字段 start 和 end，分别表示该窗口的起始、停止时间。它对 maxTimestamp()方法的实现如下：

```
public long maxTimestamp() {
    return end - 1;
}
```

TimeWindow 重写了 equals()方法和 hashCode()方法，这样 start 和 end 字段的值相等的窗口就会被视为相同的窗口：

```
public boolean equals(Object o) {
    if (this == o) {
        return true;
```

```java
    }
    if (o == null || getClass() != o.getClass()) {
        return false;
    }
    TimeWindow window = (TimeWindow) o;
    return end == window.end && start == window.start;
}
public int hashCode() {
    return MathUtils.longToIntWithBitMixing(start + end);
}
```

TimeWindow 还有一个 getWindowStartWithOffset()方法,用于计算窗口的起始时间:

```java
public static long getWindowStartWithOffset(long timestamp, long offset, long windowSize) {
    return timestamp - (timestamp - offset + windowSize) % windowSize;
}
```

从对 Window 类的观察可以了解到,Window 并不是存放窗口数据的容器,而只是一个标识。后文将会说明窗口数据被维护在 WindowOperator 的状态字段中。

### 9.3.2 WindowAssigner 类

WindowAssigner 表示窗口分配器,用来把元素分配到零个或多个窗口(Window 对象)中。它是一个抽象类,其中重要的抽象方法为 assignWindows()方法,用来给元素分配窗口。

前文介绍过 Flink 有多种类型的窗口,如 Tumbling Window、Sliding Window 等。各种类型的窗口又分为基于事件时间或处理时间的窗口。WindowAssigner 的实现类就对应着具体类型的窗口。

TumblingProcessingTimeWindows 是 WindowAssigner 的一个实现类,表示基于处理时间的 Tumbling Window。它有两个 long 类型的字段 size 和 offset,分别表示窗口的大小和窗口起始位置的偏移量,其中 offset 的作用在 TimeWindow 的 getWindowStartWithOffset()方法中可以看到。它对 assignWindows()方法的实现如下:

```java
public Collection<TimeWindow> assignWindows(Object element, long timestamp,
WindowAssignerContext context) {
    final long now = context.getCurrentProcessingTime();
    long start = TimeWindow.getWindowStartWithOffset(now, offset, size);
    return Collections.singletonList(new TimeWindow(start, start + size));
}
```

可观察到分配算法与元素本身的值以及元素自带的时间戳都没有关系。分配算法先获取了系统当前的处理时间,然后调用 TimeWindow 的 getWindowStartWithOffset()方法计算出了窗口的起始时间,接着直接通过构造方法构造出了一个 TimeWindow 对象。

SlidingEventTimeWindows 是 WindowAssigner 的另一个实现类,表示基于事件时间的 Sliding Window。它有 3 个 long 类型的字段 size、slide 和 offset,分别表示窗口的大小、滑动的步长和窗口起始位置的偏移量。它对 assignWindows()方法的实现如下:

```java
public Collection<TimeWindow> assignWindows(Object element, long timestamp,
WindowAssignerContext context) {
    if (timestamp > Long.MIN_VALUE) {
        List<TimeWindow> windows = new ArrayList<>((int) (size / slide));
        long lastStart = TimeWindow.getWindowStartWithOffset(timestamp, offset, slide);
```

```
        for (long start = lastStart;
            start > timestamp - size;
            start -= slide) {
            windows.add(new TimeWindow(start, start + size));
        }
        return windows;
    } else {
        ...
    }
}
```

代码逻辑较为简单，就是根据传入的事件时间构造出多个 TimeWindow 对象，然后将其添加到列表中一并返回。

WindowAssigner 与其主要实现类的关系如图 9-7 所示。

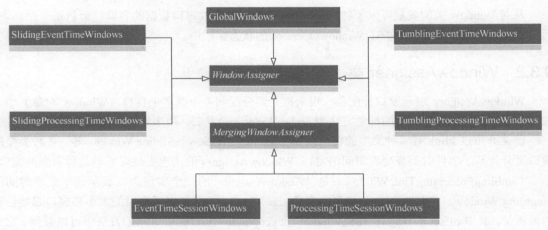

图 9-7　WindowAssigner 与其主要实现类的关系

这些类的含义分别如下。

- GlobalWindows：将所有元素分配进同一个窗口的全局窗口分配器。
- SlidingEventTimeWindows：基于事件时间的滑动窗口分配器。
- SlidingProcessingTimeWindows：基于处理时间的滑动窗口分配器。
- TumblingEventTimeWindows：基于事件时间的滚动窗口分配器。
- TumblingProcessingTimeWindows：基于处理时间的滚动窗口分配器。
- EventTimeSessionWindows：基于事件时间的会话窗口分配器。
- ProcessingTimeSessionWindows：基于处理时间的会话窗口分配器。

### 9.3.3　Trigger 类

Trigger 表示窗口触发器。它是一个抽象类，主要定义了下面 3 个方法用于确定窗口何时触发计算：

```
public abstract TriggerResult onElement(T element, long timestamp, W window, TriggerContext ctx) throws Exception;
```

```
    public abstract TriggerResult onProcessingTime(long time, W window, TriggerContext ctx)
throws Exception;
    public abstract TriggerResult onEventTime(long time, W window, TriggerContext ctx)
throws Exception;
```

这 3 个方法分别会在每个元素到来时、处理时间的定时器触发时、事件时间的定时器触发时调用。这 3 个方法的返回结果为 TriggerResult 对象。

TriggerResult 是一个枚举类，包含两个 boolean 类型的字段 fire 和 purge，分别表示窗口是否触发计算和窗口内的元素是否需要清空。其构造方法和其中的类型如下：

```
CONTINUE(false, false),
FIRE_AND_PURGE(true, true),
FIRE(true, false),
PURGE(false, true);
TriggerResult(boolean fire, boolean purge) {
    this.purge = purge;
    this.fire = fire;
}
```

窗口触发器的实现由用户根据业务需求自定义。不过 Flink 给一些内置的窗口分配器配置了默认的触发器。比如 TumblingProcessingTimeWindows 的触发器为 ProcessingTimeTrigger。它的 3 个方法的实现如下：

```
    public TriggerResult onElement(Object element, long timestamp, TimeWindow window,
TriggerContext ctx) {
        ctx.registerProcessingTimeTimer(window.maxTimestamp());
        return TriggerResult.CONTINUE;
    }
    public TriggerResult onEventTime(long time, TimeWindow window, TriggerContext ctx)
throws Exception {
        return TriggerResult.CONTINUE;
    }
    public TriggerResult onProcessingTime(long time, TimeWindow window, TriggerContext
ctx) {
        return TriggerResult.FIRE;
    }
```

由于 TumblingProcessingTimeWindows 是基于处理时间的分配器，因此其对应的触发器的 onEventTime()方法返回的一定是 CONTINUE，表示窗口不触发计算也不清空内部元素。onElement()方法同理。当处理时间的定时器触发时，onProcessingTime()方法被调用，这时窗口需要触发计算，于是会返回 FIRE。

SlidingEventTimeWindows 的默认触发器为 EventTimeTrigger，其实现思路与 ProcessingTimeTrigger 的一致，实现的方法不一一分析。下面是它的 onElement()方法：

```
    public TriggerResult onElement(Object element, long timestamp, TimeWindow window,
TriggerContext ctx) throws Exception {
        if (window.maxTimestamp() <= ctx.getCurrentWatermark()) {
            return TriggerResult.FIRE;
        } else {
            ctx.registerEventTimeTimer(window.maxTimestamp());
            return TriggerResult.CONTINUE;
        }
    }
```

这里可以看到，该方法获取了窗口的最大时间以与当前的水位线进行比较，如果水位线已经超过该时间戳，则返回 FIRE 触发计算。

Trigger 与其主要实现类的继承关系如图 9-8 所示。

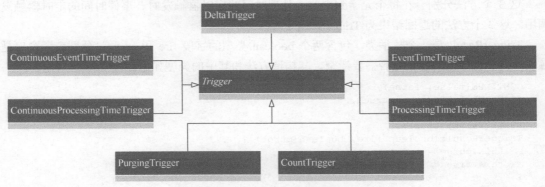

图 9-8　Trigger 与其主要实现类的继承关系

这些类的含义分别如下。

- CountTrigger：元素数达到设置的个数时触发计算的触发器。
- DeltaTrigger：基于 DeltaFunction 和设置的阈值触发计算的触发器。
- EventTimeTrigger：基于事件时间的触发器。
- ProcessingTimeTrigger：基于处理时间的触发器。
- PurgingTrigger：可包装其他触发器的清空触发器。
- ContinuousEventTimeTrigger：基于事件时间并按照一定的时间间隔连续触发计算的触发器。
- ContinuousProcessingTimeTrigger：基于处理时间并按照一定的时间间隔连续触发计算的触发器。

### 9.3.4　Evictor 类

Evictor 表示窗口剔除器，其会在窗口触发计算前或触发计算后调用其中的方法对窗口内的元素进行一定的清除。Evictor 是一个接口，其中定义了两个方法：

```
void evictBefore(Iterable<TimestampedValue<T>> elements, int size, W window, EvictorContext evictorContext);
void evictAfter(Iterable<TimestampedValue<T>> elements, int size, W window, EvictorContext evictorContext);
```

这两个方法分别在窗口中定义的计算函数被调用的前后调用。

剔除器通常使用得较少，一般的窗口分配器没有与之对应的默认剔除器。Flink 中提供了一些实现类，以 TimeEvictor 为例，它包含一个 long 类型的 windowSize 字段和一个 boolean 类型的 doEvictAfter 字段。

Evictor 对两个方法的实现如下：

```
public void evictBefore(Iterable<TimestampedValue<Object>> elements, int size, W window, EvictorContext ctx) {
```

```
        if (!doEvictAfter) {
            evict(elements, size, ctx);
        }
    }
    public void evictAfter(Iterable<TimestampedValue<Object>> elements, int size, W window,
EvictorContext ctx) {
        if (doEvictAfter) {
            evict(elements, size, ctx);
        }
    }
    private void evict(Iterable<TimestampedValue<Object>> elements, int size, EvictorContext
ctx) {
        if (!hasTimestamp(elements)) {
            return;
        }
        long currentTime = getMaxTimestamp(elements);
        long evictCutoff = currentTime - windowSize;
        for (Iterator<TimestampedValue<Object>> iterator = elements.iterator(); iterator.
hasNext(); ) {
            TimestampedValue<Object> record = iterator.next();
            if (record.getTimestamp() <= evictCutoff) {
                iterator.remove();
            }
        }
    }
```

在 evict() 方法中，完成了清除元素的工作。

如果指定了 evictor，那么在 WindowedStream 构造 StreamOperator 时，就会构造出 EvictingWindowOperator，否则会构造出 WindowOperator。

Evictor 与其实现类的关系如图 9-9 所示。

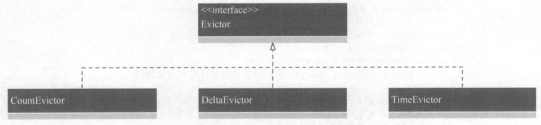

图 9-9　Evictor 与其实现类的关系

这些类的含义分别如下。

- CountEvictor：保留固定个数的元素的剔除器。
- DeltaEvictor：基于 DeltaFunction 和设置的阈值触发计算的剔除器。
- TimeEvictor：仅将元素保留一段固定时间的剔除器。

## 9.4　WindowOperator

前面分别分析了窗口分配器、窗口触发器和窗口剔除器的作用，从中可了解它们基本实现了数据流模型的核心思想。不过,目前并不知道这些组件在运行时如何协同工作。

从 WindowedStream 中的方法可以看到,这些字段被封装进了 StreamOperator 对象。定义了 evictor 时,构造的实现类是 EvictingWindowOperator;没有定义 evictor 时,构造的实现类是 WindowOperator。EvictingWindowOperator 是 WindowOperator 的子类。

下面分别介绍 WindowOperator 和 EvictingWindowOperator 的字段和方法。

对于 WindowOperator,除了窗口分配器和窗口触发器的相关字段,可以先了解下面两个字段。
- windowStateDescriptor:StateDescriptor 类型,表示窗口状态描述符。
- windowState:InternalAppendingState 类型,表示窗口的状态,窗口内的元素都在其中维护。

窗口中的元素并没有保存在 Window 对象中(从之前的示例代码中也可以看出 Window 中并没有维护元素的值),而是维护在 windowState 中。windowStateDescriptor 则是创建 windowState 所需用到的描述符。Window 对象在维护窗口内元素的过程中作为 windowState 字段的 namespace。因为 TimeWindow 等实现类重写了 equals()方法和 hashCode()方法,所以即便每个元素在分配窗口时会创建不同的 Window 对象,它们也仍然属于同一个 namespace。

当有元素到来时,会调用 WindowOperator 的 processElement()方法:

```
public void processElement(StreamRecord<IN> element) throws Exception {
    // 分配窗口
    final Collection<W> elementWindows = windowAssigner.assignWindows(
        element.getValue(), element.getTimestamp(), windowAssignerContext);
    ...
    if (windowAssigner instanceof MergingWindowAssigner) { // Session Window 的情况
        ...
    } else {
        for (W window: elementWindows) { // 非 Session Window 的情况
            ...
            // 将 Window 对象设置为 namespace 并添加元素到 windowState 中
            windowState.setCurrentNamespace(window);
            windowState.add(element.getValue());
            triggerContext.key = key;
            triggerContext.window = window;
            // 获取 TriggerResult,确定接下来是否需要触发计算或清空窗口
            TriggerResult triggerResult = triggerContext.onElement(element);
            if (triggerResult.isFire()) {
                ACC contents = windowState.get();
                if (contents == null) {
                    continue;
                }
                // 触发计算
                emitWindowContents(window, contents);
            }
            if (triggerResult.isPurge()) {
                // 清空窗口
                windowState.clear();
            }
            ...
        }
    }
    ...
}
```

重要的步骤已通过注释表示出。首先给元素分配窗口，分配的逻辑已在前文分析过。根据窗口的不同类型，会进入不同的逻辑分支。如果窗口分配器是 MergingWindowAssigner 类，那么表示窗口为 Session Window（只有 Session Window 才需要合并操作），否则进入 else 分支。这里分析非 Session Window 的情况。

接下来，将元素添加到状态中。然后调用 triggerContext 的 onElement()方法，实际上就是调用触发器的 onElement()方法。该方法的实现前文已分析过。接着，根据 TriggerResult 的值实现触发计算或清空窗口。触发计算时，会取出状态中的值进行处理；清空窗口实际上就是清空状态中的值。

其中，emitWindowContents()方法的实现如下：

```
private void emitWindowContents(W window, ACC contents) throws Exception {
    ...
    userFunction.process(triggerContext.key, window, processContext, contents,
timestampedCollector);
}
```

接下来的步骤由具体的 userFunction 来处理。注意，最后一个参数 timestampedCollector 是 Output 接口的实现类，最终会把元素发送给下游的算子。

在处理时间或事件时间的定时器触发时，会调用 WindowOperator 的 onProcessingTime()方法或 onEventTime()方法，其中的逻辑与 onElement()方法的大同小异，此处不再赘述。

EvictingWindowOperator 相比 WindowOperator，多了与 evictor 相关的几个字段。

- evictor：Evictor 类型。
- evictingWindowStateDescriptor：StateDescriptor 类型。
- evictorContext：EvictorContext 类型。
- evictingWindowState：InternalListState 类型。

与 WindowOperator 不同的是，EvictingWindowOperator 将元素维护在 evictingWindowState 中。evictorContext 的方法 evictBefore()方法和 evictAfter()方法，实际上就是调用 evictor 的对应方法。

EvictingWindowOperator 的 processElement()等方法与 WindowOperator 的逻辑一致，只是在 emitWindowContents()方法中加入了剔除器的相关逻辑：

```
private void emitWindowContents(W window, Iterable<StreamRecord<IN>> contents, ListState
<StreamRecord<IN>> windowState) throws Exception {
    ...
    evictorContext.evictBefore(recordsWithTimestamp, Iterables.size(recordsWithTimestamp));
    ...
    userFunction.process(triggerContext.key, triggerContext.window, processContext,
projectedContents, timestampedCollector);
    evictorContext.evictAfter(recordsWithTimestamp, Iterables.size(recordsWithTimestamp));
    ...
}
```

剔除器具体如何清除窗口内的元素需要由用户自己决定。

## 9.5 水位线

水位线（watermark）是选用事件时间来进行数据处理时特有的概念。它的本质就是时间戳，从上游流向下游,表示系统认为数据中的事件时间在该时间戳之前的数据都已到达。

Flink 中用水位类表示水位线。本节主要分析 Flink 中如何产生水位线、水位线如何随着数据流在算子间流转以及 Flink 如何处理多个输入流传来的水位。

### 9.5.1 产生水位线

Watermark 类是 StreamElement 类的子类,表示它是数据流中的一种元素。它有一个 long 类型的 timestamp 字段,表示它代表的时间戳,以毫秒为单位。

因为水位是事件时间下的概念,所以首先需要将配置设置为 EventTime:

```
final StreamExecutionEnvironment env = StreamExecutionEnvironment.getExecutionEnvironment();
env.setStreamTimeCharacteristic(TimeCharacteristic.EventTime);
```

产生水位线有以下两种方式:
- 直接由 source 算子产生;
- 通过 assignTimestampsAndWatermarks()方法指定 TimestampAssigner 提取时间戳产生。

**1. 直接由 source 算子产生**

如果希望由 source 算子产生水位线,那么要重写 SourceFunction 的 run()方法,通过 SourceContext 对象发送水位:

```
void run(SourceContext<T> ctx) throws Exception;
```

SourceContext 是接口,其实现类的构造在 StreamSourceContexts 类的 getSourceContext()方法中实现,由设置的时间类型决定:

```
final SourceFunction.SourceContext<OUT> ctx;
switch (timeCharacteristic) {
   case EventTime:
      ctx = new ManualWatermarkContext<>(
         output,
         processingTimeService,
         checkpointLock,
         streamStatusMaintainer,
         idleTimeout);
      break;
   case IngestionTime:
      ctx = new AutomaticWatermarkContext<>(
         output,
         watermarkInterval,
         processingTimeService,
         checkpointLock,
         streamStatusMaintainer,
         idleTimeout);
      break;
   case ProcessingTime:
      ctx = new NonTimestampContext<>(checkpointLock, output);
      break;
   default:
      throw new IllegalArgumentException(String.valueOf(timeCharacteristic));
}
```

ManualWatermarkContext 实现了 SourceContext 的 emitWatermark()方法：

```
public void emitWatermark(Watermark mark) {
    if (allowWatermark(mark)) {
        synchronized (checkpointLock) {
            ...
            processAndEmitWatermark(mark);
        }
    }
}
protected void processAndEmitWatermark(Watermark mark) {
    output.emitWatermark(mark);
}
```

output 的 emitWatermark()方法最终会调用每个子任务的最后一个输出 RecordWriterOutput 的 emitWatermark()方法：

```
public void emitWatermark(Watermark mark) {
    watermarkGauge.setCurrentWatermark(mark.getTimestamp());
    serializationDelegate.setInstance(mark);
    if (streamStatusProvider.getStreamStatus().isActive()) {
        try {
            recordWriter.broadcastEmit(serializationDelegate);
        } catch (Exception e) {
            throw new RuntimeException(e.getMessage(), e);
        }
    }
}
```

可以看到水位被广播到了每个下游子任务。

于是在重写 run()方法时会构造 Watermark 对象，调用 emitWatermark()方法就可以发送 Watermark 对象。

2. 通过 assignTimestampsAndWatermarks()方法产生

另一种产生水位线的方式是在调用 window()等方法前，调用 assignTimestampsAndWatermarks()方法添加一个算子专门用于提取时间戳。该方法有两个同名的重载方法，如下：

```
public SingleOutputStreamOperator<T> assignTimestampsAndWatermarks(
        AssignerWithPeriodicWatermarks<T> timestampAndWatermarkAssigner) {
    final int inputParallelism = getTransformation().getParallelism();
    final AssignerWithPeriodicWatermarks<T> cleanedAssigner = clean(timestampAndWatermarkAssigner);
    TimestampsAndPeriodicWatermarksOperator<T> operator =
        new TimestampsAndPeriodicWatermarksOperator<>(cleanedAssigner);
    return transform("Timestamps/Watermarks", getTransformation().getOutputType(), operator)
        .setParallelism(inputParallelism);
}
public SingleOutputStreamOperator<T> assignTimestampsAndWatermarks(
        AssignerWithPunctuatedWatermarks<T> timestampAndWatermarkAssigner) {
    final int inputParallelism = getTransformation().getParallelism();
    final AssignerWithPunctuatedWatermarks<T> cleanedAssigner = clean(timestampAndWatermarkAssigner);
    TimestampsAndPunctuatedWatermarksOperator<T> operator =
        new TimestampsAndPunctuatedWatermarksOperator<>(cleanedAssigner);
    return transform("Timestamps/Watermarks", getTransformation().getOutputType(), operator)
```

```
        .setParallelism(inputParallelism);
}
```

传入的参数的类型分别为 AssignerWithPeriodicWatermarks 和 AssignerWithPunctuatedWatermarks，它们都是继承自 TimestampAssigner 接口的子接口。TimestampAssigner 接口定义了从元素中提取时间戳的方法：

```
long extractTimestamp(T element, long previousElementTimestamp);
```

AssignerWithPeriodicWatermarks 中定义了获取 Watermark 对象的方法，根据处理时间该方法会被周期性地调用：

```
Watermark getCurrentWatermark();
```

AssignerWithPunctuatedWatermarks 中定义了构造 Watermark 对象的方法，该方法会在每条数据到来时被调用：

```
Watermark checkAndGetNextWatermark(T lastElement, long extractedTimestamp);
```

它们分别被封装进了 TimestampsAndPeriodicWatermarksOperator 和 TimestampsAndPunctuatedWatermarksOperator。

其中，TimestampsAndPeriodicWatermarksOperator 有一个 long 类型的 currentWatermark 字段，表示当前的水位代表的时间戳。

TimestampsAndPeriodicWatermarksOperator 的 processElement()方法如下：

```
public void processElement(StreamRecord<T> element) throws Exception {
    final long newTimestamp = userFunction.extractTimestamp(element.getValue(),
        element.hasTimestamp() ? element.getTimestamp() : Long.MIN_VALUE);
    output.collect(element.replace(element.getValue(), newTimestamp));
}
```

TimestampsAndPeriodicWatermarksOperator 会从元素中提取时间戳并将之封装进 StreamRecord 中。另外，它会周期性地调用 onProcessingTime()方法：

```
public void onProcessingTime(long timestamp) throws Exception {
    Watermark newWatermark = userFunction.getCurrentWatermark();
    if (newWatermark != null && newWatermark.getTimestamp() > currentWatermark) {
        currentWatermark = newWatermark.getTimestamp();
        output.emitWatermark(newWatermark);
    }
    long now = getProcessingTimeService().getCurrentProcessingTime();
    getProcessingTimeService().registerTimer(now + watermarkInterval, this);
}
```

该方法会给下游发送水位，并注册再次定时调用该方法。

TimestampsAndPunctuatedWatermarksOperator 也有一个 long 类型的 currentWatermark 字段。其 processElement()方法如下：

```
public void processElement(StreamRecord<T> element) throws Exception {
    final T value = element.getValue();
    final long newTimestamp = userFunction.extractTimestamp(value,
        element.hasTimestamp() ? element.getTimestamp() : Long.MIN_VALUE);
    output.collect(element.replace(element.getValue(), newTimestamp));
    final Watermark nextWatermark = userFunction.checkAndGetNextWatermark(value,
newTimestamp);
```

```
        if (nextWatermark != null && nextWatermark.getTimestamp() > currentWatermark) {
            currentWatermark = nextWatermark.getTimestamp();
            output.emitWatermark(nextWatermark);
        }
    }
```

该方法会在每个元素到来时提取时间戳，调用用户自己实现的 checkAndGetNextWatermark() 方法构造水位，如果其时间戳大于 currentWatermark 的时间戳，则将其发送给下游。

## 9.5.2 多个数据流传来的水位

要讨论多个数据流传来的水位，就要从接收水位的地方开始说起。与接收一般的数据元素一样，这个过程在 StreamTaskNetworkInput 的 processElement() 方法中实现：

```
private void processElement(StreamElement recordOrMark, DataOutput<T> output) throws Exception {
    if (recordOrMark.isRecord()){
        output.emitRecord(recordOrMark.asRecord());
    } else if (recordOrMark.isWatermark()) {
        statusWatermarkValve.inputWatermark(recordOrMark.asWatermark(), lastChannel);
        ...
    }
}
```

如果接收的是水位，那么调用 statusWatermarkValve 的 inputWatermark() 方法：

```
public void inputWatermark(Watermark watermark, int channelIndex) throws Exception {
    if (lastOutputStreamStatus.isActive() && channelStatuses[channelIndex].streamStatus.isActive()) {
        ...
        findAndOutputNewMinWatermarkAcrossAlignedChannels();
    }
}
private void findAndOutputNewMinWatermarkAcrossAlignedChannels() throws Exception {
    ...
    if (hasAlignedChannels && newMinWatermark > lastOutputWatermark) {
        lastOutputWatermark = newMinWatermark;
        output.emitWatermark(new Watermark(lastOutputWatermark));
    }
}
```

如果是两个输入流，那么调用的就是 StreamTwoInputProcessor.StreamTaskNetworkOutput 的 emitWatermark() 方法：

```
public void emitWatermark(Watermark watermark) throws Exception {
    synchronized (lock) {
        if (inputIndex == 0) {
            operator.processWatermark1(watermark);
        } else {
            operator.processWatermark2(watermark);
        }
    }
}
```

这里，根据输入流的索引决定调用 processWatermark1() 或 processWatermark2() 方法。

processWatermark1()和 processWatermark2()方法如下：

```java
public void processWatermark1(Watermark mark) throws Exception {
    input1Watermark = mark.getTimestamp();
    long newMin = Math.min(input1Watermark, input2Watermark);
    if (newMin > combinedWatermark) {
        combinedWatermark = newMin;
        processWatermark(new Watermark(combinedWatermark));
    }
}
public void processWatermark2(Watermark mark) throws Exception {
    input2Watermark = mark.getTimestamp();
    long newMin = Math.min(input1Watermark, input2Watermark);
    if (newMin > combinedWatermark) {
        combinedWatermark = newMin;
        processWatermark(new Watermark(combinedWatermark));
    }
}
```

总体而言，就是选择输入的水位中较小的那个，将其与当前水位进行对比，如果其比当前水位大，则调用 processWatermark()方法将其发送到下游：

```java
public void processWatermark(Watermark mark) throws Exception {
    ...
    output.emitWatermark(mark);
}
```

## 9.6 定时器

目前已经了解了不同的触发器会在定时器触发时调用其中的方法，还了解了水位的产生与传递方式。但是还不知道所谓的"定时器"是如何工作的，以及水位如何触发窗口的计算。

在 WindowOperator 中有一个 InternalTimerService 类型的 internalTimerService 字段，用于维护所有相关的定时任务。

该字段的初始化发生在 WindowOperator 的 open()方法中：

```java
internalTimerService =
    getInternalTimerService("window-timers", windowSerializer, this);
public <K, N> InternalTimerService<N> getInternalTimerService(
    String name,
    TypeSerializer<N> namespaceSerializer,
    Triggerable<K, N> triggerable) {
    ...
    return keyedTimeServiceHandler.getInternalTimerService(name, timerSerializer, triggerable);
}
public <N> InternalTimerService<N> getInternalTimerService(
    String name,
    TimerSerializer<K, N> timerSerializer,
    Triggerable<K, N> triggerable) {
    // 注册时间服务
    InternalTimerServiceImpl<K, N> timerService = registerOrGetTimerService(name, timerSerializer);
    // 启动时间服务
```

```
      timerService.startTimerService(
        timerSerializer.getKeySerializer(),
        timerSerializer.getNamespaceSerializer(),
        triggerable);
      return timerService;
    }
```

在注册时间服务时会实例化 InternalTimerServiceImpl 对象并会将其放入 InternalTimeServiceManager 的 timerServices 进行维护：

```
    <N> InternalTimerServiceImpl<K, N> registerOrGetTimerService(String name, TimerSerializer
<K, N> timerSerializer) {
        InternalTimerServiceImpl<K, N> timerService = (InternalTimerServiceImpl<K, N>)
timerServices.get(name);
        if (timerService == null) {
          timerService = new InternalTimerServiceImpl<>(
            localKeyGroupRange,
            keyContext,
            processingTimeService,
            createTimerPriorityQueue(PROCESSING_TIMER_PREFIX + name, timerSerializer),
            createTimerPriorityQueue(EVENT_TIMER_PREFIX + name, timerSerializer));
          timerServices.put(name, timerService);
        }
        return timerService;
    }
```

启动时间服务时，如果有恢复的定时任务，则直接开始定时：

```
    final InternalTimer<K, N> headTimer = processingTimeTimersQueue.peek();
    if (headTimer != null) {
      nextTimer = processingTimeService.registerTimer(headTimer.getTimestamp(), this::
onProcessingTime);
    }
```

其中，onProcessingTime()方法是 InternalTimerService 的实现类 InternalTimerServiceImpl 的方法，其实现如下：

```
    private void onProcessingTime(long time) throws Exception {
        nextTimer = null;
        InternalTimer<K, N> timer;
        while ((timer = processingTimeTimersQueue.peek()) != null && timer.getTimestamp()
<= time) {
            processingTimeTimersQueue.poll();
            keyContext.setCurrentKey(timer.getKey());
            triggerTarget.onProcessingTime(timer);
        }
        if (timer != null && nextTimer == null) {
            nextTimer = processingTimeService.registerTimer(timer.getTimestamp(), this::
onProcessingTime);
        }
    }
```

processingTimeTimersQueue 是 InternalTimerServiceImpl 中的队列，用于维护所有表示处理时间的 InternalTimer 对象。每个 InternalTimer 对象表示一个需要触发任务的时刻。如果队列中还有该对象，则将其取出然后调用 triggerTarget 的 onProcessingTime()方法并将 timer 传入。这就回到了触发器的方法。

另外，如果没有恢复的定时任务，在 WindowOperator 中也有多处会注册处理时间的定时任务并会调用 onProcessingTime()方法。比如当触发器基于处理时间触发时：

```
public TriggerResult onElement(Object element, long timestamp, TimeWindow window,
TriggerContext ctx) {
    ctx.registerProcessingTimeTimer(window.maxTimestamp());
    return TriggerResult.CONTINUE;
}
public void registerProcessingTimeTimer(long time) {
    internalTimerService.registerProcessingTimeTimer(window, time);

}
public void registerProcessingTimeTimer(N namespace, long time) {
    InternalTimer<K, N> oldHead = processingTimeTimersQueue.peek();
    if (processingTimeTimersQueue.add(new TimerHeapInternalTimer<>(time, (K) keyContext.
getCurrentKey(), namespace))) {
        long nextTriggerTime = oldHead != null ? oldHead.getTimestamp() : Long.MAX_VALUE;
        if (time < nextTriggerTime) {
            if (nextTimer != null) {
                nextTimer.cancel(false);
            }
            nextTimer = processingTimeService.registerTimer(time, this::onProcessingTime);
        }
    }
}
```

可看到 processingTimeTimersQueue 中添加了元素。

事件时间定时器的触发是从 AbstractStreamOperator 的 processWatermark()方法开始的：

```
public void processWatermark(Watermark mark) throws Exception {
    if (timeServiceManager != null) {
        timeServiceManager.advanceWatermark(mark);
    }
    output.emitWatermark(mark);
}
```

前文分析了 emitWatermark()方法，省略了上面的 advanceWatermark()方法。该方法如下：

```
public void advanceWatermark(Watermark watermark) throws Exception {
    for (InternalTimerServiceImpl<?, ?> service : timerServices.values()) {
        service.advanceWatermark(watermark.getTimestamp());
    }
}
public void advanceWatermark(long time) throws Exception {
    currentWatermark = time;
    InternalTimer<K, N> timer;
    while ((timer = eventTimeTimersQueue.peek()) != null && timer.getTimestamp() <= time) {
        eventTimeTimersQueue.poll();
        keyContext.setCurrentKey(timer.getKey());
        triggerTarget.onEventTime(timer);
    }
}
```

这里的逻辑与处理时间定时器的类似，也是从队列中取出 timer 对象，调用 triggerTarget 的 onEventTime()方法，进而调用触发器的对应方法。

eventTimeTimersQueue 中的元素可以在触发器触发操作时添加：

```
public TriggerResult onElement(Object element, long timestamp, TimeWindow window,
TriggerContext ctx) throws Exception {
    if (window.maxTimestamp() <= ctx.getCurrentWatermark()) {
        return TriggerResult.FIRE;
    } else {
        ctx.registerEventTimeTimer(window.maxTimestamp());
        return TriggerResult.CONTINUE;
    }
}
public void registerEventTimeTimer(long time) {
    internalTimerService.registerEventTimeTimer(window, time);
}
public void registerEventTimeTimer(N namespace, long time) {
    eventTimeTimersQueue.add(new TimerHeapInternalTimer<>(time, (K) keyContext.getCurrentKey(), namespace));
}
```

此外，对于 processingTimeTimersQueue 和 eventTimeTimersQueue 中的元素，还有多处可以添加，比如在 WindowOperator 的 processElement() 方法中，处理完每个元素后都会调用 registerCleanupTimer() 方法：

```
protected void registerCleanupTimer(W window) {
    long cleanupTime = cleanupTime(window);
    if (cleanupTime == Long.MAX_VALUE) {
        return;
    }
    if (windowAssigner.isEventTime()) {
        triggerContext.registerEventTimeTimer(cleanupTime);
    } else {
        triggerContext.registerProcessingTimeTimer(cleanupTime);
    }
}
```

在注册的过程中会分别给两个队列添加元素。

## 9.7 总结

本章从谷歌公司的两篇论文出发，深入讨论了流处理中窗口的产生背景和设计理念。Flink 的时间与窗口的相关实现在很大程度上参考了数据流模型。以该模型为讲解线索，本章主要介绍了 Flink 中的窗口分配器、窗口触发器、窗口剔除器等组件，并分析了这些组件如何在算子内协调工作，同时讨论了水位线、定时器在窗口模型中的实现原理与作用等。该模型为批流一体化打下了坚实的基础。

# 第 10 章

# 状态与容错

官方文档是这样描述 Flink 的：
Stateful Computations over Data Streams。
即在数据流上的状态计算。

可以说，状态计算（包括状态管理、检查点机制等）是 Flink 在设计上极为优秀、简约的特性之一，也是它主要的特点之一。

希望在学习完本章后，读者能够了解：
- 状态与容错的基本概念；
- 不同框架对状态与容错机制的考虑；
- Flink 对状态与容错机制的设计思想；
- Flink 在存储状态和恢复状态时的实现细节。

## 10.1 基本概念与设计思想

关于状态与容错机制，每个框架都有与之相关的诸多概念，这常常会令开发人员感到困惑。本节将通过对比 Hadoop、Spark、Flink 对这一机制的不同思考，深入讨论批处理系统和流处理系统如何"看待"状态与容错。

### 10.1.1 状态与容错的基本概念

本节从广义上对状态与容错进行讨论。

1. 什么是状态

什么是状态？当 Flink 初级开发人员看到这个问题时，脑海中想到的大概是官方文档中的例子：

```
// 自定义的算子
public class CountWindowAverage extends RichFlatMapFunction<Tuple2<Long, Long>, Tuple2<Long, Long>> {
    // 声明状态
    private transient ValueState<Tuple2<Long, Long>> sum;
    @Override
    public void flatMap(Tuple2<Long, Long> input, Collector<Tuple2<Long, Long>> out) throws Exception {
        ...
        // 更新状态
```

```java
            sum.update(currentSum);
            if (currentSum.f0 >= 2) {
                out.collect(new Tuple2<>(input.f0, currentSum.f1 / currentSum.f0));
                sum.clear();
            }
        }

        @Override
        public void open(Configuration config) {
            // 初始化状态
            ValueStateDescriptor<Tuple2<Long, Long>> descriptor =
                    new ValueStateDescriptor<>(
                            "average", // 状态名称
                            TypeInformation.of(new TypeHint<Tuple2<Long, Long>>() {}),
                            // 状态中保存的数据类型
                            Tuple2.of(0L, 0L)); // 状态默认值
            sum = getRuntimeContext().getState(descriptor);
        }
    }
    // 业务逻辑
    env.fromElements(Tuple2.of(1L, 3L), Tuple2.of(1L, 5L), Tuple2.of(1L, 7L), Tuple2.of(1L, 4L), Tuple2.of(1L, 2L))
            .keyBy(0)
            .flatMap(new CountWindowAverage())
            .print();
```

这是 Flink 的基本用法。开发人员会定义多个算子来表示业务逻辑，其中，CountWindowAverage 这一算子继承自 RichFlatMapFunction 这个抽象类。在 CountWindowAverage 中可以声明状态，如上面代码中的 sum 字段（ValueState 类型）。由于这个例子十分简单，这里不再赘述每一行代码的含义，重要部分均已给出注释。

对刚刚接触 Flink 的开发人员来说，状态的含义一般就是指在算子中定义表示状态的字段，如 ValueState 类型，然后在算子的 open() 方法中初始化，接着在算子的运算逻辑中对其进行更新。这样的理解是非常片面的，会让人误以为"状态"这一概念依赖于 Flink 引擎自己定义的数据类型和算子类型。这不但不利于开发人员理解 Flink 状态与容错机制的设计原理，也不能帮助开发人员横向对比 Flink 与其他计算引擎的类似特性。

从广义上来讲，任何一个程序都有状态。可以说，状态就是某一时刻程序的各个字段、变量在内存中的值。比如，程序从文件中读取数据，程序在内存中记录下文件读取到了什么位置，并将其保存在某个对象的 offset 字段中，以便接下来从该位置继续读取。这个 offset 字段的值其实就是有业务含义的"状态"值。

既然任何程序都有状态，那么对于分布式计算框架，无须特殊的设计，状态便天然地存在于其运行时的内存中。框架可以对这些状态进行维护（比如将其持久化），实现框架想要达到的目的（比如将状态用于容错机制）。那么，各个框架都是选取哪些字段、变量的值进行管理的呢？这是理解各个框架的状态与容错机制的关键。

2. 什么是容错

"容错"并不是 Flink 独有的概念，也并不是任何程序、框架都需要实现容错。在大数据计算领域常常把作业分类成流式计算或批量计算。对于批量计算，容错并不是必不可少的机制，因为

大部分批处理任务从时间和计算资源上来说是可控的。如果作业在中途异常停止，可以重新再执行一次。然而，对流处理作业来说并不是这样的。因为从业务上来说，流处理作业会"7×24 小时"不间断地执行。设想如果一个流处理作业执行了一年，突然因为一些异常而停止，或者因为发现了脏数据或逻辑问题而被手动停止，如果这时没有容错机制，则需要从一年前的数据开始从头执行。这从时间和计算成本上来说都无法接受。

如果作业需要容错，往往指的就是这样的过程：

程序在运行的过程中，在某一时刻对其状态进行落盘存储。在未来的某一时刻，程序因为某种情况而停止后，可以从之前落盘的数据重启并继续正常、稳定地运行。

通俗地说该过程就是存档、读档的过程。容错过程如图 10-1 所示。

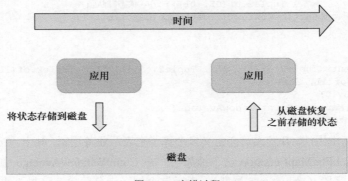

图 10-1　容错过程

**3. 状态与容错的关系**

从上面的分析可了解到，状态指的是某一时刻程序中各个字段、变量等在内存中的值，容错指的是对状态进行存储落盘、读取恢复的过程。因此，关键之处在于明确选取哪些值进行存储和恢复，以保证存储和恢复具有业务价值。对这一点的理解与取舍，便是不同框架对状态与容错机制的设计的出发点。

这里脱离具体的框架举几个例子，读者可以思考相应的设计思路接近哪个框架。

- 存储处理数据后的结果：在计算模型中，将数据按条处理。可以在处理数据的算子中定义一个字段，每处理一条数据，就按照业务逻辑对该字段进行更新。在进行状态存储时，仅存储该字段的值；在作业重启时，只需恢复该字段的值。
- 存储数据本身：在计算模型中，以数据集的方式处理数据。数据集会被多个算子处理，因此可以在数据集被某个算子处理完后将中间结果保存下来。这样在恢复时，就可以从这个完整的中间结果开始继续运行。
- 存储数据位置：计算引擎的数据一定有数据源，而某些数据源会为每条数据记录它在数据源中的位置。计算引擎可以将读取到的最新一条数据在数据源中的位置记录下来，将其作为状态进行保存和恢复。

在不同的业务和技术场景下，状态与容错的解决方案从理论上来说有无穷多种，这与每个计算框架的计算模型紧密相关。此外，框架的状态与容错机制能实现什么样的效果，还和与其对接的组件有关。比如上述第三个例子，倘若数据源并没有记录数据的位置信息，那么容错机制无法有效运行。

## 10.1.2　Hadoop 与 Spark 如何设计容错机制

一般来说，朴素的想法就是通过下面的步骤实现状态与容错机制。
（1）暂停所有数据的接收。
（2）每个任务处理当前已经接收的数据。
（3）将所有任务的状态进行持久化。
（4）恢复数据的接收和处理。

当任务出现异常时，则可以从之前持久化的地方进行恢复。Hadoop 与 Spark 的容错机制就是该思想的实现。

Hadoop 的任务可以分为 Map 任务和 Reduce 任务。这是两类分批次执行的任务，后者的输入依赖前者的输出。Hadoop 的设计思想十分简单——当任务出现异常时，重新执行该任务即可。其实，"跑"成功的任务的输出，就相当于整个任务的中间结果得到了持久化。比如 Reduce 任务出现异常需重跑时，就不必重跑它依赖的 Map 任务。

Spark 的实现也是这一想法的延续。虽然 Spark 实现的不是 Hadoop 那样的批处理，但是它仍然把"微批"当作数据处理的最小单元，整个框架实际上延续了不少批处理的思想。Spark 的容错机制相当经典，用到了其 RDD 的血统（lineage）关系。熟悉 Spark 的读者应该会了解"宽依赖""窄依赖"等概念。当 RDD 中的某个分区出现故障时，按照这种依赖关系重新计算即可。以复杂一些的宽依赖为例，Spark 会找到其父分区，经过计算重新获取结果。

如图 10-2 所示，如果 P1_0 发生故障，则 P0_0 与 P0_1 都会被重新计算，而计算 P0_0 和 P0_1 又会继续找其父分区来重新计算。按照这种血统关系来看，一直向上追溯会付出极大的代价。因此 Spark 提供了将分区计算结果持久化的方法。如果对 P0_0 与 P0_1 的数据进行了持久化，那么可以利用该结果直接恢复状态。

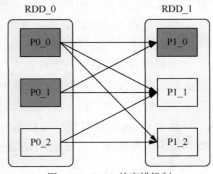

图 10-2　Spark 的容错机制

从图 10-2 的设计可以感受到，这种实现更适合用于批量计算的框架。它相当于将前一个阶段的计算结果"存档"，然后在任意时间后将该结果作为输入，来执行下一个阶段的任务。这种实现的状态存储过程显然过于复杂，并不太适用于对"低延迟"要求极高的流处理引擎。因此，Flink 设计了一套完全不同的分布式轻量级实现方式，并精巧地实现了各种一致性语义。

## 10.1.3　Flink 中容错机制的设计思想

下面从单机程序开始讨论 Flink 容错机制的设计思想。

1. 从单机程序开始

请思考对于一个运行在单个节点的进程，该如何设计容错机制。

比较容易想到的思路是，在主线程外另外开启一个线程来执行定时任务，定期地将状态数据刷写到磁盘。当作业停止后重启，则可以直接从之前刷写到磁盘的数据恢复，单机容错如图 10-3 所示。

2. 分布式容错

延续上述思路，是否可以设计一个分布式的容错机制呢？图 10-4 所示的是多节点的分布式任务，数据流的走向是从左向右的。

图 10-3　单机容错

图 10-4　多节点容错

如果给这些任务分别开启一个线程来执行定时任务，这些分布在不同物理机上的任务的确可以做到状态的存储和恢复。然而，采用这种粗暴的处理方式极容易导致业务上的异常。比如，当左边的任务处理完 a、b、c 这 3 条数据后，将数据发送至网络，在这 3 条数据还未到达中间的任务时，3 个线程同时（假设时间同步情况非常理想）触发状态存储的动作。这时左边的任务存储的状态是处理完 a、b、c 后的状态，而后两个任务（中间和右边的任务）存储的是未处理这 3 条数据时的状态。此时整个集群宕机，3 个任务被恢复后，左边的任务将从 a、b、c 这 3 条数据后的数据开始读取和处理，而后面的任务将永远无法接收到这 3 条数据。这就造成了数据的丢失。如果 3 个线程的触发时间不同步，可能会造成数据重复处理。

这个问题在流处理中被称为"一致性语义"问题。当一条数据在计算引擎中被处理"至少一次""恰好一次""最多一次"时，一致性语义分别是"at least once""exactly once""at most once"。

不同的业务场景对一致性语义有着不同的要求。举例来说，一个广告投放平台按照用户对广告的点击量进行收费，如果点击量被少算，则对平台方不利；如果点击量被多算，则对广告商不利。无论是哪种情况都不利于双方的长期合作。在这种情况下，"exactly once"语义就显得尤为重要。

基于 Flink 的计算模型与数据传输方式的设计，容错机制由栅栏（barrier）来实现。栅栏可以被理解为一条数据，会被周期性地插入数据流，随着数据一起被传输到下游。多节点容错栅栏如图 10-5 所示。

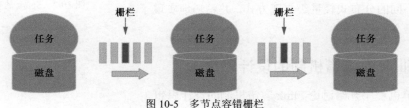

图 10-5　多节点容错栅栏

此时，每个任务将不再需要另外开启一个线程来完成定时任务，只需要在接收到栅栏时触发存储状态的动作。由于数据传输的有序性，这样的机制可以保证"exactly once"语义。

为什么这里说"可以保证"exactly once"语义，而没有说"必然"保证该语义呢？这是因为作业的拓扑可能更加复杂，多节点容错对齐如图 10-6 所示。

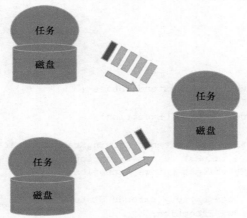

图 10-6　多节点容错对齐

如果一个进程的上游有多条数据流，那么它应该在接收到哪个栅栏时触发状态存储操作呢？

以图 10-6 为例，当右边的进程接收到下面的数据流传来的栅栏时，它可以先不触发任何操作，该数据流后面的数据也暂时不进行处理，而是将这些数据接收到缓存中。上面的数据流照常处理。当接收到上面的数据流传来的栅栏时，再触发状态存储操作。这样仍可以保证"exactly once"语义。

显然，了解了这个原理后，就可以在相应的过程中添加任何自己业务需要的策略，例如可以不让栅栏对齐就触发操作，或是每个栅栏都触发一次操作，甚至可以将部分数据丢弃，待最后一个栅栏到来时触发操作……这些不同的策略对应不同的一致性语义。Flink 实现了"exactly once"语义和"at least once"语义。

## 10.1.4　Flink 的状态与容错机制的核心概念

Flink 为需要保存在状态字段中的不同数据和位于不同算子的状态设计了不同的数据类型，为不同的持久化方式设计了不同的状态后端。这些状态的持久化依赖于检查点机制。

**1. 不同类型的状态**

Flink 为算子定义了多种可用的数据类型用于表示状态。这些状态都实现了 State 接口。

```
public interface State {
    void clear();
}
```

该接口只有一个方法需要实现。这个方法用于删除在当前键下的状态值。

继承自 State 的实现类总计有数十种，继承关系较为复杂。之所以需要这么多实现类，是因为根据算子类型、状态后端的不同，状态需要用不同的存储逻辑保存不同的信息。

所幸这些类型并不需要开发人员显式地声明。开发人员只需要记住下面 5 个接口：

- ValueState；
- ListState；
- MapState；
- ReducingState；
- AggregatingState。

这些接口的定义如下：

```
public interface ValueState<T> extends State {
   T value() throws IOException;
   void update(T value) throws IOException;
}
public interface ListState<T> extends MergingState<T, Iterable<T>> {
   void update(List<T> values) throws Exception;
   void addAll(List<T> values) throws Exception;
}
public interface MapState<UK, UV> extends State {
   UV get(UK key) throws Exception;
   void put(UK key, UV value) throws Exception;
   void putAll(Map<UK, UV> map) throws Exception;
   void remove(UK key) throws Exception;
   boolean contains(UK key) throws Exception;
   Iterable<Map.Entry<UK, UV>> entries() throws Exception;
   Iterable<UK> keys() throws Exception;
   Iterable<UV> values() throws Exception;
   Iterator<Map.Entry<UK, UV>> iterator() throws Exception;
   boolean isEmpty() throws Exception;
}
public interface ReducingState<T> extends MergingState<T, T> {}
public interface AggregatingState<IN, OUT> extends MergingState<IN, OUT> {}
```

2. 状态在不同算子中

在开发中，通常会用接口的方式声明状态：

```
private transient ListState listState;
```

接着需要在初始化方法（根据不同的接口可以是 open()方法或 initialize()方法等）中对状态进行初始化。如果是 KeyBy 前的算子，可以这样进行初始化：

```
OperatorStateStore stateStore = context.getOperatorStateStore();
listState = stateStore.getListState(new ListStateDescriptor<>(
    "myListState", // 状态名称
    TypeInformation.of(...))); // 状态中保存的数据类型
```

如果是 KeyBy 后的算子，可以这样进行初始化：

```
KeyedStateStore stateStore = context.getKeyedStateStore();
listState = stateStore.getListState(new ListStateDescriptor<>(
    "myListState", // 状态名称
    TypeInformation.of(...))); // 状态中保存的数据类型
```

虽然这两种情况都是调用 getListState() 方法进行初始化的，但是由于 stateStore 类型的不同，返回的状态类型是两个不同的实现类。这很容易理解，在一个 KeyBy 前的算子中，这个状态字段只需要保存业务上定义的状态数据。但是对于 KeyBy 后的算子，每一条来到该算子的数据，都暗含它的分区键信息。实际上，Flink 内部会在每一条数据到来时，维护它的分区键信息。在对状态数据进行存储时，会把该数据存储到对应键下的集合中。

Flink 把 KeyBy 前的算子中定义的状态称为操作符状态（operator state），把 KeyBy 后的算子中定义的状态称为键控状态（keyed state）。可以说操作符状态是一种特殊的键控状态（non-key state，即没有分区键信息的状态）。操作符状态中还有一种特殊的状态，叫作广播状态（broadcast state）。它可以被理解为一种特殊的操作符状态，它只在状态恢复时的重分配模式上与操作符状态有所区别，除此之外，在开发上和 Flink 内部对其处理的机制上与操作符状态几乎完全一致。

键控状态包括前文介绍的 5 种类型的 State：ValueState、ListState、MapState、ReducingState、AggregatingState。而操作符状态只有 ListState。

3. 状态后端

状态有多种存储媒介，也有多种存储方式，这些都由状态后端（StateBackend）来定义。Flink 支持的状态后端主要有 3 种：

- MemoryStateBackend；
- FsStateBackend；
- RocksDBStateBackend。

目前在很多大数据架构中会使用 HBase 来存储数据，因此自定义 HBaseStateBackend 成了很多技术部门的选择。还可以为以上这些状态后端定制增量存储、异步存储等特性。

4. 快照

上述机制在 Flink 中被称为"分布式快照"，它有着异步、轻量等特点。每个进程在接收到栅栏时触发的核心操作就是快照，该操作会被周期性地触发。

5. 检查点/保存点

每一次完整地结束快照后会生成多个文件，这些文件保存了该时刻集群中各节点的状态及其元信息，其被称为检查点（Checkpoint）。而保存点（Savepoint）与检查点几乎完全一致，只不过触发时机不一样。检查点是程序周期性触发并生成的，而保存点是通过命令行等方式手动生成的。通过观察源码可以发现，两者对应的绝大多数代码都是可复用的，流程和文件内容几乎完全一致。

根据官方的说法，今后可能会基于不同的状态后端对检查点和保存点进行一定的区分。

## 10.2 状态存储

状态的存储流程大致可以拆分为以下几个部分来进行理解：检查点的触发、栅栏的传输、状态数据的更新和存储、元信息的存储。从系统架构上来看，整个流程如图 10-7 所示。

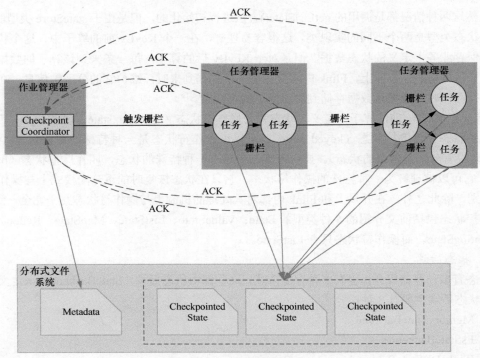

图 10-7 状态存储流程

## 10.2.1 检查点的触发

检查点的触发是从作业管理器端开始的。

### 1. CheckpointCoordinator

JobMaster 端依靠 CheckpointCoordinator 这个组件来负责检查点的触发和完成。JobMaster 端会周期性地触发检查点,对应的信息会被传递到 TaskExecutor 端,由相应的任务接收。任务完成快照操作后,会给下游的任务发送栅栏,下游任务依次执行快照操作。这些快照操作均由各个任务独立执行,将自身需要保存的状态数据进行保存,通常是将其落盘到文件,这里称该文件为"状态数据文件"。每个任务完成这些操作后,会给 JobMaster 发 ACK 信号,通知 JobMaster 该任务已完成快照操作。ACK 信号中包括状态数据文件的元信息,由 CheckpointCoordinator 负责处理。如果所有任务都发回了 ACK 信号,则 CheckpointCoordinator 会将这些元信息整合起来,写入另一个文件,这里称该文件为"状态元信息文件"。

下面对 CheckpointCoordinator 的核心字段进行介绍。

- job:JobID 类型,表示 job 的标识。
- tasksToTrigger:ExecutionVertex[]类型,表示 JobMaster 端触发检查点时信号被发往哪些任务。
- tasksToWaitFor:ExecutionVertex[]类型,表示快照操作完成后需要向 JobMaster 端发送 ACK 信号的任务集合。
- tasksToCommitTo:ExecutionVertex[]类型,表示一次完整的快照操作结束后需要触发回调函数的任务集合。
- pendingCheckpoints：Map<Long, PendingCheckpoint>类型，用于维护正在进行中的检查点。
- completedCheckpointStore：CompletedCheckpointStore 类型，用于维护已经完成的检查点。
- checkpointStorage：CheckpointStorageCoordinatorView 类型，主要用于初始化检查点的地址等元信息。
- baseInternal：表示两次检查点操作之间的基本时间间隔。
- checkpointTimeout：表示一次检查点操作所允许的最大时长。
- minPauseBetweenCheckpoints：表示一次检查点完成后距下次检查点开始的最短时间间隔。
- maxConcurrentCheckpointAttempts：表示允许同时执行的检查点的最大数量。
- jobStatusListener：JobStatusListener 类型，作业执行状态的监听器。根据作业执行的不同状态决定 CheckpointCoordinator 要执行的操作。
- lastCheckpointCompletionRelativeTime：表示上一次检查点完成的时刻。

该类与检查点触发有关的核心方法是 triggerCheckpoint()方法。

```java
public CompletableFuture<CompletedCheckpoint> triggerCheckpoint(long timestamp, boolean isPeriodic) {
    try {
        return triggerCheckpoint(timestamp, checkpointProperties, null, isPeriodic, false);
    } catch (CheckpointException e) {
        ...
    }
}
public CompletableFuture<CompletedCheckpoint> triggerCheckpoint(
    long timestamp,
    CheckpointProperties props,
    @Nullable String externalSavepointLocation,
    boolean isPeriodic,
    boolean advanceToEndOfTime) throws CheckpointException {
  ...
    synchronized (lock) {
      // 触发检查点前的检查
        preCheckBeforeTriggeringCheckpoint(isPeriodic, props.forceCheckpoint());
    }
// （1）检查是否所有 tasksToTrigger 集合中的 Task 实例的状态都是 RUNNING，将状态为 RUNNING 的
// Task 实例放入 executions 集合。如果存在状态不为 RUNNING 的 Task 实例，则终止此次检查点操作
Execution[] executions = new Execution[tasksToTrigger.length];
for (int i = 0; i < tasksToTrigger.length; i++) {
  Execution ee = tasksToTrigger[i].getCurrentExecutionAttempt();
  if (ee == null) {
    ...
  } else if (ee.getState() == ExecutionState.RUNNING) {
    executions[i] = ee;
  } else {
    …
  }
}
// （2）将 tasksToWaitFor 集合中正常实例化的 Task 实例放入 ackTasks 集合，如果存在未实例化的实
// 例，则终止此次检查点操作
Map<ExecutionAttemptID, ExecutionVertex> ackTasks = new HashMap<>(tasksToWaitFor.length);
```

```
        for (ExecutionVertex ev : tasksToWaitFor) {
            Execution ee = ev.getCurrentExecutionAttempt();
            if (ee != null) {
                ackTasks.put(ee.getAttemptId(), ev);
            } else {
                ...
            }
        }
        final CheckpointStorageLocation checkpointStorageLocation;
        final long checkpointID;
...
        // （3）实例化一个 PendingCheckpoint 对象并将其维护起来
        final PendingCheckpoint checkpoint = new PendingCheckpoint(
            job,
            checkpointID,
            timestamp,
            ackTasks,
            masterHooks.keySet(),
            props,
            checkpointStorageLocation,
            executor);
...
        try {
            synchronized (lock) {
                // （4）触发检查点前的检查
                preCheckBeforeTriggeringCheckpoint(isPeriodic, props.forceCheckpoint());
                pendingCheckpoints.put(checkpointID, checkpoint);
                ...
            }
            final CheckpointOptions checkpointOptions = new CheckpointOptions(
                props.getCheckpointType(),
                checkpointStorageLocation.getLocationReference());
            // （5）发送消息给相应的 Task 实例，触发检查点
            for (Execution execution: executions) {
                if (props.isSynchronous()) {
                    execution.triggerSynchronousSavepoint(checkpointID, timestamp, checkpointOptions,
advanceToEndOfTime);
                } else {
                    execution.triggerCheckpoint(checkpointID, timestamp, checkpointOptions);
                }
            }
            numUnsuccessfulCheckpointsTriggers.set(0);
            return checkpoint.getCompletionFuture();
        }
        catch (Throwable t) {
            ...
        }
    }
```

上述代码的核心逻辑为以下 4 个步骤。

（1）检查是否所有 tasksToTrigger 集合中的任务的状态都是 RUNNING，将状态为 RUNNING 的任务放入 executions 集合。如果存在状态不为 RUNNING 的任务，则终止此次检查点操作。

（2）将 tasksToWaitFor 集合中正常实例化的任务放入 ackTasks 集合。如果存在未实例化的任务，则终止此次检查点操作。

（3）实例化一个 PendingCheckpoint 对象并将其维护起来。注意，实例化 PendingCheckpoint 对象时，将 ackTasks 集合作为参数传入其中。

（4）发送消息给相应的任务，触发检查点。这些任务就是 executions 中的实例。

其中，preCheckBeforeTriggeringCheckpoint()方法进行了如下检查：

```
private void preCheckBeforeTriggeringCheckpoint(boolean isPeriodic, boolean forceCheckpoint)
throws CheckpointException {
    // 如果此时 CheckpointCoordinator 已经 shutdown，则终止此次检查点操作
    if (shutdown) {
        throw new CheckpointException(CheckpointFailureReason.CHECKPOINT_COORDINATOR_SHUTDOWN);
    }
    // 当周期性检查点未开启时不允许周期性地触发检查点
    if (isPeriodic && !periodicScheduling) {
        throw new CheckpointException(CheckpointFailureReason.PERIODIC_SCHEDULER_SHUTDOWN);
    }
     // 当 Checkpoint 并非强制执行时，进行下面的检查
    if (!forceCheckpoint) {
        // 当队列中还有检查点时终止此次检查点操作
        if (triggerRequestQueued) {
            throw new CheckpointException(CheckpointFailureReason.ALREADY_QUEUED);
        }
        // 如果正在 Pending 的 Checkpoint 的个数超过限制，则终止此次检查点操作，并且将
        // triggerRequestQueued 的值设置为 true
        checkConcurrentCheckpoints();
        // 检查两次检查点操作之间的时间间隔是否足够。如果不够则终止此次检查点操作，并且设置在规定
        // 时间后执行下一个检查点操作
        checkMinPauseBetweenCheckpoints();
    }
}
```

### 2. ScheduledTrigger

触发检查点的方法入口是 CheckpointCoordinator 的 triggerCheckpoint()方法，这个方法是由 ScheduledTrigger 线程触发的。

```
private final class ScheduledTrigger implements Runnable {
    @Override
    public void run() {
        try {
            triggerCheckpoint(System.currentTimeMillis(), true);
        }
        catch (Exception e) {
            ...
        }
    }
}
```

这个线程会定时执行，由此周期性地触发检查点。这个任务的开启由 CheckpointCoordinator 的 scheduleTriggerWithDelay()方法实现。

```
private ScheduledFuture<?> scheduleTriggerWithDelay(long initDelay) {
    return timer.scheduleAtFixedRate(
        new ScheduledTrigger(),
        initDelay, baseInterval, TimeUnit.MILLISECONDS);
}
```

### 3. CheckpointCoordinator 的初始化与 ScheduledTrigger 线程的执行

CheckpointCoordinator 的初始化在 ExecutionGraph 的初始化过程中实现，checkpointCoordinator 是 ExecutionGraph 中的一个字段。因为在初始化 ExecutionGraph 的过程中可能需要从之前保存的检查点文件或保存点文件中恢复数据，进而影响最终返回的 ExecutionGraph 对象，所以两者产生依赖在设计上是合理的。

在 ExecutionGraphBuilder 的 buildGraph() 方法中，调用了 executionGraph.enableCheckpointing() 方法。enableCheckpointing() 方法如下：

```
public void enableCheckpointing(
    CheckpointCoordinatorConfiguration chkConfig,
    List<ExecutionJobVertex> verticesToTrigger,
    List<ExecutionJobVertex> verticesToWaitFor,
    List<ExecutionJobVertex> verticesToCommitTo,
    List<MasterTriggerRestoreHook<?>> masterHooks,
    CheckpointIDCounter checkpointIDCounter,
    CompletedCheckpointStore checkpointStore,
    StateBackend checkpointStateBackend,
    CheckpointStatsTracker statsTracker) {
  // （1）初始化 tasksToTrigger、tasksToWaitFor、tasksToCommitTo 集合
  ExecutionVertex[] tasksToTrigger = collectExecutionVertices(verticesToTrigger);
  ExecutionVertex[] tasksToWaitFor = collectExecutionVertices(verticesToWaitFor);
  ExecutionVertex[] tasksToCommitTo = collectExecutionVertices(verticesToCommitTo);
  ...
  // （2）实例化 CheckpointCoordinator 对象
  checkpointCoordinator = new CheckpointCoordinator(
    jobInformation.getJobId(),
    chkConfig,
    tasksToTrigger,
    tasksToWaitFor,
    tasksToCommitTo,
    checkpointIDCounter,
    checkpointStore,
    checkpointStateBackend,
    ioExecutor,
    new ScheduledExecutorServiceAdapter(checkpointCoordinatorTimer),
    SharedStateRegistry.DEFAULT_FACTORY,
    failureManager);
  ...
  // （3）判断是否开启了检查点机制
  if (chkConfig.getCheckpointInterval() != Long.MAX_VALUE) {
    // （4）如果开启了检查点机制，则注册一个监听 Job 状态的监听器
    registerJobStatusListener(checkpointCoordinator.createActivatorDeactivator());
  }
  ...
}
```

上述代码的核心逻辑为以下 4 个步骤。

（1）初始化 tasksToTrigger、tasksToWaitFor、tasksToCommitTo 集合。这几个集合在接下来实例化 CheckpointCoordinator 对象时被作为参数传入，直接赋值给 CheckpointCoordinator 的相应字段。

（2）实例化 CheckpointCoordinator 对象。注意，这里传入的参数还有 chkConfig，这个参数包含关

于检查点的全部配置。CheckpointCoordinator 中的 checkpointTimeout、maxConcurrentCheckpointAttempts 等字段的值都来自该参数。

（3）判断是否开启了检查点机制。chkConfig.getCheckpointInterval()获取的就是在业务代码中开启检查点机制时设置的时间间隔。若没有开启检查点机制或者设置的时间间隔过短，则该值会在 Flink 内部被设置为 Long.MAX_VALUE，表示不开启检查点机制。

（4）如果开启了检查点机制，则注册监听器，用于监听 Job 的状态。当监听到状态发生改变时，触发相应的操作。

在第 1 步中，调用了 collectExecutionVertices()方法。该方法的逻辑比较简单，就是从参数 jobVertices 中获取其中的 taskVertices 对象。

```
private ExecutionVertex[] collectExecutionVertices(List<ExecutionJobVertex> jobVertices) {
    if (jobVertices.size() == 1) {
        ExecutionJobVertex jv = jobVertices.get(0);
        if (jv.getGraph() != this) {
            throw new IllegalArgumentException("Can only use ExecutionJobVertices of this ExecutionGraph");
        }
        return jv.getTaskVertices();
    }
    else {
        ArrayList<ExecutionVertex> all = new ArrayList<>();
        for (ExecutionJobVertex jv : jobVertices) {
            if (jv.getGraph() != this) {
                throw new IllegalArgumentException("Can only use ExecutionJobVertices of this ExecutionGraph");
            }
            all.addAll(Arrays.asList(jv.getTaskVertices()));
        }
        return all.toArray(new ExecutionVertex[all.size()]);
    }
}
```

因此，对于这个方法，关注的重点是参数值 verticesToTrigger、verticesToWaitFor 和 verticesToCommitTo 是如何得来的。在 enableCheckpointing()方法的上一层，也就是 ExecutionGraphBuilder.build()方法中，有这样一段代码：

```
List<ExecutionJobVertex> triggerVertices =
    idToVertex(snapshotSettings.getVerticesToTrigger(), executionGraph);
List<ExecutionJobVertex> ackVertices =
    idToVertex(snapshotSettings.getVerticesToAcknowledge(), executionGraph);
List<ExecutionJobVertex> confirmVertices =
    idToVertex(snapshotSettings.getVerticesToConfirm(), executionGraph);
```

其中，idToVertex()方法的逻辑与前面介绍的 collectExecutionVertices()的逻辑类似，因此重点成了 snapshotSettings 对象中的 3 个对象是如何初始化的。

```
public List<JobVertexID> getVerticesToTrigger() {
    return verticesToTrigger;
}
public List<JobVertexID> getVerticesToAcknowledge() {
    return verticesToAcknowledge;
```

```
}
public List<JobVertexID> getVerticesToConfirm() {
    return verticesToConfirm;
}
```

找到对象的初始化流程，可以很容易地在 StreamingJobGraphGenerator 类的 configure-Checkpointing()方法中发现下面的代码：

```
for (JobVertex vertex : jobVertices.values()) {
    if (vertex.isInputVertex()) {
        triggerVertices.add(vertex.getID());
    }
    commitVertices.add(vertex.getID());
    ackVertices.add(vertex.getID());
}
```

根据这里的逻辑，可以推断出 CheckpointCoordinator 中 tasksToTrigger 集合表示的就是所有 source 任务实例。而另外两个集合表示的就是包括 source 在内的所有 Task 实例。

在第 4 步中，注册了一个监听器。

```
public void registerJobStatusListener(JobStatusListener listener) {
    if (listener != null) {
        jobStatusListeners.add(listener);
    }
}
```

监听器对象通过 checkpointCoordinator.createActivatorDeactivator()方法获得。

```
public JobStatusListener createActivatorDeactivator() {
    synchronized (lock) {
        if (shutdown) {
            throw new IllegalArgumentException("Checkpoint coordinator is shut down");
        }
        if (jobStatusListener == null) {
            jobStatusListener = new CheckpointCoordinatorDeActivator(this);
        }
        return jobStatusListener;
    }
}
```

CheckpointCoordinatorDeActivator 类封装了 CheckpointCoordinator 对象，实现了 JobStatusListener 接口。其中实现的方法如下：

```
public void jobStatusChanges(JobID jobId, JobStatus newJobStatus, long timestamp,
Throwable error) {
    if (newJobStatus == JobStatus.RUNNING) {
        // 开启检查点线程
        coordinator.startCheckpointScheduler();
    } else {
        // 关闭检查点线程
        coordinator.stopCheckpointScheduler();
    }
}
```

当 JobStatus 转变为 RUNNING 时，会通知所有的监听器调用 jobStatusChanges()方法。CheckpointCoordinator 的 startCheckpointScheduler()方法会调用 scheduleTriggerWithDelay()方

法，开启前文讲的 **ScheduledTrigger** 线程，如下：

```
public void startCheckpointScheduler() {
    synchronized (lock) {
        ...
        currentPeriodicTrigger = scheduleTriggerWithDelay(getRandomInitDelay());
    }
}
```

至此，整个启动流程就串联起来了。

## 10.2.2 栅栏的传输

栅栏从 source 任务中生成，并随着消息数据向下游传输。

### 1. 将信号从 JobMaster 端发送到 TaskExecutor 端

在 JobMaster 端的 CheckpointCoordinator 组件触发检查点操作后，获取了 source 任务对应的实例。根据 10.2.1 节的分析可知，在 CheckpointCoordinator 中最后会调用 execution.triggerCheckpoint()方法。在这个方法中，会获取对应的 TaskExecutor 的位置，并将消息发送给该 TaskExecutor。

具体代码如下：

```
public void triggerCheckpoint(long checkpointId, long timestamp, CheckpointOptions checkpointOptions) {
    triggerCheckpointHelper(checkpointId, timestamp, checkpointOptions, false);
}
private void triggerCheckpointHelper(long checkpointId, long timestamp, CheckpointOptions checkpointOptions, boolean advanceToEndOfEventTime) {
    ...
    final LogicalSlot slot = assignedResource;
    if (slot != null) {
        final TaskManagerGateway taskManagerGateway = slot.getTaskManagerGateway();
        taskManagerGateway.triggerCheckpoint(attemptId, getVertex().getJobId(), checkpointId, timestamp, checkpointOptions, advanceToEndOfEventTime);
    } else {
        ...
    }
}
```

在 triggerCheckpointHelper()方法中，将 assignedResource 字段的值赋给 slot 变量，然后通过它获取 TaskManagerGateway 对象，进而将信号传递给 TaskExecutor 端。这个 assignedResource 字段的值会在部署 Task 实例时初始化，因此它包含 TaskExecutor 的信息。

### 2. 栅栏在任务间的传输

从 JobMaster 端传来的信号会让源头的任务准备快照操作，并给下游发送栅栏。下游的 Task 实例会接收到这些栅栏，着手准备快照操作，并将栅栏继续向下传递。如果一个任务需要接收多个栅栏，那么会根据不同的策略实现不同的一致性语义。

到了 TaskExecutor 端，首先会根据传来的标识找到对应的任务，调用它的方法触发快照操作。

```
public CompletableFuture<Acknowledge> triggerCheckpoint(
    ExecutionAttemptID executionAttemptID,
    long checkpointId,
    long checkpointTimestamp,
```

```
            CheckpointOptions checkpointOptions,
            boolean advanceToEndOfEventTime) {
    ...
        final Task task = taskSlotTable.getTask(executionAttemptID);
        if (task != null) {
            task.triggerCheckpointBarrier(checkpointId, checkpointTimestamp, checkpointOptions,
advanceToEndOfEventTime);
            return CompletableFuture.completedFuture(Acknowledge.get());
        } else {
            ...
        }
    }
```

接下来会进入 Task 类的 triggerCheckpointBarrier()方法：

```
public void triggerCheckpointBarrier(
        final long checkpointID,
        final long checkpointTimestamp,
        final CheckpointOptions checkpointOptions,
        final boolean advanceToEndOfEventTime) {
    final AbstractInvokable invokable = this.invokable;
    final CheckpointMetaData checkpointMetaData = new CheckpointMetaData(checkpointID,
checkpointTimestamp);
    if (executionState == ExecutionState.RUNNING && invokable != null) {
        try {
            invokable.triggerCheckpointAsync(checkpointMetaData, checkpointOptions,
advanceToEndOfEventTime);
        }
        catch (RejectedExecutionException ex) {
            ...
        }
        catch (Throwable t) {
            ...
        }
    }
    else {
        ...
    }
}
```

一般情况下此处调用的是 StreamTask 实现类的 triggerCheckpointAsync()方法：

```
public Future<Boolean> triggerCheckpointAsync(
        CheckpointMetaData checkpointMetaData,
        CheckpointOptions checkpointOptions,
        boolean advanceToEndOfEventTime) {
    return mailboxProcessor.getMainMailboxExecutor().submit(
            () -> triggerCheckpoint(checkpointMetaData, checkpointOptions, advanceToEndOf-
EventTime),
            "checkpoint %s with %s",
            checkpointMetaData,
            checkpointOptions);
}
```

从方法名能看出，这是一个异步操作。因此可以直接进入 StreamTask 的 triggerCheckpoint()方法进行分析。

```java
private boolean triggerCheckpoint(
    CheckpointMetaData checkpointMetaData,
    CheckpointOptions checkpointOptions,
    boolean advanceToEndOfEventTime) throws Exception {
    try {
        ...
        // 执行检查点
        boolean success = performCheckpoint(checkpointMetaData, checkpointOptions,
checkpointMetrics, advanceToEndOfEventTime);
        if (!success) {
            // 如果执行失败，则将失败信息回传给 JobMaster 进行处理
            declineCheckpoint(checkpointMetaData.getCheckpointId());
        }
        return success;
    } catch (Exception e) {
        ...
    }
}
```

performCheckpoint()方法会执行检查点，如果执行失败（如在 StreamTask 未正常运行时），则会调用 declineCheckpoint()方法将失败信息回传给 JobMaster，由 CheckpointCoordinator 进行处理。回传时还会带上 CheckpointException 异常对象，因为 CheckpointCoordinator 会根据异常类型进行不同的处理。这些处理主要是将检查点从 pendingCheckpoints 集合中移除，或重新启动检查点流程，或直接抛出异常等。

核心方法 performCheckpoint()如下：

```java
private boolean performCheckpoint(
    CheckpointMetaData checkpointMetaData,
    CheckpointOptions checkpointOptions,
    CheckpointMetrics checkpointMetrics,
    boolean advanceToEndOfTime) throws Exception {
    final long checkpointId = checkpointMetaData.getCheckpointId();
    if (isRunning) {
        actionExecutor.runThrowing(() -> {
            ...
            // （1）发送栅栏前的准备工作
            operatorChain.prepareSnapshotPreBarrier(checkpointId);
            // （2）将栅栏发送到下游
            operatorChain.broadcastCheckpointBarrier(
                checkpointId,
                checkpointMetaData.getTimestamp(),
                checkpointOptions);
            // （3）异步执行快照操作
            checkpointState(checkpointMetaData, checkpointOptions, checkpointMetrics);
        });
        return true;
    } else {
        ...
    }
}
```

该方法的逻辑为以下 3 个步骤。

（1）发送栅栏前的准备工作。

（2）将栅栏发送到下游。

（3）异步执行快照操作。

第 1 步在多数情况下不会进行任何操作。因为无须在算子的快照操作结束后再将栅栏向下游发送，所以可以先发送栅栏，再异步执行快照操作。这里分析栅栏的传输。

broadcastCheckpointBarrier()方法如下：

```
public void broadcastCheckpointBarrier(long id, long timestamp, CheckpointOptions checkpointOptions) throws IOException {
    // 创建 Barrier 对象
    CheckpointBarrier barrier = new CheckpointBarrier(id, timestamp, checkpointOptions);
    for (RecordWriterOutput<?> streamOutput : streamOutputs) {
        // 将栅栏向下游传输
        streamOutput.broadcastEvent(barrier);
    }
}
public void broadcastEvent(AbstractEvent event) throws IOException {
    recordWriter.broadcastEvent(event);
}
public void broadcastEvent(AbstractEvent event) throws IOException {
    // 对 CheckpointBarrier 对象进行序列化
    try (BufferConsumer eventBufferConsumer = EventSerializer.toBufferConsumer(event)) {
        for (int targetChannel = 0; targetChannel < numberOfChannels; targetChannel++) {
            tryFinishCurrentBufferBuilder(targetChannel);
            // 将 CheckpointBarrier 对象广播到下游
            targetPartition.addBufferConsumer(eventBufferConsumer.copy(), targetChannel);
        }
        if (flushAlways) {
            flushAll();
        }
    }
}
```

这时才真正创建了 Barrier 对象（CheckpointBarrier 对象），并将其向下游进行广播。

上述代码调用了 EventSerializer.toBufferConsumer()方法对事件进行序列化。相关代码如下：

```
public static BufferConsumer toBufferConsumer(AbstractEvent event) throws IOException {
    final ByteBuffer serializedEvent = EventSerializer.toSerializedEvent(event);
    MemorySegment data = MemorySegmentFactory.wrap(serializedEvent.array());
    return new BufferConsumer(data, FreeingBufferRecycler.INSTANCE, false);
}
public static ByteBuffer toSerializedEvent(AbstractEvent event) throws IOException {
    final Class<?> eventClass = event.getClass();
    if (eventClass == EndOfPartitionEvent.class) {
        return ByteBuffer.wrap(new byte[] { 0, 0, 0, END_OF_PARTITION_EVENT });
    }
    else if (eventClass == CheckpointBarrier.class) {
        // 对 CheckpointBarrier 进行序列化
        return serializeCheckpointBarrier((CheckpointBarrier) event);
    }
    else if (eventClass == EndOfSuperstepEvent.class) {
        return ByteBuffer.wrap(new byte[] { 0, 0, 0, END_OF_SUPERSTEP_EVENT });
    }
    else if (eventClass == CancelCheckpointMarker.class) {
        CancelCheckpointMarker marker = (CancelCheckpointMarker) event;
```

```
            ByteBuffer buf = ByteBuffer.allocate(12);
            buf.putInt(0, CANCEL_CHECKPOINT_MARKER_EVENT);
            buf.putLong(4, marker.getCheckpointId());
            return buf;
        }
        else {
            try {
                final DataOutputSerializer serializer = new DataOutputSerializer(128);
                serializer.writeInt(OTHER_EVENT);
                serializer.writeUTF(event.getClass().getName());
                event.write(serializer);
                return serializer.wrapAsByteBuffer();
            }
            catch (IOException e) {
                ...
            }
        }
    }
```

toSerializedEvent()方法会根据不同类型的事件对其进行序列化。

可以看出，经过这些序列化操作后，事件与数据都被放入了缓冲区。事件会与数据一起，通过相同的机制传输到下游。在网络传输层，两者没有任何区别。

以上是 source 任务的触发和传输流程。source 任务的信号是 JobMaster 端传来的，而下游算子的栅栏是上游算子传来的，因此流程上有所区别。

3. 下游算子接收栅栏后的处理

对于 StreamTask，任务会在一个循环中不断处理下一条数据，而获取下一条数据的方式会通过 InputGate。由于事件与数据都通过同样的机制传输到下游，因此下游也会通过同样的机制来获取事件。此处不赘述任务间数据传输的流程细节，仅从下游获取数据开始分析。

在 CheckpointedInputGate 类的 pollNext()方法中，会获取一个 BufferOrEvent 对象，从其类名可以看出，这个类封装的就是数据或者事件。因此在该方法中有这样的判断：

```
if (barrierHandler.isBlocked(offsetChannelIndex(bufferOrEvent.getChannelIndex()))) {
    // 如果channel被阻塞，则将数据缓存起来
    bufferStorage.add(bufferOrEvent);
    if (bufferStorage.isFull()) {
        barrierHandler.checkpointSizeLimitExceeded(bufferStorage.getMaxBufferedBytes());
        bufferStorage.rollOver();
    }
}
else if (bufferOrEvent.isBuffer()) {
    return next;
}
else if (bufferOrEvent.getEvent().getClass() == CheckpointBarrier.class) {
    // 如果是栅栏，则进行与检查点相关的处理
    CheckpointBarrier checkpointBarrier = (CheckpointBarrier) bufferOrEvent.getEvent();
    if (!endOfInputGate) {
        if (barrierHandler.processBarrier(checkpointBarrier, offsetChannelIndex
(bufferOrEvent.getChannelIndex()), bufferStorage.getPendingBytes())) {
            bufferStorage.rollOver();
        }
    }
```

```
        }
        else if (bufferOrEvent.getEvent().getClass() == CancelCheckpointMarker.class) {
            if (barrierHandler.processCancellationBarrier((CancelCheckpointMarker) bufferOrEvent.
getEvent())) {
                bufferStorage.rollOver();
            }
        }
        else {
            if (bufferOrEvent.getEvent().getClass() == EndOfPartitionEvent.class) {
                if (barrierHandler.processEndOfPartition()) {
                    bufferStorage.rollOver();
                }
            }
            return next;
        }
```

首先注意第一个 if 分支，该分支判断 channel 是否被阻塞。前文介绍栅栏原理时说过，如果一个任务需要接收多个栅栏，那么在最后一个栅栏到来前，其余的数据流的数据会先被缓存起来，不会经业务代码处理。这一缓存的处理就是在这里实现的。

接着进入 CheckpointBarrier 的相关分支。与栅栏处理和检查点有关的方法是 barrierHandler.processBarrier()方法。barrierHandler 对象是 CheckpointBarrierHandler 类型。该类有两个实现类：CheckpointBarrierAligner 和 CheckpointBarrierTracker，这两个实现类分别对应"exactly once"语义和"at least once"语义。barrierHandler 对象的初始化在 StreamTask 的初始化阶段实现，具体是在 InputProcessorUtil 类的 createCheckpointBarrierHandler()方法中实现，如下：

```
    private static CheckpointBarrierHandler createCheckpointBarrierHandler(
            CheckpointingMode checkpointMode,
            int numberOfInputChannels,
            String taskName,
            AbstractInvokable toNotifyOnCheckpoint) {
        switch (checkpointMode) {
            case EXACTLY_ONCE:
                return new CheckpointBarrierAligner(
                    numberOfInputChannels,
                    taskName,
                    toNotifyOnCheckpoint);
            case AT_LEAST_ONCE:
                return new CheckpointBarrierTracker(numberOfInputChannels, toNotifyOnCheckpoint);
            default:
                throw new UnsupportedOperationException("Unrecognized Checkpointing Mode: " +
checkpointMode);
        }
    }
```

下面逐个分析两个实现类的 processBarrier()方法。

CheckpointBarrierAligner 的 processBarrier()方法如下：

```
    public boolean processBarrier(CheckpointBarrier receivedBarrier, int channelIndex,
long bufferedBytes) throws Exception {
        final long barrierId = receivedBarrier.getId();
        // 如果只有一个 InputChannel，则直接触发检查点
        if (totalNumberOfInputChannels == 1) {
            if (barrierId > currentCheckpointId) {
```

```
            currentCheckpointId = barrierId;
            notifyCheckpoint(receivedBarrier, bufferedBytes, latestAlignmentDurationNanos);
        }
        return false;
    }
    boolean checkpointAborted = false;
    if (numBarriersReceived > 0) {
        // 本次检查点已经收到过一些栅栏的信息
        if (barrierId == currentCheckpointId) {
            // （2）正常情况下进入这里
            onBarrier(channelIndex);
        }
        else if (barrierId > currentCheckpointId) {
            ...
        }
        else {
            return false;
        }
    }
    else if (barrierId > currentCheckpointId) {
        // （1）一次新的检查点的第一个栅栏
        beginNewAlignment(barrierId, channelIndex);
    }
    else {
        ...
    }
    // （3）判断是否接收了所有的栅栏
    if (numBarriersReceived + numClosedChannels == totalNumberOfInputChannels) {
        ...
        // （4）通知开始进行与检查点相关的操作
        notifyCheckpoint(receivedBarrier, bufferedBytes, latestAlignmentDurationNanos);
        return true;
    }
    return checkpointAborted;
}
```

不考虑只有一个 InputChannel，非异常情况下的核心逻辑有以下 4 步（注意注释的顺序）：

（1）接收到第一个栅栏，执行 beginNewAlignment()方法做一些记录。

（2）接收到后续栅栏时进入 onBarrier()方法做一些记录。

（3）判断是否接收了所有的栅栏。

（4）如果接收了所有的栅栏，则开始进行与检查点相关的操作，调用 notifyCheckpoint()方法。

其中，beginNewAlignment()方法如下：

```
protected void beginNewAlignment(long checkpointId, int channelIndex) throws IOException {
    currentCheckpointId = checkpointId;
    onBarrier(channelIndex);
}
```

记录本次 checkpointId，接着调用 onBarrier()方法，如下：

```
protected void onBarrier(int channelIndex) throws IOException {
    if (!blockedChannels[channelIndex]) {
        blockedChannels[channelIndex] = true;
        numBarriersReceived++;
```

```
        }
        else {
            ...
        }
    }
```

将这个 channel 标记为阻塞,并且记录接收到的栅栏的个数。blockedChannels 在前面调用 barrierHandler.isBlocked()方法时会用到:

```
public boolean isBlocked(int channelIndex) {
    return blockedChannels[channelIndex];
}
```

对于 CheckpointBarrierTracker,因为不会有阻塞的操作,所以该类没有这样一个字段,该方法直接返回 false。

CheckpointBarrierTracker 的 processBarrier()方法如下:

```
public boolean processBarrier(CheckpointBarrier receivedBarrier, int channelIndex,
long bufferedBytes) throws Exception {
    final long barrierId = receivedBarrier.getId();
    // 如果只有一个 InputChannel,则直接触发检查点
    if (totalNumberOfInputChannels == 1) {
        notifyCheckpoint(receivedBarrier, 0, 0);
        return false;
    }
    ...
    if (barrierCount != null) {
        // (3)记录接收到一个栅栏
        int numBarriersNew = barrierCount.incrementBarrierCount();
        if (numBarriersNew == totalNumberOfInputChannels) {
            if (!barrierCount.isAborted()) {
                // (4)当接收到所有栅栏时,触发检查点
                notifyCheckpoint(receivedBarrier, 0, 0);
            }
        }
    }
    else {
        // (1)一次新的检查点的第一个栅栏
        if (barrierId > latestPendingCheckpointID) {
            latestPendingCheckpointID = barrierId;
            // (2)记录接收到一个栅栏
            pendingCheckpoints.addLast(new CheckpointBarrierCount(barrierId));
            ...
        }
    }
    return false;
}
```

此处代码的核心逻辑与前面代码的类似,在接收到所有栅栏后才触发检查点,此处不赘述。此处代码与前面代码的主要区别其实在于前面分析过的 barrierHandler.isBlocked()方法。由于 CheckpointBarrierTracker 的该方法返回 false,因此数据会被正常处理,但是这样就有可能导致在发生异常并重启时这部分数据被多处理一次,造成"at least once"语义。

无论是 CheckpointBarrierAligner 还是 CheckpointBarrierTracker,触发检查点时调用的都是父

类的 notifyCheckpoint() 方法。该方法如下：

```
protected void notifyCheckpoint(CheckpointBarrier checkpointBarrier, long bufferedBytes,
long alignmentDurationNanos) throws Exception {
    if (toNotifyOnCheckpoint != null) {
        ...
          toNotifyOnCheckpoint.triggerCheckpointOnBarrier(
             checkpointMetaData,
             checkpointBarrier.getCheckpointOptions(),
             checkpointMetrics);
    }
}
```

这里的 **toNotifyOnCheckpoint** 字段就是 **StreamTask** 对象。

```
public void triggerCheckpointOnBarrier(
     CheckpointMetaData checkpointMetaData,
     CheckpointOptions checkpointOptions,
     CheckpointMetrics checkpointMetrics) throws Exception {
   try {
      // 执行检查点核心逻辑
      if (performCheckpoint(checkpointMetaData, checkpointOptions, checkpointMetrics,
false)) {
        ...
      }
   }
   catch (CancelTaskException e) {
     ...
   }
   catch (Exception e) {
     ...
   }
}
```

到了这一步，就又回到了 performCheckpoint() 方法。其逻辑之前已经分析过，该任务会将栅栏继续向下游传输，然后触发快照操作。

### 10.2.3　状态数据的更新和存储

前文分析了栅栏的传输，现在分析每个算子接收到栅栏后触发的快照操作。

#### 1. CheckpointedFunction

在继续介绍检查点的流程前，先来介绍一个用户自定义函数可以实现的接口——CheckpointedFunction。要让算子拥有状态并能够在检查点周期中对状态进行更新、存储，可以实现 CheckpointedFunction 接口。该接口并不是唯一的选择，这里仅以它为例。

该接口有两个方法：

```
public interface CheckpointedFunction {
   void snapshotState(FunctionSnapshotContext context) throws Exception;
   void initializeState(FunctionInitializationContext context) throws Exception;
}
```

实现该接口后，需要在 initializeState() 方法中完成状态的初始化（此过程包含注册操作），在 snapshotState() 方法中完成状态的更新。

这两个方法会在任务和检查点的生命周期中被调用。

2. 更新和存储流程

performCheckpoint()的实现步骤前文已经分析过，其中最后一步就是异步执行快照操作。

状态字段是可以由业务代码的开发人员将其定义在各个算子中的，即之前介绍过的 ListState、ValueState 等类型的字段。这些字段的值的更新逻辑由业务代码维护，本部分着重分析这些字段的值是如何在检查点的生命周期中被 Flink 内部维护、更新并存储的。

从 StreamTask 的 checkpointState()方法出发，会进入 CheckpointingOperation 的 executeCheckpointing()方法，如下：

```java
public void executeCheckpointing() throws Exception {
    startSyncPartNano = System.nanoTime();
    try {
        // （1）对每个算子进行快照操作
        for (StreamOperator<?> op : allOperators) {
            checkpointStreamOperator(op);
        }
        ...
        // （2）创建异步任务并提交任务。该任务用于通知 JobMaster 本任务的检查点操作已完成
        AsyncCheckpointRunnable asyncCheckpointRunnable = new AsyncCheckpointRunnable(
            owner,
            operatorSnapshotsInProgress,
            checkpointMetaData,
            checkpointMetrics,
            startAsyncPartNano);
        owner.cancelables.registerCloseable(asyncCheckpointRunnable);
        owner.asyncOperationsThreadPool.execute(asyncCheckpointRunnable);
    } catch (Exception ex) {
        ...
    }
}
```

核心逻辑已体现在注释中。本部分会关注每个任务中状态数据的更新和存储。其中 checkpointStreamOperator()方法会进行快照操作。

```java
private void checkpointStreamOperator(StreamOperator<?> op) throws Exception {
    if (null != op) {
        OperatorSnapshotFutures snapshotInProgress = op.snapshotState(
            checkpointMetaData.getCheckpointId(),
            checkpointMetaData.getTimestamp(),
            checkpointOptions,
            storageLocation);
        operatorSnapshotsInProgress.put(op.getOperatorID(), snapshotInProgress);
    }
}
```

这里的核心逻辑是对算子调用 snapshotState()方法，返回的对象是 OperatorSnapshotFutures 对象，由 operatorSnapshotsInProgress 字段进行维护。注意，上面 executeCheckpointing()方法中第 2 步构造异步任务时将 operatorSnapshotsInProgress 字段放入其中使之作为构造方法的参数。

```java
public final OperatorSnapshotFutures snapshotState(long checkpointId, long timestamp,
CheckpointOptions checkpointOptions,
    CheckpointStreamFactory factory) throws Exception {
```

```
    // （1）获取 KeyGroupRange 对象，初始化 OperatorSnapshotFutures 对象，初始化上下文对象
    KeyGroupRange keyGroupRange = null != keyedStateBackend ?
        keyedStateBackend.getKeyGroupRange() : KeyGroupRange.EMPTY_KEY_GROUP_RANGE;
    OperatorSnapshotFutures snapshotInProgress = new OperatorSnapshotFutures();
    StateSnapshotContextSynchronousImpl snapshotContext = new StateSnapshotContext-
SynchronousImpl(
        checkpointId,
        timestamp,
        factory,
        keyGroupRange,
        getContainingTask().getCancelables());
    try {
        // （2）调用算子的 snapshotState()方法
        snapshotState(snapshotContext);
        ...
        // （3）对 OperatorState 或 KeyedState 进行快照操作，并将返回的 Future 对象设置给
snapshotInProgress
        if (null != operatorStateBackend) {
            snapshotInProgress.setOperatorStateManagedFuture(
                operatorStateBackend.snapshot(checkpointId, timestamp, factory,
checkpointOptions));
        }
        if (null != keyedStateBackend) {
            snapshotInProgress.setKeyedStateManagedFuture(
                keyedStateBackend.snapshot(checkpointId, timestamp, factory,
checkpointOptions));
        }
    } catch (Exception snapshotException) {
        ...
    }
    return snapshotInProgress;
}
```

上述代码的核心逻辑如下。

（1）初始化上下文对象等。

（2）snapshotState()方法的主要功能是调用业务代码中自定义的 snapshotState()方法。

（3）根据是否是 KeyedStream 对 KeyedState 或 OperatorState 进行快照操作。

第 2 步中的 snapshotState()方法会进入 StreamingFunctionUtils 的 snapshotFunctionState()方法，进而调用 trySnapshotFunctionState()方法。如果算子实现了 CheckpointedFunction 接口，则会进入下面的逻辑：

```
if (userFunction instanceof CheckpointedFunction) {
    ((CheckpointedFunction) userFunction).snapshotState(context);
    return true;
}
```

如果没有实现 CheckpointedFunction 而实现了其他接口，则会进入其他分支。

通常，业务代码中的 snapshotState()方法会完成状态的更新。以 FlinkKafkaConsumerBase 类为例，其 snapshotState()方法有这样的逻辑：

```
unionOffsetStates.clear();
for (Map.Entry<KafkaTopicPartition, Long> kafkaTopicPartitionLongEntry : currentOffsets.entrySet()) {
```

```
            unionOffsetStates.add(
                  Tuple2.of(kafkaTopicPartitionLongEntry.getKey(), kafkaTopicPartitionLongEntry.
getValue()));
      }
```

这里的 unionOffsetStates 字段表示该类定义并注册的状态。这是一种标准的维护方式。在其他业务逻辑中，用另一个字段或临时变量 currentOffsets 来维护当前的状态，而每一次调用 snapshotState()方法时，会先将 unionOffsetStates 的值清空，然后会将 currentOffsets 中的值全部放入 unionOffsetStates，交由 Flink 内部维护。之后的状态存储，就是将之前在 Flink 内部注册过的状态中保存的值进行存储。

状态更新完毕后，最终会调用 operatorStateBackend 或 keyedStateBackend 的 snapshot()方法。

这个 snapshot()方法的实现与算子类型（是否是 KeyBy 后的算子）、状态后端类型有关，但整体流程是一致的。这里以 DefaultOperatorStateBackendSnapshotStrategy 类为例进行分析。该类的 snapshot()方法会对 OperatorState 和 BroadcastState 一起进行处理，由于两者的处理方式一致，故略去 BroadcastState 的相关代码：

```
public RunnableFuture<SnapshotResult<OperatorStateHandle>> snapshot(
      final long checkpointId,
      final long timestamp,
      @Nonnull final CheckpointStreamFactory streamFactory,
      @Nonnull final CheckpointOptions checkpointOptions) throws IOException {
   ...
      // （1）构造 Map 用来存放此次要存储的状态的值
      final Map<String, PartitionableListState<?>> registeredOperatorStatesDeepCopies =
         new HashMap<>(registeredOperatorStates.size());
   ...
      try {
         // （2）将 Flink 内部维护的状态的值复制到新构造的 Map 中
         if (!registeredOperatorStates.isEmpty()) {
            for (Map.Entry<String, PartitionableListState<?>> entry : registeredOperatorStates.
entrySet()) {
               PartitionableListState<?> listState = entry.getValue();
               if (null != listState) {
                  listState = listState.deepCopy();
               }
               registeredOperatorStatesDeepCopies.put(entry.getKey(), listState);
            }
         }
      ...
      } finally {
         ...
      }
      // （3）构造一个异步任务
      AsyncSnapshotCallable<SnapshotResult<OperatorStateHandle>> snapshotCallable =
         new AsyncSnapshotCallable<SnapshotResult<OperatorStateHandle>>() {
            ...
         };
      final FutureTask<SnapshotResult<OperatorStateHandle>> task =
         snapshotCallable.toAsyncSnapshotFutureTask(closeStreamOnCancelRegistry);
      // （4）执行任务
      if (!asynchronousSnapshots) {
         task.run();
```

```
    }
    return task;
}
```

上述代码的逻辑十分清晰，关键点在于异步任务的逻辑。相关代码如下：

```
new AsyncSnapshotCallable<SnapshotResult<OperatorStateHandle>>() {
    @Override
    protected SnapshotResult<OperatorStateHandle> callInternal() throws Exception {
        CheckpointStreamFactory.CheckpointStateOutputStream localOut =
            streamFactory.createCheckpointStateOutputStream(CheckpointedStateScope.EXCLUSIVE);
        snapshotCloseableRegistry.registerCloseable(localOut);

        // （1）获取注册的 OperatorState 的元信息
        List<StateMetaInfoSnapshot> operatorMetaInfoSnapshots =
            new ArrayList<>(registeredOperatorStatesDeepCopies.size());
        for (Map.Entry<String, PartitionableListState<?>> entry :
            registeredOperatorStatesDeepCopies.entrySet()) {
            operatorMetaInfoSnapshots.add(entry.getValue().getStateMetaInfo().snapshot());
        }
        ...
        // （2）构造视图用来写入状态数据
        DataOutputView dov = new DataOutputViewStreamWrapper(localOut);
        OperatorBackendSerializationProxy backendSerializationProxy =
            new OperatorBackendSerializationProxy(operatorMetaInfoSnapshots,
                broadcastMetaInfoSnapshots);
        // （3）写入状态的元信息
        backendSerializationProxy.write(dov);
        // （4）写入状态数据
        int initialMapCapacity =
            registeredOperatorStatesDeepCopies.size() +
                registeredBroadcastStatesDeepCopies.size();
        final Map<String, OperatorStateHandle.StateMetaInfo> writtenStatesMetaData =
            new HashMap<>(initialMapCapacity);
        for (Map.Entry<String, PartitionableListState<?>> entry :
            registeredOperatorStatesDeepCopies.entrySet()) {
            PartitionableListState<?> value = entry.getValue();
            long[] partitionOffsets = value.write(localOut);
            OperatorStateHandle.Mode mode = value.getStateMetaInfo().getAssignmentMode();
            writtenStatesMetaData.put(
                entry.getKey(),
                new OperatorStateHandle.StateMetaInfo(partitionOffsets, mode));
        }
        ...
        OperatorStateHandle retValue = null;
        if (snapshotCloseableRegistry.unregisterCloseable(localOut)) {
            // （5）关闭输出流，返回文件描述信息，将其进行封装并返回
            StreamStateHandle stateHandle = localOut.closeAndGetHandle();
            if (stateHandle != null) {
                retValue = new OperatorStreamStateHandle(writtenStatesMetaData, stateHandle);
            }
            return SnapshotResult.of(retValue);
        } else {
            ...
        }
    }
    ...
};
```

上述代码的总体逻辑如下。

（1）获取注册的 OperatorState 的元信息。
（2）构造视图。
（3）写入状态的元信息。
（4）写入状态数据。
（5）关闭输出流，返回文件描述信息。

第 1 步中获取的元信息主要包括序列化方式、状态名称、状态类型、状态重分配策略等。在一些特殊的场景中，可能需要开发人员解析检查点过程生成的文件甚至需要对其进行修改，了解这些元信息的内容和格式对解析这些文件十分有帮助。

第 3 步就是对这些元信息进行序列化。这部分元信息序列化结束后，就开始将状态数据一条条拿出来进行序列化。在 value.write() 方法内部，会用该状态注册时指定的序列化器进行序列化，并且会将每一条数据序列化后在输出流中的位置返回，即 partitionOffsets。接着，partitionOffsets 和状态的重分配策略会被一起封装进 writtenStatesMetaData。这个重分配策略有 3 种，在介绍状态恢复时会详细讲解：

```
enum Mode {
    SPLIT_DISTRIBUTE,
    UNION,
    BROADCAST
}
```

第 5 步会关闭输出流，返回文件描述信息，其中包括文件路径等。文件描述信息与 writtenStatesMetaData 会被一起封装成 OperatorStreamStateHandle 对象并返回。

状态数据的存储至此结束。可以看出，状态数据文件（如图 10-8 所示）的前半部分是状态的描述信息，其中包含序列化器的信息，后半部分是用该序列化器进行序列化后得到的具体的状态数据。因此，理论上这个文件是自解析的，可以先反序列化前半部分的信息，然后利用序列化器对后半部分数据进行反序列化。

图 10-8　状态数据文件

## 10.2.4　元信息的存储

当算子中的状态数据都存储完成后，会在 AsyncCheckpointRunnable 定义的异步任务中通知作业管理器其快照操作已完成。

```
public void run() {
    FileSystemSafetyNet.initializeSafetyNetForThread();
    try {
        // 封装状态数据的描述信息
        TaskStateSnapshot jobManagerTaskOperatorSubtaskStates =
            new TaskStateSnapshot(operatorSnapshotsInProgress.size());
        TaskStateSnapshot localTaskOperatorSubtaskStates =
            new TaskStateSnapshot(operatorSnapshotsInProgress.size());
```

```
            for (Map.Entry<OperatorID, OperatorSnapshotFutures> entry : operatorSnapshots-
InProgress.entrySet()) {
                OperatorID operatorID = entry.getKey();
                OperatorSnapshotFutures snapshotInProgress = entry.getValue();
                OperatorSnapshotFinalizer finalizedSnapshots =
                    new OperatorSnapshotFinalizer(snapshotInProgress);
                jobManagerTaskOperatorSubtaskStates.putSubtaskStateByOperatorID(
                    operatorID,
                    finalizedSnapshots.getJobManagerOwnedState());
                localTaskOperatorSubtaskStates.putSubtaskStateByOperatorID(
                    operatorID,
                    finalizedSnapshots.getTaskLocalState());
            }
            ...
            if (asyncCheckpointState.compareAndSet(CheckpointingOperation.AsyncCheckpointState.
RUNNING,
                CheckpointingOperation.AsyncCheckpointState.COMPLETED)) {
                // 通知作业管理器已完成快照操作
                reportCompletedSnapshotStates(
                    jobManagerTaskOperatorSubtaskStates,
                    localTaskOperatorSubtaskStates,
                    asyncDurationMillis);
            } else {
                ...
            }
        } catch (Exception e) {
            ...
        } finally {
            ...
        }
    }
```

将状态数据的描述信息进行封装，然后调用 reportCompletedSnapshotStates()方法将这些信息回传给作业管理器。

```
    private void reportCompletedSnapshotStates(
        TaskStateSnapshot acknowledgedTaskStateSnapshot,
        TaskStateSnapshot localTaskStateSnapshot,
        long asyncDurationMillis) {
        TaskStateManager taskStateManager = owner.getEnvironment().getTaskStateManager();
        ...
        taskStateManager.reportTaskStateSnapshots(
            checkpointMetaData,
            checkpointMetrics,
            hasAckState ? acknowledgedTaskStateSnapshot : null,
            hasLocalState ? localTaskStateSnapshot : null);
    }
```

最终会通过 RPC 来调用 JobMaster 的对应方法：

```
public void acknowledgeCheckpoint(
        final JobID jobID,
        final ExecutionAttemptID executionAttemptID,
        final long checkpointId,
        final CheckpointMetrics checkpointMetrics,
        final TaskStateSnapshot checkpointState) {
    schedulerNG.acknowledgeCheckpoint(jobID, executionAttemptID, checkpointId,
checkpointMetrics, checkpointState);
```

```java
    }
    public void acknowledgeCheckpoint(final JobID jobID, final ExecutionAttemptID
executionAttemptID, final long checkpointId, final CheckpointMetrics checkpointMetrics,
final TaskStateSnapshot checkpointState) {
        // 将状态数据的描述信息封装成 AcknowledgeCheckpoint 对象
        final CheckpointCoordinator checkpointCoordinator = executionGraph.getCheckpoint-
Coordinator();
        final AcknowledgeCheckpoint ackMessage = new AcknowledgeCheckpoint(
            jobID,
            executionAttemptID,
            checkpointId,
            checkpointMetrics,
            checkpointState);
        final String taskManagerLocationInfo = retrieveTaskManagerLocation(executionAttemptID);
        if (checkpointCoordinator != null) {
            ioExecutor.execute(() -> {
                try {
                    // 用 CheckpointCoordinator 接收 AcknowledgeCheckpoint 对象
                    checkpointCoordinator.receiveAcknowledgeMessage(ackMessage, taskManager-
LocationInfo);
                } catch (Throwable t) {
                    ...
                }
            });
        } else {
            ...
        }
    }
```

注意，状态数据的描述信息最终就是 TaskStateSnapshot 对象，被封装成 AcknowledgeCheckpoint 对象被 CheckpointCoordinator 接收。

下面是 CheckpointCoordinator 接收 AcknowledgeCheckpoint 对象后的操作：

```java
    public boolean receiveAcknowledgeMessage(AcknowledgeCheckpoint message, String
taskManagerLocationInfo) throws CheckpointException {
        ...
        synchronized (lock) {
            final PendingCheckpoint checkpoint = pendingCheckpoints.get(checkpointId);
            if (checkpoint != null && !checkpoint.isDiscarded()) {
                // 接收 ACK 信号
                switch (checkpoint.acknowledgeTask(message.getTaskExecutionId(), message.
getSubtaskState(), message.getCheckpointMetrics())) {
                    case SUCCESS: // 成功接收 ACK 信号
                        // 如果该任务的所有并行实例都回传了 ACK 信号，则完成此次检查点操作
                        if (checkpoint.areTasksFullyAcknowledged()) {
                            completePendingCheckpoint(checkpoint);
                        }
                        break;
                    case DUPLICATE:
                        break;
                    case UNKNOWN:
                        ...
                    case DISCARDED:
                        ...
                }
```

```
            return true;
        }
        ...
    }
}
```

接收 ACK 信号时的主要逻辑如下：

```
public TaskAcknowledgeResult acknowledgeTask(
    ExecutionAttemptID executionAttemptId,
    TaskStateSnapshot operatorSubtaskStates,
    CheckpointMetrics metrics) {
  synchronized (lock) {
    if (discarded) {
        return TaskAcknowledgeResult.DISCARDED;
    }
    // 从 notYetAcknowledgedTasks 中移除该并行实例
    final ExecutionVertex vertex = notYetAcknowledgedTasks.remove(executionAttemptId);
    if (vertex == null) {
        if (acknowledgedTasks.contains(executionAttemptId)) {
            return TaskAcknowledgeResult.DUPLICATE;
        } else {
            return TaskAcknowledgeResult.UNKNOWN;
        }
    } else {
        acknowledgedTasks.add(executionAttemptId);
    }
    List<OperatorID> operatorIDs = vertex.getJobVertex().getOperatorIDs();
    int subtaskIndex = vertex.getParallelSubtaskIndex();
    long ackTimestamp = System.currentTimeMillis();
    long stateSize = 0L;
    if (operatorSubtaskStates != null) {
        for (OperatorID operatorID : operatorIDs) {
            OperatorSubtaskState operatorSubtaskState =
                operatorSubtaskStates.getSubtaskStateByOperatorID(operatorID);
            if (operatorSubtaskState == null) {
                operatorSubtaskState = new OperatorSubtaskState();
            }
            OperatorState operatorState = operatorStates.get(operatorID);
            if (operatorState == null) {
                operatorState = new OperatorState(
                    operatorID,
                    vertex.getTotalNumberOfParallelSubtasks(),
                    vertex.getMaxParallelism());
                // 将状态数据的描述信息放入 operatorStates 字段进行维护
                operatorStates.put(operatorID, operatorState);
            }
            operatorState.putState(subtaskIndex, operatorSubtaskState);
            stateSize += operatorSubtaskState.getStateSize();
        }
    }
    ...
    return TaskAcknowledgeResult.SUCCESS;
  }
}
```

这里的核心逻辑主要可概括为两步，第 1 步是从 notYetAcknowledgedTasks 中移除该并行实例，第 2 步是将状态数据的描述信息放入 operatorStates 字段进行维护。

正常情况下会返回 TaskAcknowledgeResult.SUCCESS，于是会进入 SUCCESS 分支。这里会判断是否所有并行实例都回传了 ACK 信号：

```
public boolean areTasksFullyAcknowledged() {
    return notYetAcknowledgedTasks.isEmpty() && !discarded;
}
```

可以看到这里就是用 notYetAcknowledgedTasks 来进行判断的。

如果所有并行实例都回传了 ACK 信号，则表示此次检查点操作可以结束了。最后需要整理和封装所有状态数据的描述信息，并需要将其写入元信息文件。

```
private void completePendingCheckpoint(PendingCheckpoint pendingCheckpoint) throws CheckpointException {
    ...
    try {
        try {
            // 整理和封装所有状态数据的描述信息，并将其写入元信息文件
            completedCheckpoint = pendingCheckpoint.finalizeCheckpoint();
        }
    ...
    for (ExecutionVertex ev : tasksToCommitTo) {
        Execution ee = ev.getCurrentExecutionAttempt();
        if (ee != null) {
            // 回调任务管理器端的监听器中定义的方法
            ee.notifyCheckpointComplete(checkpointId, timestamp);
        }
    }
}
```

上面方法中的主要逻辑就是调用 PendingCheckpoint 的 finalizeCheckpoint() 方法来完成元信息文件的持久化。另一处值得注意的是，最后调用了 Execution 对象的 notifyCheckpointComplete() 方法，会回调任务管理器端定义的一些关于检查点的监听器。

finalizeCheckpoint() 方法的实现如下：

```
public CompletedCheckpoint finalizeCheckpoint() throws IOException {
    synchronized (lock) {
        try {
            // 构造 Savepoint 对象
            final Savepoint savepoint = new SavepointV2(checkpointId, operatorStates.values(), masterStates);
            final CompletedCheckpointStorageLocation finalizedLocation;
            // 构造输出流
            try (CheckpointMetadataOutputStream out = targetLocation.createMetadataOutputStream()) {
                // 输出元信息到文件
                Checkpoints.storeCheckpointMetadata(savepoint, out);
                // 关闭输出流
                finalizedLocation = out.closeAndFinalizeCheckpoint();
            }
            ...
        }
        catch (Throwable t) {
```

```
            ...
        }
    }
}
```

其中，输出元信息的方法 storeCheckpointMetadata()如下：

```
public static <T extends Savepoint> void storeCheckpointMetadata(
    T checkpointMetadata,
    OutputStream out) throws IOException {
  DataOutputStream dos = new DataOutputStream(out);
  storeCheckpointMetadata(checkpointMetadata, dos);
}
public static <T extends Savepoint> void storeCheckpointMetadata(
    T checkpointMetadata,
    DataOutputStream out) throws IOException {
  // 输出 header 信息和版本信息
  out.writeInt(HEADER_MAGIC_NUMBER);
  out.writeInt(checkpointMetadata.getVersion());
  // 输出元信息
  SavepointSerializer<T> serializer = SavepointSerializers.getSerializer(checkpointMetadata);
  serializer.serialize(checkpointMetadata, out);
}
```

根据以上序列化的步骤，可以了解到状态的元信息文件如图 10-9 所示。

图 10-9　元信息文件

而 Checkpoint Metadata 的格式和内容可以从 SavepointSerializer 的 serialize()方法观察到，具体如图 10-10 所示。

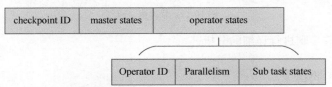

图 10-10　Checkpoint Metadata 的格式和内容

其中重要的信息就是 operator states 部分，包含之前从任务管理器端传回的状态数据的描述信息。

## 10.3　状态恢复

状态恢复的过程基本就是状态存储的逆过程。状态恢复流程大致可以分为以下几部分：元信息的读取、状态的重分配和状态数据的恢复。状态恢复流程如图 10-11 所示。

由于在恢复过程中，任务的并行度可能已经发生改变，因此在作业管理器的 CheckpointCoordinator 读取状态数据的元信息后，需要根据新的并行度，将状态数据分配给新的并行实例（具体是将文件的位置信息等分配给每个并行实例）。这就是状态的重分配。这些信息最后会被封装在 TaskDeploymentDescriptor 对象中，在调度和部署的过程中自然而然地跟随每个 TDD 对象来到任务管理器端。每个并行实例在任务管理器端初始化时，会根据这些信息到具体的状态数据文件中读取数据，这样就完成了状态的恢复。

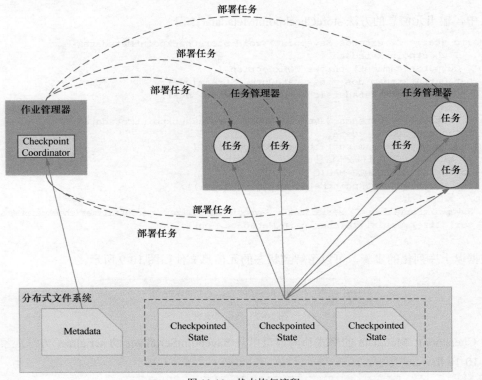

图 10-11　状态恢复流程

下面依次对这些步骤进行详细的分析。

## 10.3.1　元信息的读取

元信息的读取和状态的重分配都是在作业生成执行图的过程中进行的。从前文了解到，生成 ExecutionGraph 时，会调用 SchedulerBase 的 createAndRestoreExecutionGraph() 方法：

```
private ExecutionGraph createAndRestoreExecutionGraph(
    JobManagerJobMetricGroup currentJobManagerJobMetricGroup,
    ShuffleMaster<?> shuffleMaster,
    JobMasterPartitionTracker partitionTracker) throws Exception {
    // 构造 ExecutionGraph
    ExecutionGraph newExecutionGraph = createExecutionGraph(currentJobManagerJobMetricGroup,
shuffleMaster, partitionTracker);
    // 获取 CheckpointCoordinator 实例
    final CheckpointCoordinator checkpointCoordinator = newExecutionGraph.getCheckpoint-
Coordinator();
    if (checkpointCoordinator != null) {
        // 从检查点恢复
        if (!checkpointCoordinator.restoreLatestCheckpointedState(
```

```
            new HashSet<>(newExecutionGraph.getAllVertices().values()),
            false,
            false)) {
            // 尝试从保存点恢复
            tryRestoreExecutionGraphFromSavepoint(newExecutionGraph, jobGraph.getSavepoint
RestoreSettings());
        }
    }
    return newExecutionGraph;
}
```

当没有检查点或保存点时，显然不需要进行任何恢复。在这种情况下，可以推断上面代码的 restoreLatestCheckpointedState()方法和 tryRestoreExecutionGraphFromSavepoint()方法没有对 newExecutionGraph 对象进行任何实质性的改变。因此前文并未详细分析过这两个方法。然而，当有状态需要恢复时，这两个方法中的内容就值得探究了。

其中 restoreLatestCheckpointedState()方法如下：

```
public boolean restoreLatestCheckpointedState(
        final Set<ExecutionJobVertex> tasks,
        final boolean errorIfNoCheckpoint,
        final boolean allowNonRestoredState) throws Exception {
    synchronized (lock) {
        ...
        CompletedCheckpoint latest = completedCheckpointStore.getLatestCheckpoint
(isPreferCheckpointForRecovery);
        if (latest == null) {
            if (errorIfNoCheckpoint) {
                throw new IllegalStateException("No completed checkpoint available");
            } else {
                return false;
            }
        }
        ...
        return true;
    }
}
```

这里省略了前后的代码，只保留了中间部分。中间部分代码的逻辑是获取最近的 Completed-Checkpoint，如果值为 null，则返回 false。注意，在 createAndRestoreExecutionGraph()方法中调用 restoreLatestCheckpointedState()方法时，传入的参数 errorIfNoCheckpoint 的值为 false。

completedCheckpointStore 是维护 CompletedCheckpoint 的地方。从 10.2 节可了解到，当检查点完成时，最终会调用 CheckpointCoordinator 的 completePendingCheckpoint()方法，在该方法中执行完 pendingCheckpoint.finalizeCheckpoint()方法后会返回 CompletedCheckpoint 对象，随后该对象会被添加到 completedCheckpointStore 中。这里的 completedCheckpointStore.getLatestCheckpoint()方法就是用于获取最近一次完成的 CompletedCheckpoint。

这里讨论的场景是作业停止后，从文件中进行状态恢复，因此这里返回的值为 null。

回到 createAndRestoreExecutionGraph()方法，如果调用 restoreLatestCheckpointedState()方法返回 false，那么会调用 tryRestoreExecutionGraphFromSavepoint()方法：

```
public boolean restoreSavepoint(
        String savepointPointer,
```

```
        boolean allowNonRestored,
        Map<JobVertexID, ExecutionJobVertex> tasks,
        ClassLoader userClassLoader) throws Exception {
    // 根据元信息文件地址构造 checkpointLocation
    final CompletedCheckpointStorageLocation checkpointLocation = checkpointStorage.
resolveCheckpoint(savepointPointer);
    // 读取元信息文件，构造 CompletedCheckpoint 对象
    CompletedCheckpoint savepoint = Checkpoints.loadAndValidateCheckpoint(
        job, tasks, checkpointLocation, userClassLoader, allowNonRestored);
    // 将构造的 CompletedCheckpoint 对象添加到 completedCheckpointStore 中
    completedCheckpointStore.addCheckpoint(savepoint);
    // 完成状态的重分配
    return restoreLatestCheckpointedState(new HashSet<>(tasks.values()), true,
allowNonRestored);
}
```

这里的 savepointPointer 就是执行作业时指定的保存点地址（元信息文件地址）。resolveCheckpoint()方法会将该地址字符串进行简单的封装，形成 CompletedCheckpointStorageLocation 对象。可以将该对象理解为封装了地址的对象。

接着在 Checkpoints 的 loadAndValidateCheckpoint()方法中读取元信息文件，将其反序列化到内存中，并封装在 CompletedCheckpoint 对象中。从 10.2 节可以大致了解到，在该对象中会封装具体的状态数据的描述信息，如文件地址等。同时，该方法还会校验算子的最大并行度是否发生改变等。Flink 中的最大并行度的含义会在 10.4 节详细介绍。

构造出的 CompletedCheckpoint 对象会被添加到 completedCheckpointStore 中进行维护。

最后，又来到 restoreLatestCheckpointedState()方法。这时来分析该方法的其余部分：

```
public boolean restoreLatestCheckpointedState(
        final Set<ExecutionJobVertex> tasks,
        final boolean errorIfNoCheckpoint,
        final boolean allowNonRestoredState) throws Exception {
    synchronized (lock) {
        ...
        // 获取最近一次完成的 CompletedCheckpoint 对象
        CompletedCheckpoint latest = completedCheckpointStore.getLatestCheckpoint
(isPreferCheckpointForRecovery);
        ...
        // 状态的重分配
        final Map<OperatorID, OperatorState> operatorStates = latest.getOperatorStates();
        StateAssignmentOperation stateAssignmentOperation =
            new StateAssignmentOperation(latest.getCheckpointID(), tasks, operatorStates,
allowNonRestoredState);
        stateAssignmentOperation.assignStates();
        ...
        return true;
    }
}
```

由于先前添加了一个 CompletedCheckpoint，因此这里不会直接返回 false，而是进入后面的逻辑，即状态的重分配。

### 10.3.2 状态的重分配

状态的重分配，就是指把状态数据的描述信息重新分配给新的并行实例的过程，因此这些新

的并行实例被部署到任务管理器端后,就能知道去哪里获取它需要恢复的状态数据。

在上面的代码中,operatorStates 对象中封装的就是状态数据的描述信息。可以看到它是一个 Map,其中 key 为 OperatorID 类型,表示算子的唯一标识;value 为 OperatorState 类型,封装了一个算子的所有状态数据的描述信息,如文件地址等。后文会更详细地介绍 OperatorState 类中的内容。

既然要把状态数据的描述信息分配给各个并行实例,那么这里一定需要代表并行实例的对象,即上面代码中的 tasks 对象。它是 ExecutionJobVertex 对象的集合。

这两个对象被封装到 StateAssignmentOperation 对象中,然后调用其 assignStates()方法完成状态的重分配。

下面分析 assignStates()方法的实现逻辑:

```
public void assignStates() {
   Map<OperatorID, OperatorState> localOperators = new HashMap<>(operatorStates);
   for (ExecutionJobVertex executionJobVertex : this.tasks) {
      List<OperatorID> operatorIDs = executionJobVertex.getOperatorIDs();
      List<OperatorID> altOperatorIDs = executionJobVertex.getUserDefinedOperatorIDs();
      List<OperatorState> operatorStates = new ArrayList<>(operatorIDs.size());
      boolean statelessTask = true;
      for (int x = 0; x < operatorIDs.size(); x++) {
         OperatorID operatorID = altOperatorIDs.get(x) == null
            ? operatorIDs.get(x)
            : altOperatorIDs.get(x);
         OperatorState operatorState = localOperators.remove(operatorID);
         if (operatorState == null) {
            operatorState = new OperatorState(
               operatorID,
               executionJobVertex.getParallelism(),
               executionJobVertex.getMaxParallelism());
         } else {
            statelessTask = false;
         }
         operatorStates.add(operatorState);
      }
      if (statelessTask) {
         continue;
      }
      assignAttemptState(executionJobVertex, operatorStates);
   }
}
```

一个 ExecutionJobVertex 中包含多个算子,因此这里的两个 for 循环实际上就是匹配对应算子的过程。如果一个 ExecutionJobVertex 中的所有算子都没有定义状态,则继续遍历下一个 ExecutionJobVertex,否则会调用 assignAttemptState()方法进行状态的重分配。

```
private void assignAttemptState(ExecutionJobVertex executionJobVertex, List<OperatorState> operatorStates) {
      List<OperatorID> operatorIDs = executionJobVertex.getOperatorIDs();
      // (1)计算新的并行度
      int newParallelism = executionJobVertex.getParallelism();
      // (2)根据并行度和最大并行度计算 KeyGroupRange 的值
      List<KeyGroupRange> keyGroupPartitions = createKeyGroupPartitions(
         executionJobVertex.getMaxParallelism(),
```

```
        newParallelism);
    // (3) 计算 ExecutionJobVertex 中并行实例的总个数
    final int expectedNumberOfSubTasks = newParallelism * operatorIDs.size();
    // (4) 重分配 OperatorState
    Map<OperatorInstanceID, List<OperatorStateHandle>> newManagedOperatorStates =
        new HashMap<>(expectedNumberOfSubTasks);
    Map<OperatorInstanceID, List<OperatorStateHandle>> newRawOperatorStates =
        new HashMap<>(expectedNumberOfSubTasks);
    reDistributePartitionableStates(
        operatorStates,
        newParallelism,
        operatorIDs,
        newManagedOperatorStates,
        newRawOperatorStates);
    // (5) 重分配 KeyedState
    Map<OperatorInstanceID, List<KeyedStateHandle>> newManagedKeyedState =
        new HashMap<>(expectedNumberOfSubTasks);
    Map<OperatorInstanceID, List<KeyedStateHandle>> newRawKeyedState =
        new HashMap<>(expectedNumberOfSubTasks);
    reDistributeKeyedStates(
        operatorStates,
        newParallelism,
        operatorIDs,
        keyGroupPartitions,
        newManagedKeyedState,
        newRawKeyedState);
    // (6) 将上面分配好的结果封装到 ExecutionJobVertex 中
    assignTaskStateToExecutionJobVertices(
        executionJobVertex,
        newManagedOperatorStates,
        newRawOperatorStates,
        newManagedKeyedState,
        newRawKeyedState,
        newParallelism);
}
```

状态分为操作符状态和键控状态，在重分配时是将两者分开考虑的。核心逻辑发生在第 4 步和第 5 步。注意，在第 4 步和第 5 步中分别构造的 Map，其中的 key 类型都为 OperatorInstanceID，value 类型分别为 OperatorStateHandle 和 KeyedStateHandle，其实就表明了在重分配完成后，这些 Map 中维护了每个并行实例及其状态数据的描述信息。最后，在第 6 步中会将这些信息与 executionJobVertex 中对应的对象进行绑定。

10.4 节会介绍操作符状态和键控状态的重分配策略，这里直接介绍第 6 步的方法，观察重分配完成后，相关的信息如何与 executionJobVertex 中的对象进行绑定。

```
private void assignTaskStateToExecutionJobVertices(
    ExecutionJobVertex executionJobVertex,
    Map<OperatorInstanceID, List<OperatorStateHandle>> subManagedOperatorState,
    Map<OperatorInstanceID, List<OperatorStateHandle>> subRawOperatorState,
    Map<OperatorInstanceID, List<KeyedStateHandle>> subManagedKeyedState,
    Map<OperatorInstanceID, List<KeyedStateHandle>> subRawKeyedState,
    int newParallelism) {
    List<OperatorID> operatorIDs = executionJobVertex.getOperatorIDs();
    for (int subTaskIndex = 0; subTaskIndex < newParallelism; subTaskIndex++) {
```

```
        // 获取每个子任务的索引对应的 Execution 对象
        Execution currentExecutionAttempt = executionJobVertex.getTaskVertices()
[subTaskIndex]
            .getCurrentExecutionAttempt();
        TaskStateSnapshot taskState = new TaskStateSnapshot(operatorIDs.size());
        boolean statelessTask = true;
        for (OperatorID operatorID : operatorIDs) {
            OperatorInstanceID instanceID = OperatorInstanceID.of(subTaskIndex, operatorID);
            OperatorSubtaskState operatorSubtaskState = operatorSubtaskStateFrom(
                instanceID,
                subManagedOperatorState,
                subRawOperatorState,
                subManagedKeyedState,
                subRawKeyedState);
            if (operatorSubtaskState.hasState()) {
                statelessTask = false;
            }
            taskState.putSubtaskStateByOperatorID(operatorID, operatorSubtaskState);
        }
        if (!statelessTask) {
            // 将状态数据的描述信息封装到 Execution 对象中
            JobManagerTaskRestore taskRestore = new JobManagerTaskRestore(restoreCheckpointId,
 taskState);
            currentExecutionAttempt.setInitialState(taskRestore);
        }
    }
}
```

观察到状态数据的描述信息被封装到了对应的 Execution 对象中，通过其中的 taskRestore 字段进行维护。最终在 Execution 的 deploy() 方法中构造 TaskDeploymentDescriptor 对象时，该字段的值被直接赋给了 TaskDeploymentDescriptor 对象的同名字段。

### 10.3.3　状态数据的恢复

taskRestore 被封装进 TDD 对象后，随并行实例一起被部署到了任务管理器端。在 TaskExecutor 的 submitTask() 方法中，会将其取出并将其封装到 TaskStateManager 对象中，进而封装到 Task 对象中，在 Task 运行时调用的 doRun() 方法中，它又会被封装到 Environment 对象中。

另外，在 StreamTask 的初始化过程中，会调用其 initializeStateAndOpen() 方法，该方法会遍历每个 StreamOperator，调用其 initializeState() 方法。这些步骤在前文介绍 StreamTask 的生命周期时已提及。

StreamOperator 的 initializeState() 方法会调用 AbstractStreamOperator 实现的方法：

```
public final void initializeState() throws Exception {
    ...
    // 构造 StreamOperatorStateContext 对象，其中包含状态后端等对象，用于恢复状态到内存
    final StreamTaskStateInitializer streamTaskStateManager =
        Preconditions.checkNotNull(containingTask.createStreamTaskStateInitializer());
    final StreamOperatorStateContext context =
        streamTaskStateManager.streamOperatorStateContext(
            getOperatorID(),
            getClass().getSimpleName(),
```

```
            getProcessingTimeService(),
            this,
            keySerializer,
            streamTaskCloseableRegistry,
            metrics);
    // 获取 OperatorState 和 KeyedState 的状态后端
    this.operatorStateBackend = context.operatorStateBackend();
    this.keyedStateBackend = context.keyedStateBackend();
    ...
    try {
        // 在用户自定义函数中恢复状态等
        StateInitializationContext initializationContext = new StateInitializationContextImpl(
            context.isRestored(),
            operatorStateBackend,
            keyedStateStore,
            keyedStateInputs,
            operatorStateInputs);
        initializeState(initializationContext);
    } finally {
        ...
    }
}
```

构造 StreamTaskStateInitializer 对象的过程如下：

```
public StreamTaskStateInitializer createStreamTaskStateInitializer() {
    return new StreamTaskStateInitializerImpl(
        getEnvironment(),
        stateBackend);
}
public StreamTaskStateInitializerImpl(
    Environment environment,
    StateBackend stateBackend) {
    this.environment = environment;
    this.taskStateManager = Preconditions.checkNotNull(environment.getTaskStateManager());
    this.stateBackend = Preconditions.checkNotNull(stateBackend);
}
```

由此可以观察到，Environment 对象的 taskStateManager 字段被封装到了 StreamTaskStateInitializer 对象中并返回。StreamTaskStateInitializer 的 streamOperatorStateContext() 方法可以构造出 StreamOperatorStateContext 对象，StreamOperatorStateContext 对象中封装了状态后端，可以用于恢复状态到内存。

```
public StreamOperatorStateContext streamOperatorStateContext(
    @Nonnull OperatorID operatorID,
    @Nonnull String operatorClassName,
    @Nonnull ProcessingTimeService processingTimeService,
    @Nonnull KeyContext keyContext,
    @Nullable TypeSerializer<?> keySerializer,
    @Nonnull CloseableRegistry streamTaskCloseableRegistry,
    @Nonnull MetricGroup metricGroup) throws Exception {
    ...
    // 通过 taskStateManager 获取 PrioritizedOperatorSubtaskState 对象
    final PrioritizedOperatorSubtaskState prioritizedOperatorSubtaskStates =
        taskStateManager.prioritizedOperatorState(operatorID);
    ...
```

```
    try {
        // KeyedState 状态后端
        keyedStatedBackend = keyedStatedBackend(
            keySerializer,
            operatorIdentifierText,
            prioritizedOperatorSubtaskStates,
            streamTaskCloseableRegistry,
            metricGroup);
        // OperatorState 状态后端
        operatorStateBackend = operatorStateBackend(
            operatorIdentifierText,
            prioritizedOperatorSubtaskStates,
            streamTaskCloseableRegistry);
        ...
        // 将状态后端封装到 StreamOperatorStateContext 对象中并返回
        return new StreamOperatorStateContextImpl(
            prioritizedOperatorSubtaskStates.isRestored(),
            operatorStateBackend,
            keyedStatedBackend,
            timeServiceManager,
            rawOperatorStateInputs,
            rawKeyedStateInputs);
    } catch (Exception ex) {
        ...
    }
}
```

在上面的方法中，利用 taskStateManager 获取了 PrioritizedOperatorSubtaskState 对象。PrioritizedOperatorSubtaskState 对象可以理解为对状态数据的描述信息的又一种封装。随后利用 PrioritizedOperatorSubtaskState 对象构造了状态后端，将其封装到 StreamOperatorStateContext 对象中并返回。

回到 initializeState()方法，最后构造了 StateInitializationContext 对象，调用了 initializeState() 方法。如果 userfunction 实现了 CheckpointedFunction 接口，那么在这个方法中，就会调用用户自定义函数中的 initializeState()方法。通常，该方法会进行用户自定义函数中定义的状态的初始化，比如 FlinkKafkaConsumer 中的状态就是这样恢复的。如果是在 open()方法中恢复状态，则会在 StreamOperator 调用 open()方法时进行状态的恢复。

## 10.4 状态的重分配策略

如果一个框架中没有定义和维护状态，那么对它改变并行度是很容易的，只需在改变并行度后维护好上下游关系，如图 10-12 所示。

前文已经介绍过 Flink 启动并生成执行图的流程，可以了解到，当在代码中改变并行度后，会根据新的并行度生成新的 ExecutionGraph，上下游关系会自然而然地建立起来。

然而，如果在算子中定义了状态，那么改变并行度后，就需要考虑如何将状态分配给新的并行实例，有状态流框架如图 10-13 所示。

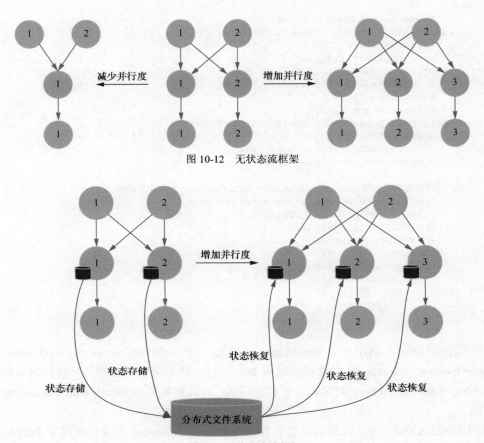

图 10-12 无状态流框架

图 10-13 有状态流框架

先前的作业将状态数据持久化到分布式文件系统中，作业重启后每个新的并行实例需要获取它应该恢复的那部分数据。

Flink 对操作符状态和键控状态有不同的重分配策略。

## 10.4.1 操作符状态的重分配

操作符状态有 3 种重分配模式，分别为 SPLIT_DISTRIBUTE、UNION 和 BROADCAST。它们在代码中为枚举类中的类型，在 OperatorStateHandle 接口中的定义如下：

```
enum Mode {
   SPLIT_DISTRIBUTE,
   UNION,
   BROADCAST
}
```

用户在定义状态时，Flink 会根据调用的方法给状态设置重分配模式。比如当用户自定义函数实现了 CheckpointedFunction 接口时，往往会在 initializeState()方法中定义状态：

```
public void initializeState(FunctionInitializationContext context) throws Exception {
    OperatorStateStore operatorStateStore = context.getOperatorStateStore();
    ListState state = operatorStateStore.getUnionListState(...);
    ...
}
```

其中调用的 getUnionListState()方法就会将状态设置成 UNION 模式：

```
public <S> ListState<S> getUnionListState(ListStateDescriptor<S> stateDescriptor) throws Exception {
    return getListState(stateDescriptor, OperatorStateHandle.Mode.UNION);
}
```

对于另外两种模式是类似的。

状态的重分配模式在状态存储时会被序列化到状态数据的描述信息中，在状态恢复时，会读取每个状态的重分配模式，进而进行不同的分配。

对于 SPLIT_DISTRIBUTE 模式，Flink 采取的策略是将所有状态汇总，然后将其均匀地分配给新的并行实例，如图 10-14 所示。

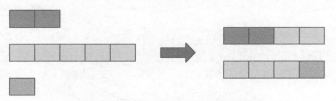

图 10-14　SPLIT_DISTRIBUTE 模式

对于 UNION 模式，Flink 采取的策略是将所有状态汇总，并且将其分发给所有并行实例，如图 10-15 所示。

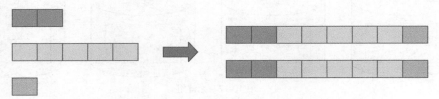

图 10-15　UNION 模式

对于 BROADCAST 模式，由于这种广播的状态在各个并行实例中本来就完全一样，因此重分配后仍然保持在各个并行实例中完全一样即可，如图 10-16 所示。

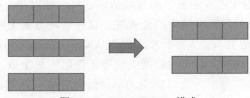

图 10-16　BROADCAST 模式

## 10.4.2 键控状态的重分配

键控状态的重分配似乎更简单一些。因为这类状态本身就是数据经过 KeyBy 操作后在对应的并行实例中维护的状态，所以如果改变了并行度，仍然可以按照 KeyBy 的规则，将这些状态读取到新的并行实例中去维护。

如图 10-17 所示，并行度从 3 改成了 4，Flink 仍然对 key 取哈希值，再对并行度取余，以此确定状态数据被读取到了哪个并行实例。但是从图 10-17 可以观察到，状态数据在存储时是顺序写入，但是在状态恢复时变成了随机读取。

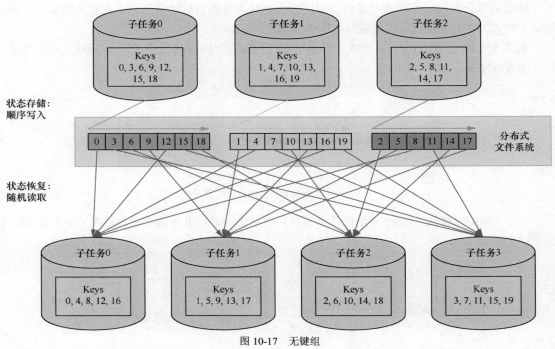

图 10-17 无键组

注： 1. hash (key) = key (identity)
  2. subtask (key) = hash (key) % parallelism

要解决这个问题，有多种思路。其中一种思路是，将之前并行实例的状态数据全部读取到新的并行实例中，再过滤掉不需要的数据。虽然这样保证了顺序读取，但是造成了大量无用数据的 I/O，分布式文件系统也会因此接收到更多的读取请求。另一种思路是在存储状态数据时，给每一条数据都记录元信息，如该数据在文件中的位置等。这种解决方案无疑会增加大量的存储成本。

Flink 提供的解决方案是引入最大并行度和键组（Key Group）的概念。最大并行度是在作业执行前指定好的，而且无法修改。最大并行度的值等于键组的总个数。每个 key 都会按照一定规则被分入某个键组。在状态存储时，会以键组为单位进行存储；在状态恢复时，会以键组为单位被整体读取到某个新的并行实例中。这意味着每个算子的并行度的值不能超过最大并行度的值。

有键组如图 10-18 所示。

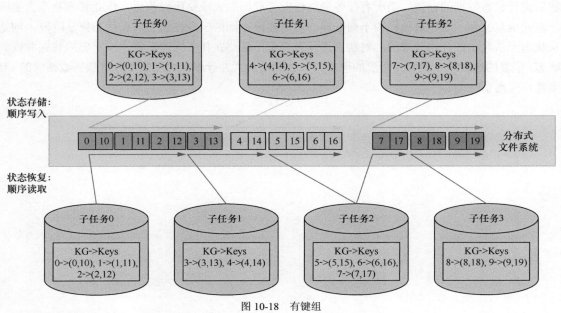

图 10-18　有键组

注：1. hash (key) = key (identity)
　　2. key_group (key) = hash (key) % number_of_key_groups
　　3. subtask (key) = key_group (key) * parallelism/number_of_key_groups

当并行度和最大并行度确定后，就可以计算出每个键组应该被分配到哪个并行实例中。比如在图 10-18 中，当并行度为 3 时，键组 0～键组 3 都在子任务 0 中，键组 4～键组 6 都在子任务 1 中，键组 7～键组 9 都在子任务 2 中。在决定某个 key 应该被分配到哪个并行实例时，首先是根据其哈希值计算出它应该属于哪个键组，然后根据所属键组确定它属于哪个并行实例。当并行度改为 4 时，只需要重新分配键组的所属情况。如图 10-18 所示，新的并行实例在读取状态数据时可以获得顺序读取在性能上的优势。另外，这种情况虽然需要多维护一些键组的信息，但数据量极小，并不会占用过多的存储空间。

读者可以进一步思考一个问题——为什么不能修改最大并行度呢？这个问题的答案很简单。假设一种简单场景，作业重启后修改了最大并行度，而没有修改并行度，那么原先属于某个并行实例的键组可能会被读取到另一个并行实例中。而在数据读取的过程中，由于并行度并未发生变化，key 的分区规则就没有发生变化，这使得原先属于某个 key 的状态并不在接收这个 key 的数据的并行实例中，会造成状态的不一致。

## 10.5　总结

本章深入讨论了 Flink 状态与容错机制的设计思想，分析了 Flink 内部的一致性语义的实现原理。从流程上来看，Flink 的容错机制可以从状态存储和状态恢复两方面来理解。状态存储的过程从检查点的触发开始，从 source 任务开始发送栅栏，每个任务收到栅栏后便开始进行快照操作，然后进行状态数据的存储。当所有任务通知作业管理器端完成快照操作后，作业管理器端会生成状态元信息文件。通过对栅栏的不同处理，可以实现不同的一致性语义。在状态恢复阶段，则是从状态元信息文件开始读取的，对状态进行重分配，最终在任务管理器端读取对应的状态并恢复状态。恢复的过程涉及最大并行度的概念。Flink 利用最大并行度减少了恢复分区状态数据时的 I/O 消耗，提高了性能。

# 第二部分　特性开发篇

# 第 11 章

# 动态调整并行度

本章介绍的动态调整并行度的方案是针对流处理作业而言的。在流处理作业的执行过程中，每个任务都需要一直不停地执行，每个物理节点、并行实例的工作负载都时刻变化着。在这种情况下，提前分配的资源可能就会在执行过程中的某些时刻显得过多或过少，造成资源浪费或系统性能低下。因此，在流处理作业的执行过程中根据工作负载的变化动态地调整并行度就成了合理的需求。

调整并行度的方式可以分为手动调整和自动调整。手动调整的方式依赖技术人员的经验，过程烦琐、低效，更加耗费人力，成本较高，因此在工程上更多的情况是实现自动调整。在以往大多数的工程实践中，无论是对类似 Flink 的流处理计算引擎进行任务并行度的调整，还是对其他分布式框架的分区调整（如 Kafka 的扩/缩容），往往会依赖吞吐量、CPU 利用率等粗粒度的指标，结果常常不尽如人意——要么是进行了完全错误的调整，要么是调整的数值过大或过小，导致需在正确值附近来回"摆动调整"，长时间才能达到收敛。

USENIX 组织的第十三届 OSDI 会议（USENIX Symposium on Operating Systems Design and Implementation）收录了一篇论文 "Three steps is all you need: fast, accurate, automatic scaling decisions for distributed streaming dataflows"，其中介绍了叫作 DS2 的模型，提供了目前优秀的动态调整 Flink 并行度的方案。

希望在学习完本章后，读者能够了解：

- 动态调整流处理系统任务并行度的设计思想和算法原理；
- 动态调整并行度的架构设计；
- 关于收集 Flink 内部统计指标的源码修改方案。

## 11.1 模型设计

本节将会归纳上述论文的主要内容，并结合 Flink 本身的实现，介绍 DS2 模型的核心设计思想。

### 11.1.1 传统模型的局限

论文 "Three steps is all you need: fast, accurate, automatic scaling decisions for distributed streaming dataflows" 首先分析了现存模型的局限。过去这些自动化的模型未能产生良好的效果，主要有下面几个原因。

- 缺乏全面的性能模型：主要体现在对资源的调整仅着眼于单个任务的并行度调整，却忽略了资源调整对其他任务的影响。一个全面的性能模型应当将流处理作业的任务形成的整个拓扑结构考虑在内。
- 依赖粗粒度的外部指标：比如使用吞吐量、CPU 利用率等外部指标，而非框架内部的指标。有一些模型使用反压作为指标，如果没有出现反压，模型就失去了判断依据。
- 使用启发式的调整：得到统计结果后，没有精准的调整策略，因此许多模型往往会进行保守的微调，导致多次调整后性能问题依然没有得到改善。

DS2 模型一方面没有考虑外部指标，而是计算出任务真实的处理和输出能力作为指标，另一方面着眼于整体的拓扑结构，将修改上游任务后对下游任务的影响纳入考虑范围，给出了更加精准、全面的调整策略。

### 11.1.2　DS2 模型的核心概念

在 Flink 中，由于每个任务的计算逻辑不同、所在物理节点的资源不同，因此可能导致这样的情况——有些并行实例负载较低，在短时间内就处理完了上游传输过来的数据，剩下的时间都处于等待输入的状态；另一些并行实例负载过高，不但自身一直处于繁忙状态，还会产生反压从而影响上游并行实例的输出。

图 11-1 所示为观察速率与真实速率。

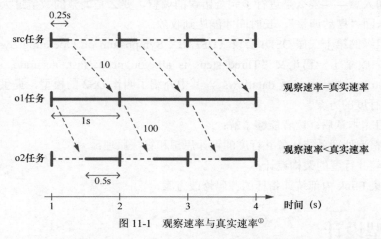

图 11-1　观察速率与真实速率[①]

这里的瓶颈发生在 o1 任务。由于 o1 任务自身处理和输出的速度较慢，因此 o2 任务有 1/2 的时间在等待 o1 任务的输出，而 src 任务受到反压的影响，只有 1/4 的时间在处理和输出数据。

这里 DS2 模型引入了"观察速率"和"真实速率"的概念。外部观察者会观察到 o1 任务每秒处理了 10 条记录，o2 任务每秒处理了 100 条记录，由此 o1 任务和 o2 任务的观察速率分别为 10recs 和 100recs。然而，如果从内部观察，会发现 o2 任务只有 0.5s 在处理数据，因此它们的真

---

① 图 11-1、图 11-2 及本章公式引自论文 *Three steps is all you need: fast, accurate, automatic scaling decisions for distributed streaming dataflows*。

实速率分别为 10recs 和 200recs。

如果希望 src 任务能够满负荷执行，那么 o1 任务的并行度必须为当前的 4 倍。这时 o1 任务的处理速率就会变为 40recs，这意味着它每秒会给 o2 任务输出 400 条记录。要保证 o2 任务刚好能够满负荷处理这些数据，它的并行度应调整为当前的 2 倍，以获得 400recs 的处理速率。

基于上述内容可知，DS2 模型主要引入了下面这些概念。

- 实用时间（useful time）：任务花费在反序列化、计算和序列化上的时间。这是数据在任务内部真正被处理所花费的时间。这部分时间不包括等待输入或输出的时间。
- 观察时间（observed time）：这部分时间是对外部观察者而言的时间，指数据在任务中滞留的所有时间，不仅包括实用时间，还包括等待输入和输出的时间。
- 真实速率（true rates）：表示单位实用时间内对数据的处理速率。这是真正衡量任务处理数据的能力的指标。
- 观察速率（observed rates）：表示单位观察时间内对数据的处理速率。这是对外部观察者而言的速率。

DS2 模型通过这些概念和全局调整的方式，理论上能一步到位地使每个任务的并行度达到最优。总体而言，该模型满足了下面这些性质。

- 对于同样的情况给予稳定的调整策略。
- 能实现准确的调整。调整后不会因调整值错误需再次调整而导致在准确值附近来回摆动调整。也不会在调整后给出过多的资源而导致资源浪费。
- 整个调整周期较短，很快就能达到最优状态。

### 11.1.3 算法原理

用 $W_u$ 表示实用时间，用 $W$ 表示观察时间，则有下面的关系：

$$\lambda_p = \frac{R_{\text{prc}}}{W_u}$$

$$\lambda_o = \frac{R_{\text{psd}}}{W_u}$$

$$\hat{\lambda}_p = \frac{R_{\text{prc}}}{W}$$

$$\hat{\lambda}_o = \frac{R_{\text{psd}}}{W}$$

其中，$R_{\text{prc}}$ 和 $R_{\text{psd}}$ 分别表示在这段观察时间内从上游拉取的数据量和向下游推送的数据量。由此 $\lambda_p$、$\lambda_o$、$\hat{\lambda}_p$ 和 $\hat{\lambda}_o$ 分别表示真实处理速率、真实输出速率、观察处理速率和观察输出速率。

用 $G$ 表示任务的逻辑拓扑，用 $A$ 表示任务的相邻关系。令任务从 $i=0$ 开始标识，一直到 $i=m-1$，同时令 $0 \leq i < j < m$。若第 $i$ 个任务和第 $j$ 个任务相邻，那么有 $A\{ij\}=1$，否则 $A\{ij\}=0$。对于第 $i$ 个任务 $o_i$，其并行度为 $p_i \geq 1$，该任务所有并行实例共同形成的真实处理速率和真实输出速率分别为：

$$O_i[\lambda_p] = \sum_{k=1}^{k=p_i} \lambda_p^k$$
$$O_i[\lambda_o] = \sum_{k=1}^{k=p_i} \lambda_o^k$$

那么，该任务的最优并行度 $pi_i$ 可以通过下面的公式计算：

$$pi_i = \left[\sum_{\forall j: j<i} A_{ji} \cdot O_j[\lambda_o] * \cdot \left(\frac{o_i[\lambda_p]}{p_i}\right)^{-1}\right], n \leqslant i < m$$

其中，$m$ 为拓扑中任务的总数，$n$ 为 source 任务的总数。$o_j[\lambda_o]^*$ 表示的则是 $o_j$ 任务的上游任务都达到最优并行度时 $o_j$ 任务的最优输出速率，具体计算公式如下：

$$o_j[\lambda_o]^* = \lambda_{\text{src}}^j, o \leqslant j < n$$

$$o_j[\lambda_o]^* = \frac{o_j[\lambda_o]}{o_j[\lambda_p]} \cdot \sum_{\forall u: u<j} A_{uj} \cdot O_u[\lambda_o]^*, n \leqslant j < m$$

其中，$\lambda_{\text{src}}^j$ 表示第 $j$ 个 source 任务的输出速率。

整体思想是，从 source 任务的输出速率开始，依次计算其下游任务的最优输出速率。在计算最优并行度时，先对其上游的输出求和，再除以当前的真实处理速率，就可得到理想并行度。

### 11.1.4 架构设计

根据以上设计思想与算法原理可知，首先需要对 Flink 源码进行修改，使其能够统计真实速率和观察速率等指标，并周期性地输出。这些指标最好能够持久化到文件系统或数据库，以便 DS2 模型中的组件进行收集和计算。

DS2 模型的组件主要包括扩展管理器（scaling manager）和扩展策略（scaling policy）。扩展管理器负责监控指标并调用扩展策略拉取统计结果进行计算。通过扩展策略计算出需要修改的并行度后，将结果返回给扩展管理器，再由扩展管理器根据该并行度重启作业。

DS2 模型架构如图 11-2 所示。

其中，扩展管理器可以设置如下内容。

- 策略间隔（policy interval）：表示扩展策略多长时间拉取一次统计结果。这个值不能设置得太短，需要保证一次拉取的数据是足够用于进行计算的。
- 预热时间（warm-up time）：表示一次调整后，要间隔多少个策略间隔再开始收集统计结果。这是因为重启作业后，需要一定的预热时间使作业的执行达到稳定状态。
- 激活时间（activation time）：表示扩展管理器多长时间调用扩展策略触发一次计算。因为 Flink 中存在窗口操作，刚开始数据会被分配到窗口中，这时处理速率很高；当窗口被触发计算时，就会进行真正的逻辑处理，这时处理速率就会变低。所以不能根据一次收集的数据来进行计算，而是需要在一段时间内收集多次数据，然后根据一定的策略选择合适的值进行并行度的调整。这个时间间隔就是激活时间。
- 目标速率比（target rate ratio）：表示当前达到的速率与目标速率的比率。增添并行度后可能会造成额外的网络通信成本，这个比率反映了真实情况和理想情况的差异。

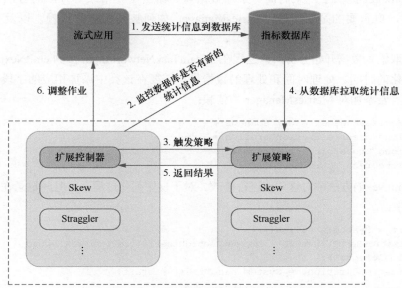

图 11-2 DS2 模型架构

### 11.1.5 使用 DS2 模型的注意事项

从图 11-2 可以观察到，扩展管理器和扩展策略只是对管理器组件和策略组件的一种实现。如果之后需要解决数据倾斜等问题，仍然可以使用 DS2 模型实现 Skew Manager 和 Skew Policy 等。只不过对当前实现的动态调整并行度来说，DS2 模型并不能解决数据倾斜等性能问题。需要注意的是，DS2 模型不但不能解决数据倾斜的问题，实际上还不能识别该问题。当发生数据倾斜时，DS2 模型仍然会对并行度进行调整，只不过调整后不会有明显的性能改善。针对这种情况，如果发现几次调整后性能没有得到明显的改善，则说明可能出现了数据倾斜等 DS2 模型无法解决的问题，这时应该对系统限制调整次数，避免继续进行不必要的调整，造成时间和资源的浪费。

另外，DS2 模型会忽略微小的调整。因为 I/O、GC 等过程都可能会影响统计结果。这些微小的影响可能会导致计算出来的结果是增加或减少一两个并行实例，这是毫无必要的。

在极少数情况下，有可能会发生调整后反而性能降低的情况。这时需要提供回退机制，以保证作业能够回退到之前的并行度。

## 11.2 指标收集

要实现上述架构，Flink 需要提供收集和输出指标的功能。

首先可以在 Flink-runtime 模块下的 org.apache.Flink.runtime.util 包中添加一个 MetricsManager 类来管理指标。MetricsManager 应当是 Task 级的，单独负责每一个并行实例的指标统计与输出。

MetricsManager 提供了记录时间、累加数据量、输出统计结果等方面的方法。在了解这些方法的实现前，更重要的是要了解 MetricsManager 如何被初始化和依赖，以及其在何处发挥作用。

可以从获取输入数据开始分析，该过程在 StreamTaskNetworkInput 类的 emitNext() 方法中实现。要记录反序列化的时间、处理时间和处理的数据量，需要在该类中添加相应的字段来对这些值进行维护，同时需要添加对 MetricsManager 的依赖：

```
private MetricsManager metricsManager;
private long deserializationDuration = 0;
private long processingDuration = 0;
private long recordsProcessed = 0;
```

接着在 emitNext() 方法中对这些值进行修改。对于反序列化过程，将时间记录在 deserializationDuration 字段中：

```
long start = System.nanoTime();
DeserializationResult result = currentRecordDeserializer.getNextRecord(deserializationDelegate);
deserializationDuration += System.nanoTime() - start;
```

数据被处理时，将处理时间和处理的数据量记录在 processingDuration 和 recordsProcessed 中：

```
long processingStart = System.nanoTime();
processElement(deserializationDelegate.getInstance(), output);
processingDuration += System.nanoTime() - processingStart;
recordsProcessed++;
```

当获取新的 BufferOrEvent 对象时，会调用 processBufferOrEvent() 方法：

```
Optional<BufferOrEvent> bufferOrEvent = checkpointedInputGate.pollNext();
if (bufferOrEvent.isPresent()) {
    processBufferOrEvent(bufferOrEvent.get());
} else {
    ...
}
```

在 processBufferOrEvent() 方法中通知 MetricsManager 获取了一个新的 BufferOrEvent 对象：

```
if (bufferOrEvent.isBuffer()) {
    lastChannel = bufferOrEvent.getChannelIndex();
    currentRecordDeserializer = recordDeserializers[lastChannel];
    currentRecordDeserializer.setNextBuffer(bufferOrEvent.getBuffer());
    metricsManager.newInputBuffer(System.nanoTime());
}
```

最后，需要在获取 BufferOrEvent 对象前，通知 MetricsManager 输出统计结果：

```
if (deserializationDuration > 0) {
    metricsManager.inputBufferConsumed(System.nanoTime(), deserializationDuration, processingDuration, recordsProcessed);
    deserializationDuration = 0;
    processingDuration = 0;
    recordsProcessed = 0;
```

```
}
Optional<BufferOrEvent> bufferOrEvent = checkpointedInputGate.pollNext();
```
并且需要将相应的字段的值清零。

在输出数据时，会调用 RecordWriter 的 emit()方法。因此需要在 RecordWriter 中添加对 MetricsManager 的依赖，在 emit()方法中利用 MetricsManager 进行输出数据和序列化所花费时间的统计：

```
protected void emit(T record, int targetChannel) throws IOException, InterruptedException {
    metricsManager.incRecordsOut();
    long start = System.nanoTime();
    serializer.serializeRecord(record);
    long end = System.nanoTime();
    metricsManager.addSerialization(end - start);
    if (copyFromSerializerToTargetChannel(targetChannel)) {
        serializer.prune();
    }
}
```

在 ChannelSelectorRecordWriter 的 broadcastEmit()方法中需要进行类似的操作：

```
public void broadcastEmit(T record) throws IOException, InterruptedException {
    metricsManager.incRecordsOut();
    long start = System.nanoTime();
    serializer.serializeRecord(record);
    long end = System.nanoTime();
    metricsManager.addSerialization(end - start);
    boolean pruneAfterCopying = false;
    for (int targetChannel = 0; targetChannel < numberOfChannels; targetChannel++) {
        if (copyFromSerializerToTargetChannel(targetChannel)) {
            pruneAfterCopying = true;
        }
    }
    if (pruneAfterCopying) {
        serializer.prune();
    }
}
```

上面的 emit()和 broadcastEmit()方法都调用了 copyFromSerializerToTargetChannel()方法，在该方法中也需要记录序列化所花费的时间，或者通知缓冲区已满。

```
metricsManager.outputBufferFull(System.nanoTime());
if (flushAlways) {
    long start = System.nanoTime();
    flushTargetPartition(targetChannel);
    long end = System.nanoTime();
    metricsManager.addSerialization(end - start);
    metricsManager.outputBufferFull(System.nanoTime());
}
```

在 BroadcastRecordWriter 和 ChannelSelectorRecordWriter 的 requestNewBufferBuilder()方法中，需要统计申请资源时等待的时间，实现方式如下：

```
long bufferStart = System.nanoTime();
BufferBuilder builder = targetPartition.getBufferBuilder();
long bufferEnd = System.nanoTime();
if (bufferEnd - bufferStart > 0) {
    metricsManager.addWaitingForWriteBufferDuration(bufferEnd - bufferStart);
}
```

## 11.3 指标管理

了解了 MetricsManager 如何在 Flink 原有逻辑中收集指标后，现在来了解其内部如何维护、管理这些指标。

MetricsManager 包含下面这些字段。

- taskId：任务 id。
- instanceId：并行实例的 id。
- recordsIn：自上一次输出统计结果后接收的总数据量。
- recordsOut：自上一次输出统计结果后输出的总数据量。
- usefulTime：自上一次输出统计结果后的实用时间。
- waitingTime：自上一次输出统计结果后的等待时间。
- currentWindowStart：本次统计的开始时间。
- status：ProcessingStatus 类型，处理的状态，其中封装了序列化所花费的时间、等待申请资源所花费的时间等。
- windowSize：通过配置给出该字段的值，收集指标的周期。当超过这个周期后可以输出结果。
- ratesPath：输出文件的路径。
- epoch：每次输出的唯一标识。

其中的 status 字段的类型为 ProcessingStatus，其也是 runtime 模块下 util 包中新添加的类。在 11.2 节中调用 MetricsManager 中的方法进行指标收集时，实际上都是将这些信息维护在 status 中。

上面有关统计指标的方法，实现分别如下：

```
public void newInputBuffer(long timestamp) {
    status.setProcessingStart(timestamp);
    status.setWaitingForReadBufferDuration(timestamp - status.getProcessingEnd());
}
public void addSerialization(long serializationDuration) {
    status.addSerialization(serializationDuration);

}
public void incRecordsOut() {
    status.incRecordsOut();
}
public void addWaitingForWriteBufferDuration(long duration) {
    status.addWaitingForWriteBuffer(duration);
}
```

输出统计结果的方法的实现如下:

```java
public void inputBufferConsumed(long timestamp, long deserialization, long processing,
long numRecords) {
    synchronized (status) {
        if (currentWindowStart == 0) {
            currentWindowStart = timestamp;
        }
        status.setProcessingEnd(timestamp);
        // 累加统计值
        recordsIn += numRecords;
        recordsOut += status.getNumRecordsOut();
        usefulTime += processing + status.getSerializationDuration() + deserialization
            - status.getWaitingForWriteBufferDuration();
        // 清空 status 中的状态
        status.clearCounters();
        if (timestamp - currentWindowStart > windowSize) {
            // 计算速率
            long duration = timestamp - currentWindowStart;
            double trueProcessingRate = (recordsIn / (usefulTime / 1000.0)) * 1000000;
            double trueOutputRate = (recordsOut / (usefulTime / 1000.0)) * 1000000;
            double observedProcessingRate = (recordsIn / (duration / 1000.0)) * 1000000;
            double observedOutputRate = (recordsOut / (duration / 1000.0)) * 1000000;
            // 整理统计结果
            String ratesLine = workerName + ","
                + instanceId  + ","
                + numInstances  + ","
                + currentWindowStart + ","
                + trueProcessingRate + ","
                + trueOutputRate + ","
                + observedProcessingRate + ","
                + observedOutputRate;
            List<String> rates = Arrays.asList(ratesLine);
            // 输出统计结果
            ...
            // 清空之前的状态
            recordsIn = 0;
            recordsOut = 0;
            usefulTime = 0;
            currentWindowStart = 0;
            epoch++;
        }
    }
}
```

上述代码的核心思路就是如果这段时间超过了设置的 windowSize 的值,那么会计算各项速率,将结果整理并输出。输出后清空之前的状态,进入下一次的统计。outputBufferFull()方法的逻辑是类似的,不再赘述。其中输出的逻辑可以自定义,可以输出到文件,也可以输出到数据库。

当 Flink 拥有了统计和输出这些指标的能力后,就可以在外部实现扩展管理器和扩展策略来进行并行度的动态调整了。

## 11.4 总结

本章介绍了一种动态调整流处理作业任务并行度的方案。该方案的底层设计思想来自 DS2 模型，本章详细介绍了其中的算法原理。从架构上来说，这种方案需要 Flink 提供指标给外部系统，依靠扩展管理器和扩展策略来进行计算和调整。本章提供了修改 Flink 源码的思路和代码示例，读者可以以此为基础对整个架构进行实现。

# 第 12 章

# 自适应查询执行

"自适应查询执行"（Adaptive Query Execution）是一种在运行过程中，根据运行时中间结果产生统计信息来调整和优化后续执行计划，从而提高整体执行效率的优化方式。传统的基于代价的优化方式虽然会获取数据集的统计信息，但是这种统计不一定精确和及时，而且很多时候缺乏对中间计算结果的统计。自适应查询执行则可以解决这一问题。这种技术在一些成熟的数据库查询引擎中已经有了多年应用。随着 Spark 3.0 的发布，这种技术更多地被关注开源分布式计算引擎的人们所熟知。通过阅读 Spark 3.0 的源码，人们可以更加直观地理解该技术的实现原理，并将其应用在其他计算引擎中。据笔者所知，阿里巴巴集团内部的计算引擎产品早已对类似技术有所实现。

第 11 章介绍的动态调整并行度是针对流处理作业而言的，而本章介绍的自适应查询执行是针对批处理作业的优化。在批处理作业的执行过程中，当数据发生分区后，很有可能产生数据倾斜的现象，改变并行度往往并不能彻底解决该问题。这时，可以考虑在改变下游任务并行度的同时，改变上下游的输入输出关系。比如，让数据量明显偏多的上游分区被下游多个消费者将数据拆分开进行消费，这样下游每个并行实例所要读取的数据量就相对少了。如果上游多个分区的数据量都较少，还可以用同样的思路减少下游任务并行度，让其中一些下游并行实例消费多个上游分区数据，以节省资源成本。

本章所要讨论的自适应查询执行优化方案，主要就是根据上游任务的统计结果，调整与它直接连接的下游任务的并行度和上下游的输入输出关系，以达到改善数据倾斜或实现分区合并的要求。由于在 Flink 的批处理作业调度过程中，可以先部署上游任务，等待上游任务执行完后，再通知下游任务的并行实例进行调度和部署，因此在批处理作业中就有了自适应查询执行的优化空间。

本章会先回顾和梳理 Flink 批处理作业的调度、部署、执行流程，然后分析实现自适应查询执行需要在源码中做哪些修改。

希望在学习完本章后，读者能够了解：

- 自适应查询执行优化策略的实现原理；
- Flink 中每个任务执行完后如何收集分区数据量等统计信息；
- 如何在运行过程中修改下游未部署任务的并行度；
- 如何修改上下游的输入输出关系。

## 12.1 Flink 框架下的自适应查询执行

本节将讨论 Flink 框架下自适应查询执行的实现思路。

### 12.1.1 执行阶段的划分

自适应查询执行的优化可以建立在 Flink 批处理作业的分步调度和部署之上。回顾 Flink 的 ExecutionGraph，如图 12-1 所示。

在批处理作业中，每一个 ExecutionJobVertex 可以看作一个执行阶段。以图 12-1 为例，只有当 Source 这个 ExecutionJobVertex 中的所有 ExecutionVertex 对应的任务执行完后（这里只有一个并行实例），才会判断是否可以调度和部署下游任务，即 Flat Map 中的两个并行实例。同理，只有当 Flat Map 中的两个并行实例执行完后，才会部署该 ExecutionJobVertex 的下游任务，即 Keyed Aggregation --> Sink 任务中的两个并行实例。

这就意味着，当上游任务的并行实例执行完后，在判断是否可以调度和部署下游任务的过程中，可以对整个 ExecutionGraph 进行优化，改变接下来要部署的任务的并行度，甚至改变图 12-1 中 IntermediateResultPartition 和 ExecutionEdge 的连接关系，以此解决数据倾斜问题或实现分区合并。

### 12.1.2 优化流程

Flink 中的 ExecutionGraph 表示执行图，它被调度器 DefaultScheduler 依赖。第 6 章介绍过 Flink 会根据执行图生成调度拓扑 SchedulingTopology 对象，该对象是调度过程中具体被操作的对象，被调度器的调度策略依赖。可以认为，调度器管理并维护着 Flink 作业的整个调度过程。

Flink 本身的调度器和其支持的调度策略都并未实现重优化执行图的功能。为了支持自适应查询执行的优化，就需要对现有调度器进行修改，可以通过实现新的调度策略来完成重优化。在新的调度策略下，重优化过程如图 12-2 所示。

从部署上游任务到重优化后部署下游任务的整个过程大致可以分为以下 4 步。

（1）调度器会先部署所有的 source 任务，即任务 1 的所有并行实例。这些并行实例被部署到任务管理器中执行。

（2）执行完后，每一个并行实例都会通知调度器其状态变为"已完成"。这个过程同时伴随着统计信息的回传。

（3）当调度器了解到上游任务的每一个并行实例都执行完后，聚合统计信息，对执行图进行重优化。如图 12-2 所示，任务 2 的并行度在优化过程中发生了改变。

（4）部署优化后的任务 2 中的每个并行实例。

同理，当任务 2 执行完后，也会通过同样的步骤对任务 3 进行优化。

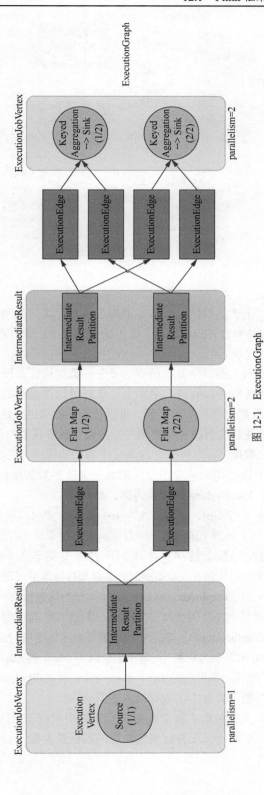

图 12-1 ExecutionGraph

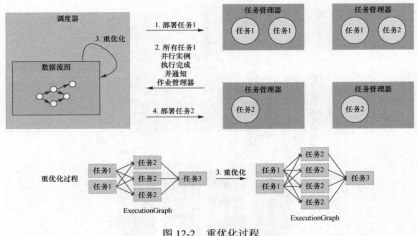

图 12-2　重优化过程

## 12.1.3　优化策略

如果将并行度调大，上游的数据可以被更多的下游并行实例分别消费，这自然就能改善数据倾斜问题。这是一个非常朴素且正确的想法。这里的关键在于，如何利用 Flink 自身的机制对这个优化策略进行相对简易的实现。

直观来说，似乎需要对上游发生数据倾斜的子分区数据进行分段，让下游每个并行实例去消费对应的分段数据。这个想法并没有错。事实上，这样可以在最细的粒度上对数据进行切分和消费。然而，基于对 Flink 数据交换机制的了解，可以想象，如果进行这样的实现，很可能需要在上游或下游的某处记录下游并行实例应当从哪里开始消费、消费到了何处、消费到何处时读取结束，并且由于需要分段地读取数据，很有可能会深入反序列化的代码，这是一件复杂且有风险的事情。

那么，有没有可能不通过切分单个子分区的数据来改善数据倾斜问题呢？这需要回顾 Flink 的数据交换机制。以一个 MapReduce 的过程为例，如图 12-3 所示。

假设上游是 5 个 map 并行实例，下游是 3 个 reduce 并行实例。根据前文对 ResultSubpartition 和 InputChannel 的介绍，可以了解到上游的每个并行实例会根据下游并行度将数据写入对应子分区。在图 12-3 中，每个 map 并行实例都将数据写入了 3 个子分区。这些子分区的索引与下游 reduce 并行实例的索引一一对应。下游每个并行实例在拉取数据时，会根据自身索引去每个上游并行实例中拉取对应的子分区。因此，可以用二元组（mapIndex, reduceIndex）对子分区进行唯一标识。比如，map4 并行实例中的 subpartition1，其标识就是（4,1）。对于每一个 reduce 并行实例，可以用（startMapIndex, endMapIndex）和（startReducerIndex, endReducerIndex）表示它拉取的子分区的范围（左闭右开）。比如，reduce2 的 startMapIndex 和 endMapIndex 分别为 0 和 5，表示它拉取了 map0 到 map4 的数据。由于它拉取的都是 subpartition2 这个子分区的数据，因此它的 startReducerIndex 和 endReducerIndex 分别为 2 和 3。

基于图 12-3，如何判断是否发生了数据倾斜呢？

当某一个 reduce 并行实例拉取了每个 map 并行实例中对应子分区的数据后，这些子分区的数据加在一起的数据量比另外的 reduce 并行实例拉取的数据量大很多，这时判断它发生了数据倾斜。

比如，可能 reduce2 拉取了 map0 到 map4 的 subpartition2 后，发现数据发生了倾斜。这时就可以考虑只拉取部分 map 并行实例的子分区数据，而对于另一些数据则可以新添加一个 reduce 并行实例去拉取，如图 12-4 所示。

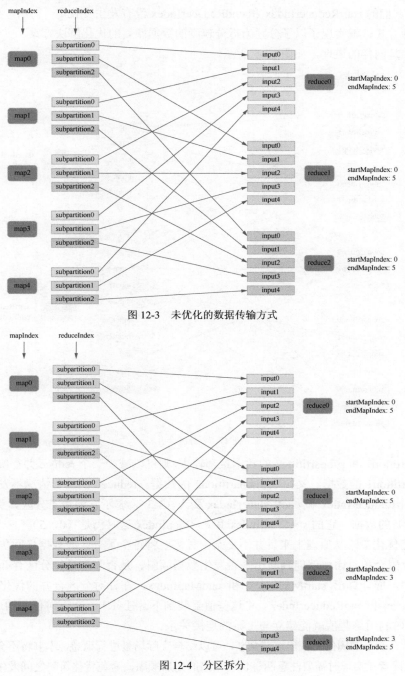

图 12-3　未优化的数据传输方式

图 12-4　分区拆分

在图 12-4 中，reduce0 和 reduce1 的消费情况没有发生变化，但是增添了 reduce3 分担了 reduce2 原有的消费压力。现在，reduce2 只拉取 map0 到 map2 的数据，reduce3 拉取 map3 和 map4 的数据。reduce2 的 startMapIndex 和 endMapIndex 变为了（0,3），reduce3 的 startMapIndex 和 endMapIndex 则为（3,5）。它们的 startReducerIndex 和 endReducerIndex 没有发生变化。

通过这种方式，就实现了以子分区为拆分粒度的数据倾斜的优化解决方案。

分区合并是同样的道理，如图 12-5 所示。

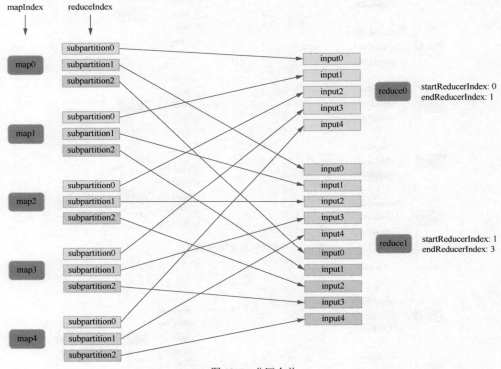

图 12-5　分区合并

若 subpartition1 和 subpartition2 的数据量都较少，则可以减少一个 reduce 并行实例，让 reduce1 既拉取 subpartition1 的数据，又拉取 subpartition2 的数据。reduce0 的消费情况不发生任何变化。reduce1 的 startReducerIndex 和 endReducerIndex 变为（1,3），表示它消费子分区索引为 1 到 3（包括 1 不包括 3）的数据。它的 startMapIndex 和 endMapIndex 这时仍是（0, 5）。

实现上述优化策略从原理上来讲并不复杂，只需要将每个子分区的数据量回传给作业管理器端，统一进行聚合和对比，就能判断出是否发生数据倾斜、是否需要进行分区合并。在针对数据倾斜的优化中，应计算出 startMapIndex 和 endMapIndex；在针对分区合并的优化中，应计算出 startReducerIndex 和 endReducerIndex。将这些值封装到下游任务中，在下游任务构造 InputGate 或 InputChannel 时利用这些值就能建立与上游的连接关系。

至于在什么水平的阈值下做出这些判断，可以结合实际场景进行调整，本书将不会过多讨论算法的实现细节。读者在实现时需要注意两种优化策略的先后顺序，避免优化策略之间发生冲突。在做判断时，不仅要考虑分区的相对数据量，还要考虑每个分区的绝对数据量，以避免不必要的优化。

## 12.2　统计信息的收集

获取每个子分区的数据量是优化的基础。Flink 本身并不具备这样的功能。不过在批处理作业中，在每个任务输出数据时，Flink 会对数据量进行记录。

数据输出时最终会被写入 ResultSubpartition，在本章所讨论的优化场景中，其实现类一定为 BoundedBlockingSubpartition。该类有一个 BoundedData 类型的 data 字段，可以认为该字段维护了数据量的信息。数据输出时，会调用 BoundedBlockingSubpartition 的 writeAndCloseBufferConsumer() 方法，在该方法中会将缓冲区中的数据写入 data 字段。

```
private void writeAndCloseBufferConsumer(BufferConsumer bufferConsumer) throws IOException
{
    try {
        final Buffer buffer = bufferConsumer.build();
        try {
            if (canBeCompressed(buffer)) {
                final Buffer compressedBuffer = parent.bufferCompressor.
compressToIntermediateBuffer(buffer);
                // 将数据写入 data 字段
                data.writeBuffer(compressedBuffer);
                if (compressedBuffer != buffer) {
                    compressedBuffer.recycleBuffer();
                }
            } else {
                // 将数据写入 data 字段
                data.writeBuffer(buffer);
            }
            ...
}
```

FileChannelBoundedData 是 BoundedData 的默认实现类，它的 writeBuffer() 方法如下：

```
public void writeBuffer(Buffer buffer) throws IOException {
    size += BufferReaderWriterUtil.writeToByteChannel(fileChannel, buffer,
headerAndBufferArray);
}
```

可以观察到，该类中用 size 字段维护了写入数据量的大小。在 Flink 的设计中，这个字段的值并没有任何实际的用处，但是 BoundedBlockingSubpartition 提供了获取该值的方法：

```
protected long getTotalNumberOfBytes() {
    return data.getSize();
}
```

结合任务的生命周期，可以在任务结束时，将这个值获取出来进行封装并回传给作业管理器。

在 Task 的 doRun() 方法的最后阶段会调用 ResultPartitionWriter 的 finish() 方法，表示分区结束生产数据。进而会调用 ResultSubpartition 的 finish() 方法：

```
for (ResultPartitionWriter partitionWriter : consumableNotifyingPartitionWriters) {
    if (partitionWriter != null) {
        partitionWriter.finish();
    }
}
public void finish() throws IOException {
```

```
        for (ResultSubpartition subpartition : subpartitions) {
            subpartition.finish();
        }
        isFinished = true;
}
```

在 BoundedBlockingSubpartition 的 finish()方法中，会最后一次将数据写入 data 字段：

```
public void finish() throws IOException {
    isFinished = true;
    flushCurrentBuffer();
writeAndCloseBufferConsumer(EventSerializer.toBufferConsumer(EndOfPartitionEvent.INSTANCE));
    data.finishWrite();
}
```

这意味着，在运行完 ResultSubpartition 的 finish()方法后，就可以调用上面的 getTotalNumberOfBytes()方法来获取数据量大小。因为每个 ResultPartition 都有多个 ResultSubpartition，所以可以在 ResultPartition 中添加一个数组字段，用来封装多个 ResultSubpartition 的 size 字段的值。

Task 的 doRun()方法最后会进入 finally 代码块，调用 notifyFinalState()方法：

```
private void notifyFinalState() {
    checkState(executionState.isTerminal());
    taskManagerActions.updateTaskExecutionState(new TaskExecutionState(jobId, executionId, executionState, failureCause));
}
```

正是在这里，任务的状态被回传给了作业管理器，由此，作业管理器可以知道该并行实例已经执行完成。可以在 notifyFinalState()方法中获取先前封装在 ResultPartition 中的数组字段，即每个子分区的数据量大小，将其封装在 TaskExecutionState 对象中并回传给作业管理器端（需要给 TaskExecutionState 添加字段并修改构造方法）。通过这样的修改，就可以使得作业管理器端在得知任务执行完成的同时获取其中每个子分区的数据量统计信息。

在作业管理器端，这个数据量统计信息会被调度器获取，进而会进入 DefaultScheduler 的 updateTaskExecutionStateInternal()方法：

```
protected void updateTaskExecutionStateInternal(final ExecutionVertexID executionVertexId,
final TaskExecutionState taskExecutionState) {
        schedulingStrategy.onExecutionStateChange(executionVertexId, taskExecutionState.getExecutionState());
        maybeHandleTaskFailure(taskExecutionState, executionVertexId);
}
```

在调度策略的 onExecutionStateChange()方法中，会获取接下来可以调度的任务，并判断是否可以部署。根据前面介绍的优化方案了解到，可以实现一个新的调度策略将重优化的操作封装在 onExecutionStateChange()方法中。不过需要注意的是，该方法的参数并不是 TaskExecutionState 类型，因此可以重载一个同名方法，将整个 TaskExecutionState 对象传入方法，主要目的是获取封装在其中的数据量。调度器在对数据量的维护和处理上，需要清晰地表示 ExecutionJobVertex、ResultPartition、ResultSubpartition 之间的关系。

## 12.3　执行图与调度拓扑的修改

这里提供一个在新的调度策略中重载 onExecutionStateChange()方法的示例：

```
public void onExecutionStateChange(ExecutionVertexID executionVertexId, ExecutionState
executionState, TaskExecutionState taskExecutionState) {
    if (!FINISHED.equals(executionState)) {
        return;
    }
    // 保存统计信息,并根据条件判断是否要重优化ExecutionGraph
    saveStatisticsAndMaybeReOptimize(taskExecutionState);
    final Set<SchedulingExecutionVertex<?, ?>> verticesToSchedule = IterableUtils
        .toStream(schedulingTopology.getVertexOrThrow(executionVertexId).getProducedResults())
        .filter(partition -> partition.getResultType().isBlocking())
        .flatMap(partition -> inputConstraintChecker.markSchedulingResultPartitionFinished
(partition).stream())
        .flatMap(partition -> IterableUtils.toStream(partition.getConsumers()))
        .collect(Collectors.toSet());
    allocateSlotsAndDeployExecutionVertices(verticesToSchedule);
}
```

重载的方法增加了 TaskExecutionState 类型的参数。在流程上除了增加了 saveStatisticsAndMaybeReOptimize()方法,其余的部分未发生变化。而在 saveStatisticsAndMaybeReOptimize()方法中,需要做如下事情。

- 保存当前并行实例传回的统计信息。
- 判断是否该任务的所有并行实例都达到"已完成"状态并传回了统计信息。
    ◆ 如果不是,则保存完统计信息后直接跳转出该方法,继续执行后面原有的流程。
    ◆ 如果是,则将统计信息进行聚合,利用具体的数据倾斜和分区合并的优化策略计算出新的并行度以及每个并行实例的 startMapIndex、endMapIndex、startReducerIndex 和 endReducerIndex。
- 构建新的 ExecutionGraph。
- 根据新的 ExecutionGraph 构建新的调度拓扑 SchedulingTopology。

其中 saveStatisticsAndMaybeReOptimize()方法示例如下:

```
private void saveStatisticsAndMaybeReOptimize(TaskExecutionState taskExecutionState) {
    if (saveStatisticsAndMaybeAccumulate(taskExecutionState)) { // 在保存统计信息的过程
// 中判断是否需要聚合统计信息
        reOptimizeExecutionGraph(); // 重优化执行图
        updateSchedulingTopology(); // 更新调度拓扑
    }
}
```

其中 saveStatisticsAndMaybeAccumulate()方法会保存该并行实例传回的统计信息,并且会判断是否所有并行实例都达到"已完成"状态。这个判断的实现可以直接参考 Flink 调度过程中的判断方式。如果获得了全部的统计信息,则可以对这些信息进行聚合,然后在 reOptimizeExecutionGraph()方法中,通过数据倾斜和分区合并的优化策略计算出新的并行度和每个并行实例的 startMapIndex、endMapIndex、startReducerIndex 和 endReducerIndex,并基于此重新构建 ExecutionGraph。

在重新构建 ExecutionGraph 的时候,主要需修改 3 处,如图 12-6 所示。

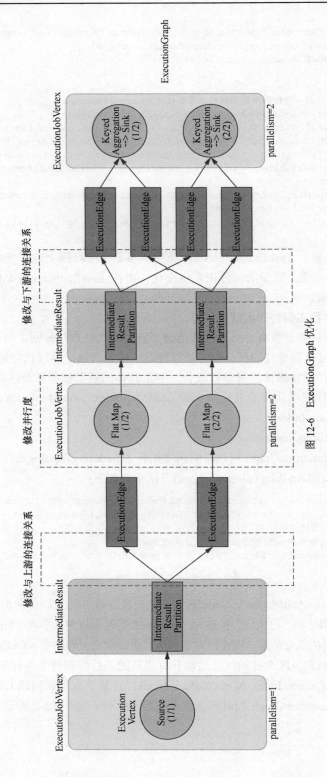

图 12-6 ExecutionGraph 优化

上游的 source 任务执行完后，需要对中间的 Flat Map 任务进行优化。直观的改变就是修改了该任务的并行度。回顾执行图的构建逻辑，ExecutionEdge 和 IntermediateResultPartition 对象是和 ExecutionVertex 对象一同构建出来的。因此，当并行度发生改变后，必然会产生新的 ExecutionEdge 和 IntermediateResultPartition 对象，这就意味着新的 ExecutionEdge 对象需要与上游的 IntermediateResultPartition 对象建立连接（否则就无法找到上游的子分区），而新的 IntermediateResultPartition 对象也需要与下游的 ExecutionEdge 建立连接。

构建 ExecutionGraph 的逻辑可以参考 attachJobGraph()方法的流程以实现重优化。

最后，由于改变了原来的 ExecutionGraph，对应的调度拓扑 SchedulingTopology 也应发生相应的变化。这部分内容在 updateSchedulingTopology()方法中实现。参考 Flink 原有逻辑，给调度器中的 schedulingTopology 字段重新赋值即可。还需要注意的是，Flink 在检查上下游依赖时，用到了 inputConstraintChecker 对象。该对象中的集合同样需要更新。可参考 LazyFromSourcesSchedulingStrategy 的 startScheduling()方法，该方法中包含 schedulingTopology 与 inputConstraintChecker 的关系，前文已详细介绍。

## 12.4 上下游关系的建立

在 Flink 原有逻辑中，下游并行实例是如何知道它应该去哪一个上游并行实例的对应子分区获取数据的呢？

在 SingleInputGate 类中有一个 int 类型的 consumedSubpartitionIndex 字段，它表示的就是上游并行实例中子分区的索引。在 InputChannel 中有一个 ResultPartitionID 类型的 partitionId 字段，它表示上游 ResultPartition 的唯一标识。这两个字段的值共同确定了 InputChannel 对象应该去哪一个上游并行实例的哪一个子分区获取数据。在 InputChannel 的 requestSubpartition()方法中可以看到对这两个变量的使用。InputChannel 要去哪个上游并行实例的 ResultPartition 获取数据是在构建 ExecutionGraph 的过程中就已经确定了的，因此关键在于要去哪些子分区获取数据。

追溯 consumedSubpartitionIndex 字段的值的由来可知，其是在作业管理器端构建 TaskDeploymentDescriptor 的过程中计算出来的。具体在 TaskDeploymentDescriptorFactory 的 createInputGateDeploymentDescriptors()方法中实现：

```
private List<InputGateDeploymentDescriptor> createInputGateDeploymentDescriptors() {
    List<InputGateDeploymentDescriptor> inputGates = new ArrayList<>(inputEdges.length);
    for (ExecutionEdge[] edges : inputEdges) {
        int numConsumerEdges = edges[0].getSource().getConsumers().get(0).size();
        // queueToRequest 的值最终会赋给 consumedSubpartitionIndex 字段
        int queueToRequest = subtaskIndex % numConsumerEdges;
        IntermediateResult consumedIntermediateResult = edges[0].getSource().
getIntermediateResult();
        IntermediateDataSetID resultId = consumedIntermediateResult.getId();
        ResultPartitionType partitionType = consumedIntermediateResult.getResultType();
        inputGates.add(new InputGateDeploymentDescriptor(
            resultId,
            partitionType,
            queueToRequest,
```

```
            getConsumedPartitionShuffleDescriptors(edges)));
    }
    return inputGates;
}
```

可以观察到，在原有逻辑中是通过取余的方式获取索引。在进行了优化后，显然不能再这样计算。这时可以利用先前计算出的 startReducerIndex 和 endReducerIndex 来确定下游并行实例要读取的子分区索引。原有逻辑是将 queueToRequest 传入 InputGateDeploymentDescriptor，现在需要将这个读取的范围传入描述符（可以实现一个新的 InputGate 描述符）。在任务管理器端对 InputGate 进行初始化时，需要对这种情况单独处理，以获取这个读取的范围，在与上游建立连接时依次与这些子分区建立连接即可。

## 12.5 总结

本章介绍了 Flink 自适应查询执行优化的原理，并提供了简单的代码示例。自适应查询执行优化是针对批处理作业而言的，它可以在执行过程中收集上游任务产生的统计信息，从而优化下游任务的执行方式，如改变并行度、修改上下游输入输出关系等。通过这种优化，可以大大改善数据倾斜带来的问题，并且在分区数据量较少时可以实现分区合并。

# 第 13 章

# Flink Sort-Merge Shuffle

在 Flink 的批处理作业中,上下游在数据交换时采用的是哈希混洗的方式。在哈希混洗过程中,上游任务的每个并行实例会根据下游任务的并行度,将数据写入相应的中间文件。下游任务的并行实例会找到上游任务的并行实例中对应的文件进行数据拉取。如果上游任务并行度为 $M$,下游任务并行度为 $R$,那么混洗过程产生的文件数就是 $M \times R$。在生产环境中,这是一个相当大的数。这么多的小文件不仅会占用大量的文件句柄数,而且在磁盘 I/O 上会消耗巨大性能。当数据量大到一定规模时,可能整个作业根本无法正常执行。

为了解决这个问题,Flink 源码工程的 Blink 分支对混洗过程进行了优化,添加了类似于 Spark 基于排序的混洗的数据交换机制,使该过程的稳定性得到了极大改善、效率得到了极大提高。本章主要对这部分内容进行介绍。

希望在学习完本章后,读者能够了解:
- Sort-Merge Shuffle 的基本原理;
- Flink 原有混洗过程与 Blink 分支混洗过程的区别;
- Sort-Merge Shuffle 的代码实现。

## 13.1 混洗机制的对比

要了解 Flink 的混洗机制如何改进,可以先回顾 Hadoop 和 Spark 的混洗机制以及 Flink 当前的哈希混洗过程。

Hadoop MapReduce 的 Map 任务产生输出数据时会先将数据写入内存中的缓冲区,等到缓冲区的数据达到某个阈值时,这些数据就会被溢写到磁盘。溢写磁盘的线程会根据数据最终要传输到的 Reduce 任务的并行实例,将数据写入相应的分区。每个分区会进行内排序。缓冲区一般会多次发生溢写,因此会产生多个文件。最后这些溢写文件会被合并成一个已分区且分区内有序的文件。合并文件的过程会发生归并排序。当 Map 端的数据写完后,Reduce 任务就开始拉取数据。每个 Reduce 实例都需要到各个 Map 实例的输出文件中拉取自己分区的数据。拉取数据的过程也会发生归并排序。当所有数据在 Reduce 端被排好序后就可以开始进行聚合处理。Hadoop 混洗如图 13-1 所示。

Spark 主要有两种混洗方式,一种是哈希混洗,另一种是与 Hadoop 的 MapReduce 过程类似的基于排序的混洗。哈希混洗没有在 Map 端对数据进行排序,而仅仅是将数据分区输出。下游根据分区的唯一标识拉取对应的数据。总体而言，哈希混洗在实现上较为简单，在运行时上游不需要进行任何排序和合并，因此速度非常快。哈希混洗的缺点在于中间文件较多，而且对需要排序的逻辑来说不太合适。基于排序的混洗可以有效解决中间文件过多的问题。Spark 的两种混洗方式如图 13-2 和图 13-3 所示。

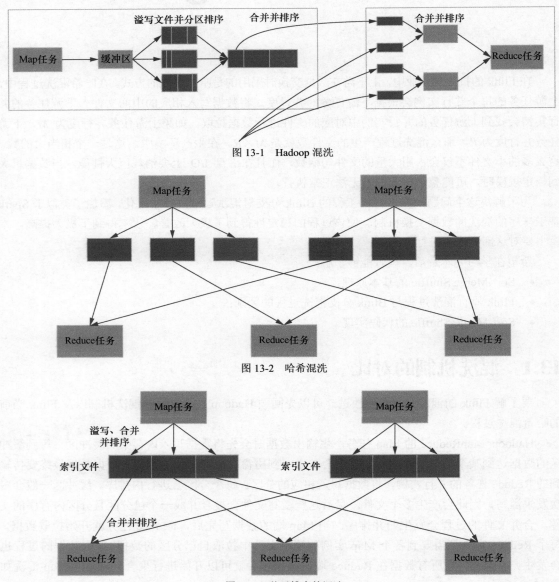

图 13-1　Hadoop 混洗

图 13-2　哈希混洗

图 13-3　基于排序的混洗

Flink 由于自身的定位，采用了更加通用、更加适合流处理的哈希混洗。不过，在需要对大规模数据进行排序的批处理任务中，哈希混洗很多时候不是最好的选择。Flink 工程中的 Blink 分支就对这个过程做了优化,实现了类似于 Hadoop MapReduce 混洗和 Spark 基于排序的混洗的数据交换过程,该过程在 Flink 中被称为 Sort-Merge Shuffle。

Sort-Merge Shuffle 会在上游输出数据时,先将数据写入缓冲区并按分区进行排序,然后会将数据溢写到文件中,同时对文件进行合并。最后会生成一个或多个大文件和对应的索引文件。

现在可以回过头来思考这样一个问题——实际上许多下游任务不必排序就可以正常地处理数据,为什么 MapReduce、Spark 等计算引擎的混洗过程都默认排序呢?如果仅仅是为了解决中间小文件过多的问题,可以在哈希混洗的基础上对文件进行合并,这样能节省排序所消耗的资源。要回答这个问题,就需要深入下游的实现层面。比如,当两个数据集进行连接操作时,如果没有对两个数据集进行排序,那么往往需要先将一个数据集读入内存,用哈希结构对其进行维护,再遍历另一个数据集,从该哈希结构中找到匹配的数据进行连接。如果是对两个有序的数据集进行连接操作,则不需要维护这样的数据结构,可以节省不少的内存空间。结合 Flink 等计算引擎的内存管理机制,这部分节省下来的内存就可以被应用到网络 I/O 的过程中,以进一步提高数据交换的效率。

这便是许多计算引擎从哈希混洗演进到 Sort-Merge Shuffle(基于排序的混洗)的另一深层原因。

## 13.2　Flink 混洗机制

Flink 进行批处理时,原有的混洗过程会将数据按分区写入对应的 ResultSubpartition。在本章的讨论范围内,ResultSubpartition 的实现类就是 BoundedBlockingSubpartition。输出数据时会调用 writeAndCloseBufferConsumer()方法:

```
private void writeAndCloseBufferConsumer(BufferConsumer bufferConsumer) throws IOException {
    try {
        final Buffer buffer = bufferConsumer.build();
        try {
            if (canBeCompressed(buffer)) {
                final Buffer compressedBuffer = parent.bufferCompressor.compressToIntermediate-
Buffer(buffer);
                data.writeBuffer(compressedBuffer);
                if (compressedBuffer != buffer) {
                    compressedBuffer.recycleBuffer();
                }
            } else {
                data.writeBuffer(buffer);
            }
            ...
        }
```

该方法会调用 writeBuffer()方法将数据写入 BoundedData 类型的 data 字段。BoundedData 有多种实现类,常用的是 FileChannelBoundedData 类,它会利用 FileChannel(java.nio.channels.FileChannel)将数据写入临时文件。

```
final class FileChannelBoundedData implements BoundedData {
    // 在读取数据时,会利用这个文件路径创建用于读取数据的 FileChannel
    private final Path filePath;
```

```
    // 利用 FileChannel 对象将数据进行写入
    private final FileChannel fileChannel;
```

每个子分区都对应一个 FileChannel 对象。它的路径从配置中获得，在 ResultSubpartition 对象的初始化过程中会由 FileChannelManager 创建扩展名为 .channel 的文件对象。如果上游并行度为 $M$，下游并行度为 $R$，那么这个混洗过程总共就会有 $M \times R$ 个 .channel 文件。下游在读取数据时，会利用文件路径 filePath 创建用于读取数据的 FileChannel 对象，定位到对应的文件进行数据读取。

## 13.3　Blink 混洗的数据流转

Flink 源码工程的 Blink 分支是在 Flink 1.5.1 上做的优化。1.5.1 版本在实现细节上与本书所介绍的 1.10 版本有一定差别，不过代码的整体结构、核心逻辑并未发生大的变动。

相比 Flink 原有的混洗机制，Blink 分支所做的优化主要体现在上游任务输出数据的阶段。在 Blink 分支中，上游任务输出数据时，会先将数据写入缓冲区，然后按照分区对其进行排序，并最终落盘。在落盘的过程中，还会按照一定的策略对文件进行合并，最终形成一个或多个数据文件以及对应的索引文件。

从数据流转角度来看，整个过程如图 13-4 所示。

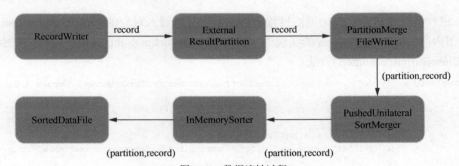

图 13-4　数据流转过程

在 Flink 原有逻辑中，RecordWriter 会将数据写入 ResultPartition，进而按照分区将其写入对应的 ResultSubpartition。在优化的 Blink 分支中，数据被写入了 ExternalResultPartition，进而到 PartitionMergeFileWriter 中进行排序、合并。PartitionMergeFileWriter 依赖 PushedUnilateralSortMerger 组件，数据写入该组件时，被封装成了二元组，第一位表示分区索引，第二位表示数据本身。PushedUnilateralSortMerger 将数据写入了 InMemorySorter，同时 PushedUnilateralSortMerger 中有多个线程，分别以 InMemorySorter 为操作对象，对数据进行排序、溢写的操作，最后数据会被写入 SortedDataFile。可以认为 SortedDataFile 对象就对应一个文件，在 PushedUnilateralSortMerger 的合并线程中这些文件会进行合并，最终形成合并后的文件和对应的索引文件。

ExternalResultPartition 等类之间的关系如图 13-5 所示。

下面对部分组件进行介绍。

## 13.3 Blink 混洗的数据流转

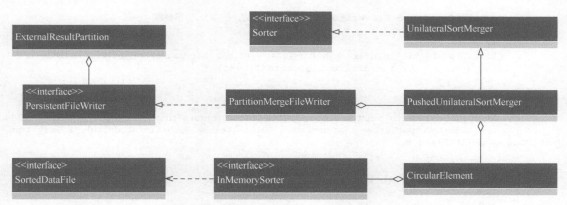

图 13-5　ExternalResultPartition 等类之间的关系

### 13.3.1　ExternalResultPartition

ExternalResultPartition 是 ResultPartitionWriter 的实现类，与 Flink 原有的 ResultPartition 是同一层级的概念。Blink 分支关于混洗过程的优化正是从这里开始的。ExternalResultPartition 之所以有 External 前缀，是因为 Blink 分支还对混洗过程设置了外部的管理器，由外部组件对混洗过程的资源（如文件）进行管理、维护，这部分内容在必要时会在下文进行相关介绍，但本章会将重点放在排序、合并的逻辑上。

在写入数据时，会调用 ExternalResultPartition 的 emitRecord() 方法，该方法的核心逻辑是对 ExternalResultPartition 进行初始化并将数据写入它依赖的 PartitionMergeFileWriter：

```
public void emitRecord(
    T record,
    int[] targetChannels,
    boolean isBroadcast,
    boolean flushAlways) throws IOException, InterruptedException {
  if (!initialized) {
    // 若没有进行初始化，则对其进行初始化
    initialize();
  }
  try {
    // 将数据写入 PersistentFileWriter
    fileWriter.add(record, targetChannels);
  } catch (Throwable e) {
    ...
  }
}
```

其中，初始化的逻辑主要是获取最终要写入的文件的根路径并实例化 fileWriter 字段。在 Sort-Merge Shuffle 中，fileWriter 的类型为 PartitionMergeFileWriter。

ExternalResultPartition 的另一个重要方法是 finish() 方法，它的调用时机与原有的 ResultPartition 的一致，是在任务生命周期的最后阶段调用。该方法的实现如下：

```
public void finish() throws IOException {
  try {
    ...
```

```
            // 调用 PersistentFileWriter 的 finish()方法
            fileWriter.finish();
            // 写索引文件
            List<List<PartitionIndex>> indicesList = fileWriter.generatePartitionIndices();
            for (int i = 0; i < indicesList.size(); ++i) {
               ...
            }
            // 写 finish 文件
            String finishedPath = ExternalBlockShuffleUtils.generateFinishedPath(partitionRootPath);
            try (FSDataOutputStream finishedOut = fs.create(new Path(finishedPath), FileSystem.WriteMode.OVERWRITE)) {
               ...
            }
        } catch (Throwable e) {
        ...
        } finally {
        ...
        }
        isFinished = true;
    }
```

从上面的过程可以观察到，该方法主要是在文件写完后进行一些整理和收尾的工作，以及生成索引文件和 finish 文件。

## 13.3.2　PartitionMergeFileWriter

PartitionMergeFileWriter 持有一个 PushedUnilateralSortMerger 类型的 sortMerger 字段。PartitionMergeFileWriter 的主要功能是将消息封装成(partition, record)这样的二元组并将其传入 sortMerger 进行处理。

下面是它的一些重要字段。

- **partitionDataRootPath**：输出文件的根路径。
- **typeSerializer**：TypeSerializer 类型，消息的序列化器。
- **allMemory**：<MemorySegment>类型，可使用的内存段。
- **reuse**：Tuple2 类型，其中 T 是泛型，表示消息的数据类型。第一位为 Integer 类型，表示分区号。
- **sortMerger**：PushedUnilateralSortMerger 类型，消息写入其中后会进行排序、溢写等操作。

它的构造方法主要就是用于给字段赋值。其中 partitionDataRootPath、typeSerializer 和 allMemory 的值由外部传入，可以直接获得，剩余的过程都是在给 sortMerger 的构造方法构造参数：

```
public PartitionMergeFileWriter(
    int numPartitions,
    String partitionDataRootPath,
    int mergeFactor,
    boolean enableAsyncMerging,
    boolean mergeToOneFile,
    MemoryManager memoryManager,
    List<MemorySegment> memory,
    IOManager ioManager,
    TypeSerializer<T> serializer,
    SerializerManager<SerializationDelegate<T>> serializerManager,
    AbstractInvokable parentTask,
```

```
            Counter numBytesOut,
            Counter numBuffersOut) throws IOException, MemoryAllocationException {
        this.partitionDataRootPath = partitionDataRootPath;
        this.typeSerializer = serializer;
        this.allMemory = memory;
        // 构造 sortMerger 所需的参数
        Class<Tuple2<Integer, T>> typedTuple = (Class<Tuple2<Integer, T>>) (Class<?>) Tuple2.class;
        TypeSerializer<?>[] serializers = new TypeSerializer[]{IntSerializer.INSTANCE, serializer.duplicate()};
        TypeSerializer<Tuple2<Integer, T>> tuple2Serializer = new TupleSerializer<>(typedTuple, serializers);
        DuplicateOnlySerializerFactory<Tuple2<Integer, T>> serializerFactory = new DuplicateOnlySerializerFactory<>(tuple2Serializer);
        int[] keyPositions = new int[]{0};
        TypeComparator<?>[] comparators = new TypeComparator<?>[]{new IntComparator(true)};
        TupleComparator<Tuple2<Integer, T>> tuple2Comparator = new TupleComparator<>(
            keyPositions, comparators, serializers);
        BufferSortedDataFileFactory<T> sortedDataFileFactory = new BufferSortedDataFileFactory<>(
            partitionDataRootPath, typeSerializer, ioManager, serializerManager, numBytesOut, numBuffersOut);
        PartitionedBufferSortedDataFileFactory<T> partitionedBufferSortedDataFileFactory =
            new PartitionedBufferSortedDataFileFactory<T>(sortedDataFileFactory, numPartitions);
        SortedDataFileMerger<Tuple2<Integer, T>> merger = new ConcatPartitionedFileMerger<T>(
            numPartitions, partitionDataRootPath, mergeFactor, enableAsyncMerging, mergeToOneFile, ioManager);
        sortMerger = new PushedUnilateralSortMerger<>(partitionedBufferSortedDataFileFactory, merger,
            memoryManager, allMemory, ioManager, parentTask, serializerFactory, tuple2Comparator,
            0, mergeFactor, false, 0,
            false, true, true, enableAsyncMerging);
    }
```

注意，传入 sortMerger 的序列化器、比较器等都将 Tuple2 类型的数据视为操作对象。

PartitionMergeFileWriter 的 add()方法如下：

```
public void add(T record, int targetPartition) throws IOException {
    reuse.f1 = record;
    reuse.f0 = targetPartition;
    sortMerger.add(reuse);
}
```

在这里，消息被封装成了（partition, record）二元组。

根据前文介绍的 ExternalResultPartition 的相关内容可知，它的 finish()方法会调用 fileWriter 的 finish()方法和 generatePartitionIndices()方法，这两个方法的实现如下：

```
public void finish() throws IOException, InterruptedException {
    // 调用 sortMerger 的相关方法进行任务结束后的处理
    sortMerger.finishAdding();
    // 通过 sortMerger 获取最后产生的溢写文件
    List<SortedDataFile<Tuple2<Integer, T>>> remainFiles = sortMerger.getRemainingSortedDataFiles();
    int nextFileId = 0;
    FileSystem localFileSystem = FileSystem.getLocalFileSystem();
    for (SortedDataFile<Tuple2<Integer, T>> file : remainFiles) {
        // 重命名文件
        localFileSystem.rename(
            new Path(file.getChannelID().getPath()),
```

```
            new Path(ExternalBlockShuffleUtils.generateDataPath(partitionDataRootPath,
nextFileId++)));
        }
    }
    public List<List<PartitionIndex>> generatePartitionIndices() throws IOException,
InterruptedException {
        List<List<PartitionIndex>> partitionIndices = new ArrayList<>();
        List<SortedDataFile<Tuple2<Integer, T>>> remainFiles = sortMerger.getRemaining
SortedDataFiles();
        for (SortedDataFile<Tuple2<Integer, T>> file : remainFiles) {
            if (!(file instanceof PartitionedSortedDataFile)) {
                throw new IllegalStateException("Unexpected file type.");
            }
            partitionIndices.add(((PartitionedSortedDataFile<T>) file).getPartitionIndexList());
        }
        return partitionIndices;
    }
```

finish()方法通过 sortMerger 对最后的溢写文件进行一些收尾处理，generatePartitionIndices()方法会通过 sortMerger 获取索引信息。由此可以了解到，核心的逻辑都在 sortMerger 中实现，它不仅能实现对消息进行排序、溢写，还能实现对溢写文件进行合并，生成索引信息。最后在任务收尾阶段，是通过它获取文件信息和索引信息对文件进行重命名并且生成索引文件和 finish 文件。

## 13.4 Blink 混洗的 Sort-Merge 过程

消息的排序、溢写以及文件的合并都在 PushedUnilateralSortMerger 中实现。PushedUnilateralSortMerger 除了主线程，还有排序线程（SortingThread）、溢写线程（SpillingThread）和合并线程（MergingThread），排序、溢写、合并的操作同步进行，由此提高了效率，线程如图 13-6 所示。

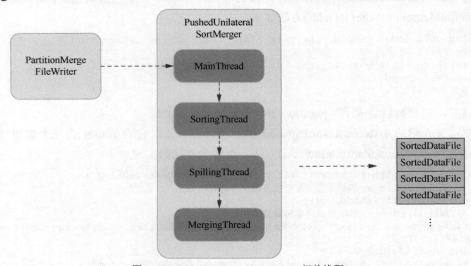

图 13-6　PushedUnilateralSortMerger 相关线程

在主线程中，消息被写入了 InMemorySorter，排序线程会以 InMemorySorter 为操作对象对消息进行排序。InMemorySorter 被封装在 CircularElement 中，这些线程通过几个队列共享

CircularElement 对象。

## 13.4.1 PushedUnilateralSortMerger

PushedUnilateralSortMerger 继承自 UnilateralSortMerger 类。UnilateralSortMerger 是 Flink 原有逻辑中就有的类。Blink 分支对它的主要改变是增加了合并线程。

在 UnilateralSortMerger 类中，有这样几个线程：

```
// 读取线程
private final ThreadBase<E> readThread;
// 排序线程
private final ThreadBase<E> sortThread;
// 溢写线程
private final ThreadBase<E> spillThread;
// 合并线程
private final ThreadBase<E> mergingThread;
```

它们的逻辑定义在内部类 ReadingThread、SortingThread、SpillingThread 和 MergingThread 中。PushedUnilateralSortMerger 重写了获取 ReadingThread 的方法，使之返回 null，因此 PushedUnilateralSortMerger 中没有读取的逻辑（消息直接从主线程写入了排序线程）。

在合并阶段，合并的对象为 SortedDataFile。由此，Blink 分支中多了 SortedDataFileFactory、SortedDataFileMerger 等组件，用于文件的创建、合并等。在 PartitionMergeFileWriter 初始化 sortMerger 时能看到构建出了这些对象并将它们作为参数传入了 PushedUnilateralSortMerger 的构造方法。

UnilateralSortMerger 有一个重要的 circularQueues 字段，它封装了几个队列，被各个线程持有。消息就是在这些队列中被放入和取出的，由此被多个线程共享。同时，几个线程还会根据条件向队列中放入各种信号，通知另外的线程进行特定的操作，大致如图 13-7 所示。

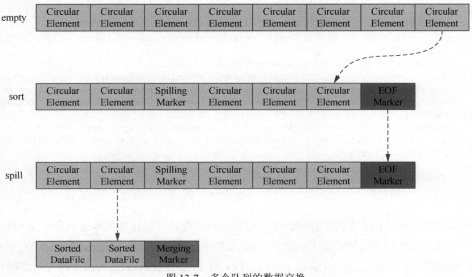

图 13-7　多个队列的数据交换

**UnilateralSortMerger** 的构造方法如下：

```
protected UnilateralSortMerger(SortedDataFileFactory<E> sortedDataFileFactory,
SortedDataFileMerger<E> merger,
        MemoryManager memoryManager, List<MemorySegment> memory, IOManager ioManager,
        MutableObjectIterator<E> input, AbstractInvokable parentTask,
        TypeSerializerFactory<E> serializerFactory, TypeComparator<E> comparator,
        int numSortBuffers, int maxNumFileHandles, boolean inMemoryResultEnabled,
        float startSpillingFraction, boolean noSpillingMemory, boolean handleLargeRecords,
        boolean objectReuseEnabled, boolean enableAsyncMerging)
    throws IOException {
  ...
    for (int i = 0; i < numSortBuffers; i++)
    {
      ...
      // 构造 InMemorySorter 对象
      if (comp.supportsSerializationWithKeyNormalization() &&
          serializer.getLength() > 0 && serializer.getLength() <= THRESHOLD_FOR_IN_
PLACE_SORTING)
      {
        buffer = new FixedLengthRecordSorter<E>(serializerFactory.getSerializer(),
comp, sortSegments);
      } else {
        buffer = new NormalizedKeySorter<E>(serializerFactory.getSerializer(), comp,
sortSegments);
      }
      // 利用 buffer 构造 CircularElement 对象，将其添加进 circularQueues 的 empty 队列
      CircularElement<E> element = new CircularElement<E>(i, buffer, sortSegments);
      circularQueues.empty.add(element);
    }
    ...
    // 构造溢写文件的队列，由溢写线程和合并线程共享
    BlockingQueue<SortedDataFileElement<E>> spilledFiles = new LinkedBlockingQueue<>();

    ...
    // 构造线程
    this.readThread = getReadingThread(exceptionHandler, input, circularQueues,
largeRecordHandler,
        parentTask, serializer, startSpillingBytes);
    this.sortThread = getSortingThread(exceptionHandler, circularQueues, parentTask);
    this.spillThread = getSpillingThread(sortedDataFileFactory, spilledFiles, merger,
        exceptionHandler, circularQueues, parentTask, memoryManager, ioManager,
serializerFactory,
        comparator, this.sortReadMemory, this.writeMemoryForSpilling, maxNumFileHandles);
    this.mergingThread = getMergingThread(merger, memoryManager, exceptionHandler,
parentTask, spilledFiles);
    ...
    // 启动线程
    startThreads();
}
```

上面的代码省略了内存分配的逻辑，剩下部分的核心逻辑是根据资源构造出多个 InMemorySorter 实例，即 buffer 对象，进而构造出 CircularElement 对象并将其添加到 circularQueues 的 empty 队列。可以简单地理解为 CircularElement 就是一个内存段，用于承载消息。最后构造线程并启动线程。注意，这个过程中还构造了溢写文件的队列 spilledFiles，它被溢写线程和合并线程共享。circularQueues 对象

被读取线程（读取线程此时对应的值为 null，表示不启动）、排序线程、溢写线程共享。由此就了解了消息如何被多个线程共享，并最终按分区顺序被写入文件中。

其中 SortedDataFileElement 就是对 SortedDataFile 的简单封装：

```java
protected static final class SortedDataFileElement<E> {
    final SortedDataFile<E> sortedDataFile;
    public SortedDataFileElement() {
        sortedDataFile = null;
    }
    public SortedDataFileElement(SortedDataFile<E> sortedDataFile) {
        this.sortedDataFile = sortedDataFile;
    }
}
```

CircularElement 中主要封装了 InMemorySorter 对象，即其中的 buffer 字段。

```java
protected static final class CircularElement<E> {
    final int id;
    final InMemorySorter<E> buffer;
    final List<MemorySegment> memory;
    public CircularElement() {
        this.id = -1;
        this.buffer = null;
        this.memory = null;
    }
    public CircularElement(int id, InMemorySorter<E> buffer, List<MemorySegment> memory) {
        this.id = id;
        this.buffer = buffer;
        this.memory = memory;
    }
}
```

CircularQueues 则封装了 3 个队列，用于存放 CircularElement 元素：

```java
protected static final class CircularQueues<E> {
    final BlockingQueue<CircularElement<E>> empty;
    final BlockingQueue<CircularElement<E>> sort;
    final BlockingQueue<CircularElement<E>> spill;
    public CircularQueues() {
        this.empty = new LinkedBlockingQueue<CircularElement<E>>();
        this.sort = new LinkedBlockingQueue<CircularElement<E>>();
        this.spill = new LinkedBlockingQueue<CircularElement<E>>();
    }
    public CircularQueues(int numElements) {
        this.empty = new ArrayBlockingQueue<CircularElement<E>>(numElements);
        this.sort = new ArrayBlockingQueue<CircularElement<E>>(numElements);
        this.spill = new ArrayBlockingQueue<CircularElement<E>>(numElements);
    }
}
```

PushedUnilateralSortMerger 在 UnilateralSortMerger 的基础上主要增加了一个 CircularElement 类型的 currentBuffer 字段，并且提供了 add()方法：

```java
public synchronized void add(E current) throws IOException {
    try {
        ...
```

```
      while (true) {
        ...
        final InMemorySorter<E> buffer = currentBuffer.buffer;
        // 将消息写入 InMemorySorter
        if (!buffer.write(current)) {
          if (buffer.isEmpty()) {
            ...
          } else {
            ...
            // 将 currentBuffer 添加到 sort 队列中
            circularQueues.sort.add(currentBuffer);
            currentBuffer = null;
          }
        } else {
          ...
        }
      }
    } catch (Throwable e) {
      ...
    }
}
```

currentBuffer 从 empty 队列中获得。核心逻辑是将消息写入 currentBuffer 持有的 buffer, 然后在一定条件下将 currentBuffer 转移到 sort 队列中。

下面将分析 InMemorySorter 如何维护写入的消息, 以及排序线程如何对消息进行排序。

### 13.4.2　NormalizedKeySorter

本节以实现类 NormalizedKeySorter 为例分析消息如何被维护。

NormalizedKeySorter 维护消息的基本思路是将消息与消息存放的位置 offset 分开维护, offset 相当于消息的指针, 只要知道了 offset, 就知道了消息的位置。在排序时, 只需要移动 offset 的位置, 而不需要移动消息本身。

在 NormalizedKeySorter 中, 消息被存放在 recordBufferSegments 字段中, offset 被存放在 sortIndex 字段中。它们的数据类型都是 ArrayList。MemorySegment 可以被理解为一个内存段。

另外, 以下 3 个字段需要注意:

recordCollector 表示输出消息时的视图, recordBuffer 和 recordBufferForComparison 表示读取消息时的视图。它们实际上都指向 recordBufferSegments。之所以读取消息时需要两个视图, 是因为需进行比较从而实现排序。

```
private final SimpleCollectingOutputView recordCollector;
private final RandomAccessInputView recordBuffer;
private final RandomAccessInputView recordBufferForComparison;
```

下面的代码实现了这几个字段的初始化:

```
this.sortIndex = new ArrayList<MemorySegment>(16);
this.recordBufferSegments = new ArrayList<MemorySegment>(16);
this.recordCollector = new SimpleCollectingOutputView(this.recordBufferSegments,
    new ListMemorySegmentSource(this.freeMemory), this.segmentSize);
this.recordBuffer = new RandomAccessInputView(this.recordBufferSegments, this.segmentSize);
```

```
this.recordBufferForComparison = new RandomAccessInputView(this.recordBufferSegments,
this.segmentSize);
```

NormalizedKeySorter 的 write()方法如下：

```java
public boolean write(T record) throws IOException {
    if (this.currentSortIndexOffset > this.lastIndexEntryOffset) {
        if (memoryAvailable()) {
            // 获取下一个可用的内存段用于写 offset
            this.currentSortIndexSegment = nextMemorySegment();
            // 将内存段放入 sortIndex 字段
            this.sortIndex.add(this.currentSortIndexSegment);
            this.currentSortIndexOffset = 0;
            this.sortIndexBytes += this.segmentSize;
        } else {
            return false;
        }
    }
    // 将消息序列化到 recordBufferSegments 中的内存段中
    try {
        this.serializer.serialize(record, this.recordCollector);
    }
    catch (EOFException e) {
        return false;
    }
    // 获取 recordBufferSegments 中当前的 offset
    final long newOffset = this.recordCollector.getCurrentOffset();
    final boolean shortRecord = newOffset - this.currentDataBufferOffset < LARGE_RECORD_THRESHOLD;
    // 将 offset 写入 currentSortIndexSegment
    this.currentSortIndexSegment.putLong(this.currentSortIndexOffset, shortRecord ?
        this.currentDataBufferOffset : (this.currentDataBufferOffset | LARGE_RECORD_TAG));
    if (this.numKeyBytes != 0) {
        // 随即将 key 写入 currentSortIndexSegment
        this.comparator.putNormalizedKey(record, this.currentSortIndexSegment, this.currentSortIndexOffset + OFFSET_LEN, this.numKeyBytes);
    }
    // 更新状态值
    this.currentSortIndexOffset += this.indexEntrySize;
    this.currentDataBufferOffset = newOffset;
    this.numRecords++;
    return true;
}
```

此时传入的消息为（partition, record）二元组，因此序列化器、比较器都是对应 Tuple2 的实例化对象。整个过程会将消息写入 recordBufferSegments 的内存段，消息写入后 recordCollector 会维护当前的 offset。接着，会获取该 offset，然后将 offset 的值连同消息中 partition 的值写入 currentSortIndexSegment 内存段，如图 13-8 所示。

NormalizedKeySorter 中定义了排序的逻辑：

```java
public int compare(int segmentNumberI, int segmentOffsetI, int segmentNumberJ, int segmentOffsetJ) {
    // 根据 MemorySegment 的编号从 sortIndex 中获取对应的内存段
    final MemorySegment segI = this.sortIndex.get(segmentNumberI);
    final MemorySegment segJ = this.sortIndex.get(segmentNumberJ);
```

```
        // 对比内存段中存放的 partition 的值
        int val = segI.compare(segJ, segmentOffsetI + OFFSET_LEN, segmentOffsetJ + OFFSET_LEN,
this.numKeyBytes);
        if (val != 0 || this.normalizedKeyFullyDetermines) {
            return this.useNormKeyUninverted ? val : -val;
        }
        // 如果 partition 的值相同,再获取 offset 的值
        final long pointerI = segI.getLong(segmentOffsetI) & POINTER_MASK;
        final long pointerJ = segJ.getLong(segmentOffsetJ) & POINTER_MASK;
        // 根据 offset 的值寻找具体的数据进行比较
        return compareRecords(pointerI, pointerJ);
    }
    private int compareRecords(long pointer1, long pointer2) {
        this.recordBuffer.setReadPosition(pointer1);
        this.recordBufferForComparison.setReadPosition(pointer2);
        try {
            return this.comparator.compareSerialized(this.recordBuffer, this.recordBuffer-
ForComparison);
        } catch (IOException ioex) {
            throw new RuntimeException("Error comparing two records.", ioex);
        }
    }
```

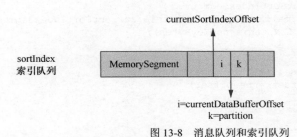

图 13-8　消息队列和索引队列

根据比较结果,调换索引数据的位置:

```
    public void swap(int segmentNumberI, int segmentOffsetI, int segmentNumberJ, int
segmentOffsetJ) {
        final MemorySegment segI = this.sortIndex.get(segmentNumberI);
        final MemorySegment segJ = this.sortIndex.get(segmentNumberJ);
        segI.swapBytes(this.swapBuffer, segJ, segmentOffsetI, segmentOffsetJ, this.
indexEntrySize);
    }
```

从上面的方法可以看到,调整的是从 sortIndex 中取出的内存段。

### 13.4.3 排序线程

PushedUnilateralSortMerger 中不启动读取线程，消息直接从主线程写入了 circularQueues 的 sort 队列中，当排序线程启动后，会从该队列中获取数据。

排序线程的核心方法如下：

```
public void go() throws IOException {
   boolean alive = true;
   while (isRunning() && alive) {
      CircularElement<E> element = null;
      try {
         // 从 sort 队列中获取 CircularElement 对象
         element = this.queues.sort.take();
      }
      catch (InterruptedException iex) {
        ...
      }
      if (element != EOF_MARKER && element != SPILLING_MARKER) {
         ...
         // 对消息进行排序
         this.sorter.sort(element.buffer);
      }
      ...
      // 将 CircularElement 对象添加到 spill 队列中
      this.queues.spill.add(element);
   }
}
```

其核心逻辑是从 sort 队列中获取 CircularElement，并对消息进行排序，再将其添加到 spill 队列中。这里用的排序算法是快速排序，调用的方法则是上面介绍过的 compare() 方法和 swap() 方法等。

### 13.4.4 溢写线程

消息被排序后，被添加到了 spill 队列中。执行溢写线程后，会从该队列中获取元素，对数据进行溢写。

之前介绍过的 sortedDataFileFactory、merger、spilledFiles 等对象被传入了溢写线程，用于给它的对应字段赋值。

溢写线程的核心逻辑如下：

```
public void go() throws IOException {
   final Queue<CircularElement<E>> cache = new ArrayDeque<CircularElement<E>>();
   CircularElement<E> element;
   while (isRunning()) {
      try {
         // 从 spill 队列中获取 CircularElement 对象
         element = this.queues.spill.take();
      }
      ...
   }
   ...
   while (isRunning()) {
```

```
        ...
        // 通过工厂对象创建 SortedDataFile 对象
        SortedDataFile<E> output = sortedDataFileFactory.createFile(writeMemory);
        ...
        // 将数据溢写到文件中
        element.buffer.writeToOutput(output, largeRecordHandler);
        // 完成溢写
        output.finishWriting();
        if (output.getBytesWritten() > 0) {
            // 将 SortedDataFile 封装并将其添加到 spilledFiles 队列中
            spilledFiles.add(new SortedDataFileElement<>(output));
            ++numSpilledFiles;
        }
        // 清空后放回到 empty 队列中
        element.buffer.reset();
        this.queues.empty.add(element);
    }
    ...
    spilledFiles.add(mergingMarker());
}
```

上面的代码省略了内存中的合并操作部分以及对所占空间较大的消息的处理部分。核心逻辑是从 spill 队列中获取 CircularElement 对象，然后构建 SortedDataFile 对象，将数据写入其中。在写入的过程中会产生一些索引信息，会通过 finishWriting() 方法对这些信息进行整理。最后，将 SortedDataFile 对象封装并将其添加到 spilledFiles 队列中，等待合并线程将其取出进行合并。

注意，在将消息写入文件时，调用的仍是 InMemorySorter 的 writeToOutput() 方法。NormalizedKeySorter 的实现如下：

```
public void writeToOutput(SortedDataFile<T> output, LargeRecordHandler<T> largeRecordsOutput)
    throws IOException {
    final int numRecords = this.numRecords;
    int currentMemSeg = 0;
    int currentRecord = 0;
    while (currentRecord < numRecords) {
        // 获取保存 offset 信息的对应内存段
        final MemorySegment currentIndexSegment = this.sortIndex.get(currentMemSeg++);
        for (int offset = 0; currentRecord < numRecords && offset <= this.lastIndexEntryOffset;
currentRecord++, offset += this.indexEntrySize) {
            // 获取具体数据的位置
            final long pointer = currentIndexSegment.getLong(offset);
            if (pointer >= 0 || largeRecordsOutput == null) {
                // 设置到指定的位置
                this.recordBuffer.setReadPosition(pointer);
                // 将数据从内存中写入文件
                output.copyRecord(this.recordBuffer);
            }
            else {
                ...
            }
        }
    }
}
```

整个过程仍是先找到 offset，然后找到对应的数据，将其写入文件。因为之前已经排过序，所以这里数据会按分区被有序写入文件。

copyRecord()方法的实现如下：

```
public void copyRecord(DataInputView serializedRecord) throws IOException {
    // 获取 partition 的值
    int partitionIndex = serializedRecord.readInt();
    if (partitionIndex != currentPartition) {
        backendFile.flush();
        currentPartition = partitionIndex;
    }
    // 更新分区信息
    partitionIndexGenerator.updatePartitionIndexBeforeWriting(
        partitionIndex, backendFile.getBytesWritten(), numRecordWritten);
    // 将数据写入文件
    backendFile.copyRecord(serializedRecord);
    numRecordWritten++;
}
```

因为消息是二元组形式，所以可以先获取 partition 信息。该信息会被维护在 partitionIndexGenerator 中，然后将数据写入文件。

finishWriting()方法的实现如下：

```
public void finishWriting() throws IOException {
    backendFile.finishWriting();
    partitionIndexGenerator.finishWriting(backendFile.getBytesWritten(), numRecordWritten);
}
```

该方法主要用于关闭输出，并整理索引信息。

## 13.4.5　合并线程

合并线程的主要逻辑是从 spilledFiles 队列中取出 SortedDataFileElement 对象，利用 SortedDataFileMerger 类型的 merger 对象对先前生成的 SortedDataFile 进行合并。

核心方法如下：

```
protected void go() throws IOException {
    List<SortedDataFile<E>> sortedDataFiles = new ArrayList<>();
    while (isRunning()) {
        try {
            // 获取 SortedDataFileElement 对象
            SortedDataFileElement<E> sortedDataFile = spilledFiles.poll(200, TimeUnit.MILLISECONDS);
            ...
            if (sortedDataFile == MERGING_MARKER) {
                // 完成最后的合并
                sortedDataFiles = merger.finishMerging(
                    writeMemory, mergeReadMemory, channelDeleteRegistry, getRunningFlag());
                break;
            } else if (sortedDataFile != null) {
                // 继续进行合并操作
                merger.notifyNewSortedDataFile(
```

```
                    sortedDataFile.sortedDataFile, writeMemory, mergeReadMemory, channel-
DeleteRegistry, getRunningFlag());
            }
        } catch (InterruptedException e) {
            ...
        }
    }
    ...
        // 利用最后的sortedDataFiles列表，构造迭代器对象并给remainingSortedDataFiles和iterator字
        // 段赋值
        MutableObjectIterator<E> finalResultIterator =
            merger.getMergingIterator(sortedDataFiles, mergeReadMemory, largeRecords,
channelDeleteRegistry);
        setResult(sortedDataFiles, finalResultIterator);
    }
}
```

核心逻辑是将取出的对象通过 merger 的 notifyNewSortedDataFile()方法进行合并，最后通过 finishMerging()方法完成最后的合并。finishMerging()方法的返回值 sortedDataFiles 表示的就是最终的文件列表。这些文件列表会构造出一个迭代器对象 finalResultIterator，通过 setResult()方法给 UnilateralSortMerger 中的字段赋值。

```
protected final void setResult(List<SortedDataFile<E>> mergedDataFiles, MutableObject-
Iterator<E> iterator) {
    synchronized (this.iteratorLock) {
        if (this.unhandledException == null) {
            this.remainingSortedDataFiles = mergedDataFiles;
            this.iterator = iterator;
            this.iteratorLock.notifyAll();
        }
    }
}
```

因此，最后是由 remainingSortedDataFiles 字段表示最终合并后的文件列表。

在 PushedUnilateralSortMerger 中，merger 的实现类是 ConcatPartitionedFileMerger。它的 notifyNewSortedDataFile()方法如下：

```
public void notifyNewSortedDataFile(SortedDataFile<Tuple2<Integer, T>> sortedDataFile,
                    List<MemorySegment> writeMemory,
                    List<MemorySegment> mergeReadMemory,
                    ChannelDeleteRegistry<Tuple2<Integer, T>> channelDeleteRegistry,
                    AtomicBoolean aliveFlag) throws IOException {
    DataFileInfo<SortedDataFile<Tuple2<Integer, T>>> dataFileInfo = new DataFileInfo<>(
        sortedDataFile.getBytesWritten(), 0, numberOfSubpartitions, sortedDataFile);
    mergePolicy.addNewCandidate(dataFileInfo);
    mergeIfPossible(mergeReadMemory, channelDeleteRegistry, aliveFlag);
}
```

这里主要要注意，SortedDataFile 被封装成了 DataFileInfo 对象，而该对象有一个 mergeRound 字段。这里在构造方法中给该字段赋值为 0。在合并策略中，文件是分层级合并的。这个值表示的就是文件的层级。合并策略是先将文件依次放入第 0 层，当满足某个条件时将这些文件合并，放入下一层，以此类推（如图 13-9 所示）。

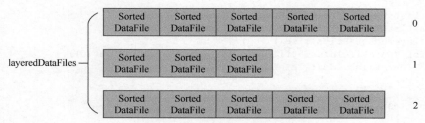

图 13-9 分层级的文件

mergePolicy 的 addNewCandidate()方法如下：

```
public void addNewCandidate(DataFileInfo<T> dataFileInfo) {
    if (isFinalMergeStarted) {
        layeredDataFiles.get(0).addLast(dataFileInfo);
    } else {
        int mergeRound = dataFileInfo.getMergeRound();
        if (layeredDataFiles.size() == mergeRound) {
            LinkedList<DataFileInfo<T>> dataFiles = new LinkedList<>();
            layeredDataFiles.add(dataFiles);
        }
        layeredDataFiles.get(mergeRound).addLast(dataFileInfo);
    }
}
```

这个 layeredDataFiles 就表示一个层级结构，其数据类型为 LinkedList<DataFileInfo<T>>。当接收到完成合并的信号时，会调用 merger 的 finishMerging()方法：

```
public List<SortedDataFile<Tuple2<Integer, T>>> finishMerging(List<MemorySegment> writeMemory,
                                        List<MemorySegment> mergeReadMemory,
                                        ChannelDeleteRegistry<Tuple2<Integer,T>> channelDeleteRegistry,
                                        AtomicBoolean aliveFlag) throws IOException {
    mergePolicy.startFinalMerge();
    mergeIfPossible(mergeReadMemory, channelDeleteRegistry, aliveFlag);
    return mergePolicy.getFinalMergeResult();
}
```

在这个方法中，会将所有剩下的文件全部放入第 0 层进行最后的合并。mergePolicy 的 startFinalMerge()方法如下：

```
public void startFinalMerge() {
    isFinalMergeStarted = true;
    for (int i = 1; i < layeredDataFiles.size(); ++i) {
        layeredDataFiles.get(0).addAll(layeredDataFiles.get(i));
    }
    if (layeredDataFiles.size() > 1) {
        layeredDataFiles = layeredDataFiles.subList(0, 1);
    }
}
```

最后调用 mergePolicy 的 getFinalMergeResult()方法从 DataFileInfo 中取出 SortedDataFile 对象并形成列表：

```
public List<T> getFinalMergeResult() {
   ArrayList<T> dataFiles = new ArrayList<>();
   if (layeredDataFiles.size() > 0) {
      for (DataFileInfo<T> fileInfo : layeredDataFiles.get(0)) {
         dataFiles.add(fileInfo.getDataFile());
      }
   }
   return dataFiles;
}
```

## 13.5 文件的读取和元信息管理

由于上游任务输出数据时的逻辑发生了变化，没有了之前的子分区，因此下游读取数据时的逻辑需要进行相应的调整。要了解读取数据时的逻辑应当如何调整，首先要知道输出文件的现状。

在合并文件结束后，生成一个或多个数据文件，同时还包括内存中的索引信息。在任务生命周期的最后阶段，会调用 ExternalResultPartition 的 finish()方法，进而运行下面这段代码：

```
// 完成合并工作并对最终生成的数据文件重命名
fileWriter.finish();
// 生成索引文件
List<List<PartitionIndex>> indicesList = fileWriter.generatePartitionIndices();
for (int i = 0; i < indicesList.size(); ++i) {
    String indexPath = ExternalBlockShuffleUtils.generateIndexPath(partitionRootPath, i);
    try (FSDataOutputStream indexOut = fs.create(new Path(indexPath), FileSystem.WriteMode.OVERWRITE)) {
        DataOutputView indexView = new DataOutputViewStreamWrapper(indexOut);
        ExternalBlockShuffleUtils.serializeIndices(indicesList.get(i), indexView);
    }
}
// 生成 finish 文件
String finishedPath = ExternalBlockShuffleUtils.generateFinishedPath(partitionRootPath);
try (FSDataOutputStream finishedOut = fs.create(new Path(finishedPath), FileSystem.WriteMode.OVERWRITE)) {
    DataOutputView finishedView = new DataOutputViewStreamWrapper(finishedOut);
    finishedView.writeInt(ExternalBlockResultPartitionMeta.SUPPORTED_PROTOCOL_VERSION);
    String externalFileType = fileWriter.getExternalFileType().name();
    finishedView.writeInt(externalFileType.length());
    finishedView.write(externalFileType.getBytes());
    finishedView.writeInt(indicesList.size());
    finishedView.writeInt(numberOfSubpartitions);
}
```

数据文件有对应的索引文件和 finish 文件。通过读取 finish 文件和索引文件，就能读取相应的数据。

### 13.5.1 ExternalBlockResultPartitionManager

以 RemoteInputChannel 请求数据的过程（LocalInputChannel 请求数据的过程没有太大区别）为例，下游会向上游发送 PartitionRequest，然后上游需要创建读取相应数据的视图。这个过程没有区别。区别在于创建视图的对象由 ResultPartitionManager 变成了 ExternalResultPartitionManager。

在原有逻辑中，ResultPartitionManager 会维护分区和子分区的信息，由于现在没有了子分区，显然利用 ResultPartitionManager 来创建视图不太合适。

ExternalResultPartitionManager 中有一个 resultPartitionMetaMap 字段，用于维护 ResultPartition 的元信息。它的数据类型为 ConcurrentHashMap。

创建视图的方法 createSubpartitionView() 的实现如下：

```java
public ResultSubpartitionView createSubpartitionView(
    ResultPartitionID resultPartitionId,
    int index,
    BufferAvailabilityListener availabilityListener) throws IOException {
    // 从 map 中获取文件描述信息
    ExternalBlockResultPartitionMeta resultPartitionMeta = resultPartitionMetaMap.get
(resultPartitionId);
    if (resultPartitionMeta == null) {
        // 从指定路径获取文件信息（主要是文件的根路径）
        LocalResultPartitionResolver.ResultPartitionFileInfo fileInfo = resultPartition-
Resolver.getResultPartitionDir(
            resultPartitionId);
        // 将文件的描述信息封装成 ExternalBlockResultPartitionMeta 对象
        resultPartitionMeta = new ExternalBlockResultPartitionMeta(
            resultPartitionId,
            shuffleServiceConfiguration.getFileSystem(),
            fileInfo);
        ExternalBlockResultPartitionMeta prevResultPartitionMeta =
            resultPartitionMetaMap.putIfAbsent(resultPartitionId, resultPartitionMeta);
        if (prevResultPartitionMeta != null) {
            resultPartitionMeta = prevResultPartitionMeta;
        }
    }
    // 通过 ExternalBlockResultPartitionMeta 对象创建视图
    ExternalBlockSubpartitionView subpartitionView = new ExternalBlockSubpartitionView(
        resultPartitionMeta,
        index,
        dirToThreadPool.get(resultPartitionMeta.getRootDir()),
        resultPartitionId,
        bufferPool,
        shuffleServiceConfiguration.getWaitCreditDelay(),
        availabilityListener);
    resultPartitionMeta.notifySubpartitionStartConsuming(index);
    return subpartitionView;
}
```

因为数据文件、索引文件、finish 文件的根路径是通过配置而来的，是固定的路径，所以这里可以直接获取该路径。这些文件的描述信息被封装成了 ExternalBlockResultPartitionMeta 对象，并被放入了 resultPartitionMetaMap 进行维护。下一次再创建视图时，就可以直接从 map 中获取该对象。

最后还可以观察到，不仅创建视图的过程发生了变化，视图对象也变成了新实现的类 ExternalBlockSubpartitionView。后文会对该视图进行分析。

### 13.5.2　ExternalBlockResultPartitionMeta

从上面将文件的描述信息封装成 ExternalBlockResultPartitionMeta 的过程可以观察到，其构造方法中传入了表示文件信息的 fileInfo 对象，其中包含文件的根路径等信息。

ExternalBlockResultPartitionMeta 中的重要字段是 subpartitionMetas，表示每个子分区的元信息（子分区的数据在某个文件中的起止位置），其数据类型为<ExternalSubpartitionMeta>数组。这个字段的初始化在 ExternalBlockResultPartitionMeta 的 initialize()方法中实现，整个过程大致是根据 fileInfo 中的文件路径依次读取 finish 文件和索引文件，将数据文件的元信息读取到 ExternalSubpartitionMeta 对象中。

ExternalSubpartitionMeta 类的字段如下：

```
static final class ExternalSubpartitionMeta {
   private final Path dataFile;
   private final long offset;
   private final long length;
```

initialize()方法的实现如下：

```
synchronized void initialize() throws IOException {
   ...
   // 读取 finish 文件进行初始化
   initializeByFinishFile();
   // 读取索引文件进行初始化
   initializeByIndexFile();
   ...
}
```

在 initializeByFinishFile()方法中，会读取 finish 文件以获取一些元信息，主要包括子分区的个数、溢写文件的个数等，并会将元信息赋值到 ExternalBlockResultPartitionMeta 的 subpartitionNum 字段和 spillCount 字段中。

在 initializeByIndexFile()方法中，会遍历索引文件，依次读取其中的索引信息，并会将其添加到 subpartitionMetas 字段中。之后在读取数据文件时，就可以根据其中的信息，找到具体的数据文件中的偏移量和读取范围进行读取。

### 13.5.3　ExternalBlockSubpartitionView

通过前面的分析可以了解，在创建视图对象时，将 ExternalBlockResultPartitionMeta 对象作为构造方法的参数传入，而 ExternalBlockResultPartitionMeta 中包含每个子分区的索引信息，因此视图对象知道去哪里读取相应的数据。

ExternalBlockSubpartitionView 实现了 Runnable 接口，它在创建以后就会被提交给线程池开始运行。它会根据索引信息创建读取器，不断将数据读取到缓冲区。在之后进入 Flink 原有逻辑调用视图对象的 getNextBuffer()方法时，它会从缓冲区读取 buffer。

ExternalBlockSubpartitionView 的核心方法如下：

```
public void run() {
   ...
   try {
      if (metaIterator == null) {
         // 初始化元信息
         initializeMeta();
      }
      ...
```

```
        while (true) {
            while (isAvailableForReadUnsafe()) {
                // 读取下一个buffer,并将其放入缓冲区
                Buffer buffer = readNextBuffer();
                enqueueBuffer(buffer);
            }
            ...
        }
    } catch (Throwable t) {
        ...
    } finally {
        ...
    }
}
```

上述代码省略了资源管理和流量控制的部分,只保留了读取数据的核心逻辑。

这里的 initializeMeta() 方法会调用 externalResultPartitionMeta 字段的初始化方法,将索引信息读取到内存进行维护。

```
private void initializeMeta() throws IOException {
    if (!externalResultPartitionMeta.hasInitialized()) {
        externalResultPartitionMeta.initialize();
    }
    List<ExternalBlockResultPartitionMeta.ExternalSubpartitionMeta> subpartitionMetas =
externalResultPartitionMeta.getSubpartitionMeta(subpartitionIndex);
    metaIterator = subpartitionMetas.iterator();
    for (ExternalBlockResultPartitionMeta.ExternalSubpartitionMeta meta : subpartitionMetas) {
        totalLength += meta.getLength();
    }
}
```

这里读取了所有属于该子分区的索引信息,初始化了迭代器,并且累加了总的数据量。

在 readNextBuffer() 方法中,会根据迭代器获取下一个读取器(对应一个数据文件),然后将数据读取到 buffer 中。

```
private Buffer readNextBuffer() throws IOException, InterruptedException {
    if (currFsIn == null) {
        // 获取下一个读取器
        currFsIn = getNextFileReader();
    }
    ...
    try {
        long lengthToRead = Math.min(currRemainLength, buffer.getMaxCapacity());
        // 将数据读取到buffer中
        currFsIn.readInto(buffer, lengthToRead);
        currRemainLength -= lengthToRead;
    }
    ...
    return buffer;
}
private SynchronousBufferFileReader getNextFileReader() throws IOException {
    SynchronousBufferFileReader nextFsIn = null;
    ExternalBlockResultPartitionMeta.ExternalSubpartitionMeta nextMeta;
    while (metaIterator.hasNext()) {
```

```
            nextMeta = metaIterator.next();
            currRemainLength = nextMeta.getLength();
            if (currRemainLength > 0) {
                FileIOChannel.ID fileChannelID = new FileIOChannel.ID(nextMeta.getDataFile().
getPath());
                nextFsIn = new SynchronousBufferFileReader(fileChannelID, false, false);
                // 根据偏移量直接移动到属于该子分区数据的位置
                nextFsIn.seekToPosition(nextMeta.getOffset());
                break;
            }
        }
    }
    return nextFsIn;
}
```

而在 enqueueBuffer()方法中,buffer 会被添加到 buffers 队列中:

```
private void enqueueBuffer(Buffer buffer) throws IOException {
    synchronized (lock) {
        ...
        buffers.add(buffer);
        ...
    }
}
```

需要读取数据时,会调用 Flink 原有逻辑的 getNextBuffer()方法,该方法的实现如下:

```
public ResultSubpartition.BufferAndBacklog getNextBuffer() {
    synchronized (lock) {
        Buffer buffer = buffers.poll();
        Buffer nextBuffer = buffers.peek();
        if (buffer != null) {
            return new ResultSubpartition.BufferAndBacklog(buffer, nextBuffer != null,
buffers.size(),
                    nextBuffer != null && !nextBuffer.isBuffer());
        } else {
            return null;
        }
    }
}
```

该方法会从 buffers 队列中获取 buffer 并返回。

## 13.6 总结

本章深入探讨了分布式计算引擎的混洗过程,回顾了 Hadoop MapReduce 和 Spark 的混洗机制,分析了为什么 Flink 需要为批处理引入 Sort-Merge Shuffle。Blink 分支对该过程已经实现了优化,本章对它的实现细节进行了详细的介绍。

# 第 14 章

# 修改检查点的状态

在含状态的流处理作业中，一般会启用检查点机制周期性地生成检查点（或手动生成保存点），一旦作业因为业务原因或技术原因必须停止，那么可以通过这些检查点重新启动，以保证之前的状态不丢失。前文已经详细分析过 Flink 如何保证数据的一致性。虽然在 Flink 的版本迭代过程中，检查点机制已经越来越完善，且效率越来越高，但是状态数据最终是以序列化后的方式存储在文件中的，这就使得状态数据对于外界（用户）十分不透明。

对于一个"7×24 小时"不断执行的含状态的流处理作业，了解这些状态数据的值是十分必要的。举例来说，从业务角度来看，流处理作业在执行过程中可能会因为一些脏数据或者之前对业务理解有偏差的问题，导致状态数据的值并不是当前的期望值，从而影响整个业务。如果能够读取状态数据文件然后将状态数据可视化，并能够对其进行分析、修改，那么可以从一个任选的检查点对状态进行修改并让作业从该处重启，从而避免解决业务问题后从最初的时间点重新开始执行，能大大节省时间成本和硬件资源。

理论上来说，既然 Flink 自身可以读取检查点文件并对其进行反序列化，那么用户应该可以引用其中用到的模块对这些文件进行读取、反序列化，甚至对其进行修改。Flink 1.9 引入的 State Processor API 就是基于这样的原理的。State Processor API 提供了一套给用户使用的接口，用于读取检查点中的状态数据。利用这套 API，不仅可以读取状态数据，还能对其进行修改并生成新的检查点文件用于作业重启（或用于启动其他应用并对其状态进行初始化）。此外还可以根据需求进行其他与状态相关的修改，如将状态后端从 Heap State Backend 改成 RocksDB State Backend 等。

本章会以上面提到的业务需求为出发点对检查点状态的读取、修改进行深入探讨。

希望在学习完本章后，读者能够了解：

- 检查点文件中状态数据及相关元信息修改的原理；
- State Processor API 的原理和应用；
- 状态数据可视化和修改的架构实现；
- 修改最大并行度的原理和方式。

## 14.1 状态修改的原理

想要修改状态，就需要了解状态数据和相应的状态元信息的序列化和反序列化的方式。这部分内容在前文已详细介绍。这里重点回顾 Flink 读取状态元信息并将状态数据读取到内存的过程。

## 14.1.1 状态元信息的读取

下面以让 Flink 作业从一个检查点恢复为例进行介绍。

在 Flink 构造执行图时，会进行状态重分配，从该过程可以观察到状态元信息的读取。

```
public boolean restoreSavepoint(
    String savepointPointer,
    boolean allowNonRestored,
    Map<JobVertexID, ExecutionJobVertex> tasks,
    ClassLoader userClassLoader) throws Exception {
    // 根据检查点的地址构造 checkpointLocation 对象
    final CompletedCheckpointStorageLocation checkpointLocation = checkpointStorage.resolveCheckpoint(savepointPointer);
    // 根据 checkpointLocation 对象解析状态元信息并将其读取到内存中，封装成 savepoint 对象
    CompletedCheckpoint savepoint = Checkpoints.loadAndValidateCheckpoint(
        job, tasks, checkpointLocation, userClassLoader, allowNonRestored);
    ...
}
```

如果想要编写程序从外部读取检查点的数据，首先要做的就是上述代码所示的操作——根据检查点的地址将状态元信息读取到内存中进行封装。

这里在获得 checkpointLocation 对象时，调用的是 checkpointStorage 对象的 resolveCheckpoint() 方法。观察 resolveCheckpoint() 方法的实现，可发现它最终调用的是 AbstractFsCheckpointStorage 的公有静态方法 resolveCheckpointPointer()：

```
public CompletedCheckpointStorageLocation resolveCheckpoint(String checkpointPointer) throws IOException {
    return resolveCheckpointPointer(checkpointPointer);
}

public static CompletedCheckpointStorageLocation resolveCheckpointPointer(String checkpointPointer) throws IOException {
    ...
}
```

当获得 checkpointLocation 对象后，再调用一个公有静态方法构造 savepoint 对象。该方法的实现如下：

```
public static CompletedCheckpoint loadAndValidateCheckpoint(
    JobID jobId,
    Map<JobVertexID, ExecutionJobVertex> tasks,
    CompletedCheckpointStorageLocation location,
    ClassLoader classLoader,
    boolean allowNonRestoredState) throws IOException {
    final StreamStateHandle metadataHandle = location.getMetadataHandle();
    final String checkpointPointer = location.getExternalPointer();
    // 获取 Savepoint 对象
    final Savepoint rawCheckpointMetadata;
    try (InputStream in = metadataHandle.openInputStream()) {
        DataInputStream dis = new DataInputStream(in);
        rawCheckpointMetadata = loadCheckpointMetadata(dis, classLoader);
    }
    final Savepoint checkpointMetadata = rawCheckpointMetadata.getTaskStates() == null ?
        rawCheckpointMetadata :
        SavepointV2.convertToOperatorStateSavepointV2(tasks, rawCheckpointMetadata);
```

```
        ...
        // 填充 operatorStates,将 OperatorState 与算子的唯一标识进行映射
        HashMap<OperatorID, OperatorState> operatorStates = new HashMap<>(checkpointMetadata.
getOperatorStates().size());
        for (OperatorState operatorState : checkpointMetadata.getOperatorStates()) {
            ...
        }
        return new CompletedCheckpoint(
                jobId,
                checkpointMetadata.getCheckpointId(),
                0L,
                0L,
                operatorStates,
                checkpointMetadata.getMasterStates(),
                props,
                location);
    }
```

上面代码的关键步骤是根据 location 对象中的信息将状态元信息读取到内存中构造出 Savepoint 对象。保存点中的 operatorStates 字段包含所有的状态元信息。因此,想要在外部程序中实现读取状态元信息的读取器,按照下面的方式实现即可,无须再构造 CompletedCheckpoint 对象:

```
public class StateReader {
    public static Collection<OperatorState> readMetadata(String savepointPath) throws
IOException {
        CompletedCheckpointStorageLocation location
                = AbstractFsCheckpointStorage.resolveCheckpointPointer(savepointPath);
        Savepoint savepoint;
        try (DataInputStream stream = new DataInputStream(location.getMetadataHandle().
openInputStream())) {
            savepoint = Checkpoints.loadCheckpointMetadata(stream, Thread.currentThread().
getContextClassLoader());
        }
        return savepoint.getOperatorStates();
    }
}
```

接下来,可以直接利用返回值(OperatorState 对象的集合)找到具体的状态数据文件进行读取。

## 14.1.2 状态数据的读取

在 Flink 的机制中,状态元信息在作业管理器端读取,状态数据在任务管理器端读取,读取过程利用了许多任务管理器内部运行时的对象和方法。在外部程序中实现状态元信息读取器的方法利用了 Flink 提供的公有方法,而没有手动实现反序列化的过程。那么读取状态数据可直接利用一些 Flink 提供的现成方法。下面以操作符(注意这里指的不是 OperatorState 类,而是与键控状态对应的操作符状态)和 FsStateBackend 状态后端为例,分析从外部程序读取状态数据的可行性。

在 Flink 原有逻辑中,状态数据读取的关键环节在 StreamTaskStateInitializerImpl 的 streamOperatorStateContext()方法中。状态元信息在该方法中被读取,在该方法中还通过如下逻辑构造了 OperatorStateBackend 对象:

```
operatorStateBackend = operatorStateBackend(
    operatorIdentifierText,
```

```
    prioritizedOperatorSubtaskStates,
    streamTaskCloseableRegistry);
```

operatorStateBackend()方法则是通过状态后端读取状态元信息来构造该对象的：

```
protected OperatorStateBackend operatorStateBackend(
    String operatorIdentifierText,
    PrioritizedOperatorSubtaskState prioritizedOperatorSubtaskStates,
    CloseableRegistry backendCloseableRegistry) throws Exception {
    ...
    BackendRestorerProcedure<OperatorStateBackend, OperatorStateHandle> backendRestorer =
        new BackendRestorerProcedure<>(
            // 生成 OperatorStateBackend 对象
            (stateHandles) -> stateBackend.createOperatorStateBackend(
                environment,
                operatorIdentifierText,
                stateHandles,
                cancelStreamRegistryForRestore),
            backendCloseableRegistry,
            logDescription);
    try {
        return backendRestorer.createAndRestore(
            prioritizedOperatorSubtaskStates.getPrioritizedManagedOperatorState());
    } finally {
        ...
    }
}
```

注意，createOperatorStateBackend()方法的第一个参数为 environment，其类型为 Environment（org.apache.Flink.runtime.execution.Environment）。这是 Flink 的任务在运行时的环境对象。由此可以想到一种策略，就是将先前读取到的状态元信息作为 Flink 作业的输入，将其分布到各个算子上，在各个并行实例中去获取 environment 等 Flink 运行时的对象，将 Flink 运行时的对象作为方法的参数去构造 operatorStateBackend。

比如，可以构建图 14-1 所示的工作流。

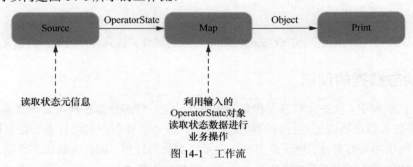

图 14-1　工作流

因为最终的目的并不是将 OperatorState 进行映射并输出到某处，而只是用一个并行实例去接收该对象，所以 Map 算子中的核心逻辑是模仿 Flink 本身获取状态数据的方式去获取这些数据。获取这些数据后，可以将其传回给作业管理器端统一处理，也可以在 Map 算子中将它们输出到外部的文件系统或数据库。至于要将什么数据输出到 Print 算子，在这里并不重要，Print 算子的作用仅仅是添加一个 Sink 使执行图能够构建成功，换作另外的 Sink 亦可。

OperatorState 类中有一个 operatorSubtaskStates 字段，其类型为 Map，其中的 OperatorSubtaskState

封装了每个并行实例的所有状态元信息。针对这里需要读取的操作符状态，可以遍历每一个 OperatorSubtaskState，获取其中的 managedOperatorState 字段，该字段的类型为 StateObjectCollection。得到 managedOperatorState 对象后，就可以利用 StateBackend 的 createOperatorStateBackend()方法获得 OperatorStateBackend 对象。

这部分逻辑在 map()方法中可以这样实现：

```
Collection<OperatorSubtaskState> states = ...        // 获取 OperatorSubtaskState 集合
for (OperatorSubtaskState subtaskState : states) {   // 遍历每个并行实例中的状态元信息
        // 获取一个并行实例包含的 managedOperatorState
        StateObjectCollection<OperatorStateHandle> managedOperatorState = subtaskState.getManagedOperatorState();
        // 根据检查点文件的状态后端定义对应的状态后端实例
        FsStateBackend stateBackend = new FsStateBackend(...);
        // 从状态数据文件中恢复数据到 OperatorStateBackend 对象
        OperatorStateBackend operatorStateBackend
                = stateBackend.createOperatorStateBackend(...);
    ...
```

OperatorStateBackend 的唯一实现类为 DefaultOperatorStateBackend，它包含一个 Map<String, PartitionableListState<?>>类型的 registeredOperatorStates 字段。这是维护状态数据的地方。其中的 key 为每个状态的名称，value 中包含具体的状态的值。继续观察 PartitionableListState 类的字段，它包含一个 RegisteredOperatorStateBackendMetaInfo 类型的字段 stateMetaInfo 和一个 ArrayList 类型的字段 internalList。这个 internalList 就包含具体的状态数据。RegisteredOperatorStateBackendMetaInfo 类维护了一些状态的元信息，其中 assignmentMode 字段的类型为 OperatorStateHandle.Mode，表示操作符状态的重分配策略，partitionStateSerializerProvider 字段的类型为 StateSerializerProvider，包含序列化器的信息。

上述一些字段无法通过公有方法获取，可以通过反射的方式获取。于是，在这个过程中就得到了状态名称、状态重分配策略、序列化信息、具体的状态的值等信息。将这些信息输出到外部的文件系统或数据库，就可以对其进行可视化、分析、修改等操作。

DefaultOperatorStateBackend 还包含一个 snapshot()方法。当对状态相关的信息进行修改后，可以将修改后的内容通过 snapshot()方法生成快照，参考 Flink 原有快照逻辑，就可以生成新的状态数据文件。这些状态数据文件的描述信息会回传给作业管理器端（如利用 Accumulator），利用同样的方式即可生成元信息文件。

## 14.2　状态处理器 API

14.1 节从细节上回顾了状态元信息和状态数据的读取过程，并简单论证了编写外部程序读取这些信息的可行性。事实上，为了满足前面说到的业务需求，Flink 从 1.9 版本开始就引入了 State Processor API，专门提供了接口让用户能够读取与状态相关的信息，并能够进行灵活的修改。这套 API 主要可以用于读取状态数据的值并对其进行分析、修改，并能够生成新的检查点文件用于启动其他应用。在修改状态的过程中，不仅可以修改状态的值，还能修改状态在序列化/反序列化过程中的序列化器（修改状态数据在 Flink 类型系统中对应的数据类型）、最大并行度等。

图 14-2 所示为一个应用的执行图。

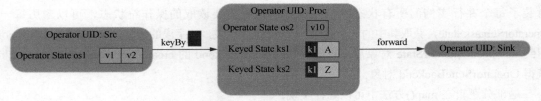

图 14-2　应用的执行图

在 source 任务中，定义了一些操作符状态。经过重分区后，在中间的算子中定义了操作符状态和键控状态。基于对这些状态的基本理解，应用所包含的状态（检查点中的数据）可以被抽象成数据库，如图 14-3 所示。

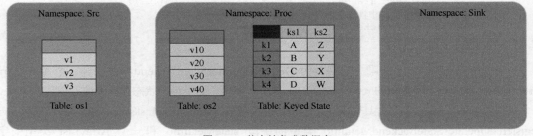

图 14-3　状态抽象成数据库

由此，基于 State Processor API 的功能，可以实现图 14-4 所示的架构，将状态数据进行可视化和修改。

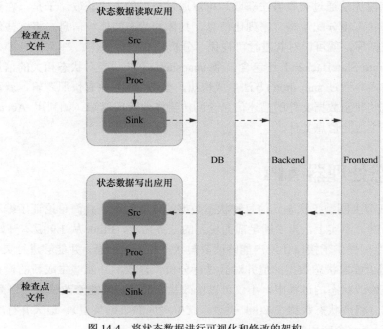

图 14-4　将状态数据进行可视化和修改的架构

图 14-4 状态数据读取应用和状态数据写出应用都是利用 State Processor API 实现的。状态数据读取应用以一个现成的检查点文件为输入，最终可以将其输出到数据库，并通过后端代码的读取将其展示到前端。用户可以在前端对数据进行业务上的修改，并通过后端代码将其保存到数据库。状态数据写出应用则通过读取数据库中修改过的状态信息，利用 State Processor API 生成新的检查点文件。

本节将对 State Processor API 的读取和写出接口进行简单的介绍。

## 14.2.1 数据的读取

State Processor API 读取状态数据的思路与 14.1 节给出的相关思路非常相似。它首先会读取状态数据的元信息，然后根据元信息中的文件地址将状态数据读取到内存中，接着分布式地处理这些状态数据。整个任务属于一个 Flink 批处理作业。

在利用 State Processor API 进行开发时，会先读取状态的元信息：

```
final String checkpointPath = ...;
final ExecutionEnvironment env = ExecutionEnvironment.getExecutionEnvironment();
StateBackend stateBackend = ...;
ExistingSavepoint load = Savepoint.load(...);
```

这里的 Savepoint 类就是 State Processor API 中定义的类。它的 load() 方法的实现如下：

```
public static ExistingSavepoint load(ExecutionEnvironment env, String path, StateBackend stateBackend) throws IOException {
    // 读取状态的元信息
    org.apache.Flink.runtime.checkpoint.savepoint.Savepoint savepoint = SavepointLoader.loadSavepoint(path);
    // 获取最大并行度
    int maxParallelism = savepoint
        .getOperatorStates()
        .stream()
        .map(OperatorState::getMaxParallelism)
        .max(Comparator.naturalOrder())
        .orElseThrow(() -> new RuntimeException("Savepoint must contain at least one operator state."));
    SavepointMetadata metadata = new SavepointMetadata(maxParallelism, savepoint.getMasterStates(), savepoint.getOperatorStates());
    return new ExistingSavepoint(env, metadata, stateBackend);
}
```

其中读取状态元信息的 SavepointLoader 的 loadSavepoint() 方法与 14.1 节实现的读取器的思路完全一致：

```
public static Savepoint loadSavepoint(String savepointPath) throws IOException {
    CompletedCheckpointStorageLocation location = AbstractFsCheckpointStorage
        .resolveCheckpointPointer(savepointPath);
    try (DataInputStream stream = new DataInputStream(location.getMetadataHandle().openInputStream())) {
        return Checkpoints.loadCheckpointMetadata(stream, Thread.currentThread().getContextClassLoader());
    }
}
```

在 State Processor API 中，将这些信息读取到内存后会经过层层封装，然后会变成 Existing Savepoint 对象。接着就可以对该对象添加逻辑计算，读取真正的状态数据到算子中进行定制化操作。

State Processor API 为操作符状态和键控状态分别提供了不同的接口用于进行状态数据的读取和写出。

对于操作符状态，可以调用 ExistingSavepoint 中的以下方法实现状态数据的读取：

```
// 读取重分配策略为 SPLIT_DISTRIBUTE 的状态
public <T> DataSet<T> readListState(String uid, String name, TypeInformation<T> typeInfo)
throws IOException {
    // 根据 uid 从元信息中找到对应的算子
    OperatorState operatorState = metadata.getOperatorState(uid);
    // 根据名称和类型构造描述符
    ListStateDescriptor<T> descriptor = new ListStateDescriptor<>(name, typeInfo);
    // 定义输入逻辑
    ListStateInputFormat<T> inputFormat = new ListStateInputFormat<>(operatorState, descriptor);
    return env.createInput(inputFormat, typeInfo);
}
// 读取重分配策略为 UNION 的状态
public <T> DataSet<T> readUnionState(String uid, String name, TypeInformation<T> typeInfo) throws IOException {
    OperatorState operatorState = metadata.getOperatorState(uid);
    ListStateDescriptor<T> descriptor = new ListStateDescriptor<>(name, typeInfo);
    UnionStateInputFormat<T> inputFormat = new UnionStateInputFormat<>(operatorState, descriptor);
    return env.createInput(inputFormat, typeInfo);
}
// 读取重分配策略为 BROADCAST 的状态
public <K, V> DataSet<Tuple2<K, V>> readBroadcastState(
    String uid,
    String name,
    TypeInformation<K> keyTypeInfo,
    TypeInformation<V> valueTypeInfo) throws IOException {
    OperatorState operatorState = metadata.getOperatorState(uid);
    MapStateDescriptor<K, V> descriptor = new MapStateDescriptor<>(name, keyTypeInfo, valueTypeInfo);
    BroadcastStateInputFormat<K, V> inputFormat = new BroadcastStateInputFormat<>(operatorState, descriptor);
    return env.createInput(inputFormat, new TupleTypeInfo<>(keyTypeInfo, valueTypeInfo));
}
```

根据上面几个方法的方法签名可以观察到，用户需要知道想要读取的状态的名称、重分配策略、算子的 uid、状态的序列化方式（状态在 Flink 类型系统中对应的类型）等，返回的数据类型为 DataSet。

以 readListState()方法为例，它构造了 ListStateInputFormat 来定义输入的逻辑，传入的 operatorState 对象为特定算子中的状态，descriptor 为其中特定状态的描述符。

ListStateInputFormat 的构造方法如下：

```
public ListStateInputFormat(OperatorState operatorState, ListStateDescriptor<OT> descriptor) {
    super(operatorState, false);
```

```
        this.descriptor = Preconditions.checkNotNull(descriptor, "The state descriptor must
not be null");
    }
```

operatorState 对象被赋值给了父类中的字段。

ListStateInputFormat 中定义了 getElements()方法,该方法的参数为 OperatorStateBackend 对象:

```
    protected final Iterable<OT> getElements(OperatorStateBackend restoredBackend) throws
Exception {
        return restoredBackend.getListState(descriptor).get();
    }
```

基于之前的了解可知,这里是可以直接返回状态数据的。需要进入 ListStateInputFormat 的父类观察何时调用 getElements()方法。

ListStateInputFormat 的父类为抽象类 OperatorStateInputFormat,其包含 OperatorState 类型的 operatorState 字段和 OperatorStateBackend 类型的 restoredBackend 字段。在其 open()方法中,实现了状态数据的读取:

```
    public void open(OperatorStateInputSplit split) throws IOException {
        registry = new CloseableRegistry();
        final BackendRestorerProcedure<OperatorStateBackend, OperatorStateHandle> backendRestorer =
            new BackendRestorerProcedure<>(
                (handles) -> createOperatorStateBackend(getRuntimeContext(), handles, registry),
                registry,
                operatorState.getOperatorID().toString()
            );
        try {
            restoredBackend = backendRestorer.createAndRestore(split.getPrioritizedManagedOperatorState());
        } catch (Exception exception) {
            throw new IOException("Failed to restore state backend", exception);
        }
        try {
            elements = getElements(restoredBackend).iterator();
        } catch (Exception e) {
            throw new IOException("Failed to read operator state from restored state backend", e);
        }
    }
```

可以观察到,restoredBackend 对象的构造与 14.1 节回顾的方式一致。有了 restoredBackend 对象,就可以调用 getElements()方法,根据描述符将状态数据取出形成迭代器。在这里,迭代器对象被赋值给了 elements 字段,随后就可以直接从该字段依次取出数据进行处理并将数据输出到下一个算子了。

对于键控状态,是类似的步骤,只不过键控状态需要用户自定义输出的数据类型。readKeyedState()方法如下:

```
    public <K, OUT> DataSet<OUT> readKeyedState(String uid, KeyedStateReaderFunction<K,
OUT> function) throws IOException {
        TypeInformation<K> keyTypeInfo;
        TypeInformation<OUT> outType;
        try {
            keyTypeInfo = TypeExtractor.createTypeInfo(
                KeyedStateReaderFunction.class,
```

```
            function.getClass(), 0, null, null);
    } catch (InvalidTypesException e) {
        ...
    }
    try {
        outType = TypeExtractor.getUnaryOperatorReturnType(
            function,
            KeyedStateReaderFunction.class,
            0,
            1,
            TypeExtractor.NO_INDEX,
            keyTypeInfo,
            Utils.getCallLocationName(),
            false);
    } catch (InvalidTypesException e) {
        ...
    }
    return readKeyedState(uid, function, keyTypeInfo, outType);
}
```

与操作符状态的状态数据的读取不同，这里的输出类型 outType 要从参数 function 中获得，其类型为 KeyedStateReaderFunction。其中的泛型 OUT 就是输出类型，需要用户自定义。

比如，在原算子中定义这样一个状态：

```
private transient MapState<String, Integer> mapState;
```

那么用户可以定义一个封装类：

```
static class KeyedMapState {
    String key;
    Map<String, Integer> value;
}
```

其中的 key 表示分区的字段，value 表示状态的具体值，其类型 Map 对应 M 状态的数据类型 MapState。

接着，用户可实现 KeyedStateReaderFunction 类，定义数据读取和输出的逻辑：

```
static class ReaderFunction extends KeyedStateReaderFunction<String, KeyedMapState> {
    // 定义与原算子相同的状态用于接收文件中的数据
    private transient MapState<String, Integer> mapState;
    @Override
    public void open(Configuration parameters) {
        MapStateDescriptor<String, Integer> mapStateDescriptor =
            new MapStateDescriptor<>(...);
        mapState = getRuntimeContext().getMapState(mapStateDescriptor);
    }
    @Override
    public void readKey(String key, Context ctx, Collector<KeyedMapState> out) throws Exception {
        // 将从文件中读取出来的数据放入 map 并发送到下一个算子
        Map<String, Integer> map = new HashMap<>();
        Iterator<Map.Entry<String, Integer>> iterator = mapState.iterator();
        while (iterator.hasNext()) {
            Map.Entry<String, Integer> next = iterator.next();
            map.put(next.getKey(), next.getValue());
        }
```

```java
            KeyedMapState keyedMapState = new KeyedMapState();
            keyedMapState.key = key;
            keyedMapState.value = map;
            out.collect(keyedMapState);
        }
    }
```

虽然整个逻辑看起来好像仅实现了操作符状态的状态数据读取过程中 Flink 已经实现好的逻辑，但实际上，原算子中可能有多个状态，这些状态都可以通过自定义的 KeyedStateReaderFunction 进行读取、封装和传输。这就是自定义 KeyedStateReaderFunction 和自定义输出的数据类型的意义。

### 14.2.2 数据的写出

State Processor API 提供了现成的输出状态数据到新的检查点文件的方法。方法大致如下：

```java
int maxParallelism = ...;
Savepoint
    .create(new RocksDBStateBackend(), maxParallelism) // 指定新的状态后端和最大并行度
    .withOperator("uid1", transformation1) // 指定要输出的算子的 uid 和对应的状态数据
    .withOperator("uid2", transformation2)
    .write(savepointPath); // 指定输出地址
```

从这里可以观察到，在输出状态数据时可以指定新的状态后端、最大并行度、要输出的算子的 uid 等。上面的 transformation 对象中定义了生成状态数据的逻辑，于是这里只需要关注 transformation 对象是如何产生的。

这个过程仍然需要将操作符状态和键控状态分开讨论。

对于操作符状态，是调用下面的接口：

```java
BootstrapTransformation transformation = OperatorTransformation
    .bootstrapWith(data)
    .transform(new SimpleBootstrapFunction());
```

这里的 data 对象可以是任意一个 DataSet 对象，可以是自定义的数据，也可以是之前利用 State Processor API 从状态数据文件中读取的数据。SimpleBootstrapFunction 是用户自定义的类，其继承了 StateBootstrapFunction 类，它是用户自定义生成状态数据的地方：

```java
public class SimpleBootstrapFunction extends StateBootstrapFunction<Integer> {
    private ListState<Integer> state;
    @Override
    public void processElement(Integer value, Context ctx) throws Exception {
        state.add(value);
    }
    @Override
    public void snapshotState(FunctionSnapshotContext context) throws Exception {
    }
    @Override
    public void initializeState(FunctionInitializationContext context) throws Exception {
        state = context.getOperatorState().getListState(new ListStateDescriptor<>("state", Types.INT));
    }
}
```

在 processElement() 方法中，传进的数据被放入了自定义的状态。这个状态的名称、类型可以

是自定义的。

对于键控状态，则需要指定分区器：

```
BootstrapTransformation transformation = OperatorTransformation
    .bootstrapWith(data)
    .keyBy(acc -> acc.id) // 指定分区器
    .transform(new KeyedBootstrapper());
```

这里的 KeyedBootstrapper 类是自定义的类，继承自 KeyedStateBootstrapFunction 类。它的实现与 StateBootstrapFunction 的实现思路一致，如：

```
public class KeyedBootstrapper extends KeyedStateBootstrapFunction<Integer, Account> {
    ValueState<Double> state;
    @Override
    public void open(Configuration parameters) {
        ValueStateDescriptor<Double> descriptor = new ValueStateDescriptor<>("total", Types.DOUBLE);
        state = getRuntimeContext().getState(descriptor);
    }
    @Override
    public void processElement(Account value, Context ctx) throws Exception {
        state.update(value.amount);
    }
}
```

在 StateBootstrapFunction 和 KeyedStateBootstrapFunction 的实现类中自定义了状态并接收了数据，当整个 State Processor API 定义的作业执行完时，这些数据会被输出到指定的路径形成新的检查点文件。

前文专门提及过最大并行度是无法修改的。然而，在 State Processor API 中，这些状态数据被读取到了 Flink 任务中，并基于指定的分区器重新进行了分区，因此在保存检查点时，这些状态数据就已包含了以新的最大并行度为基础的分区信息。这样，这些状态数据文件就可以用来启动最大并行度与之相等的应用。

## 14.3 总结

本章以修改检查点状态这一业务需求为出发点，分析了修改检查点文件的可行性，读者可以通过 Flink 中提供的方法实现检查点文件的反序列化并生成新的检查点文件。Flink 1.9 中新增了 State Processor API，该 API 正是利用这种思路对状态数据文件中的各种信息进行读取和修改的。通过这种方式，状态数据和相关元信息对于用户变得更加透明，这样在业务上就有了更加灵活的操作空间。